Advances in Experimental Medicine and Biology

For further volumes:
http://www.springer.com/series/5584

Denis Corbeil
Editor

Prominin-1 (CD133): New Insights on Stem & Cancer Stem Cell Biology

Springer

Editor
Denis Corbeil
Biotechnological Centre (BIOTEC)
Dresden University of Technology
Dresden, Germany

ISSN 0065-2598
ISBN 978-1-4899-8692-4 ISBN 978-1-4614-5894-4 (eBook)
DOI 10.1007/978-1-4614-5894-4
Springer New York Heidelberg Dordrecht London

Printed on acid-free paper

Springer is part of Springer Science+Business Media (www.springer.com)

To my son Xavier

Preface

This book was conceived to celebrate the fifteenth anniversary of the discovery of prominin-1, also known in the medical field as CD133. Since its original description in 1997 in the murine system by Anja Weigmann, myself, Andrea Hellwig and Wieland B. Huttner (Proc Natl Acad Sci USA, 94; 12425–12430), and independently in the human system by David W. Buck and his team (Blood, 90; 5002–5012; 5013–5021), this cholesterol-binding pentaspan membrane glycoprotein has emerged as the object of great attention worldwide. This coincides with the identification and isolation of stem cells from different types of tissue and organ for which prominin-1 has become one of the most valuable cell surface markers with clinical value. Expectations for the development of novel therapies through the replacement and regeneration of damaged or diseased tissues based on isolated stem cells has made this field one of the leading edges on the frontiers of modern medicine. The expression of prominin-1 by putative cancer stem cells has also brought new horizon in cancer treatments, and this molecule may be regarded as a potential target in the eradication of cancers. The medical significance of prominin-1 is also highlighted in the visual system where mutations in the *PROM1* gene cause retinal degeneration.

As a biochemist, my long-standing interest in prominin-1 has brought me in the field of tissue engineering and cellular regeneration to understand the biological basis of tissue formation. The study of the molecular cell biology of prominin-1 in diverse cell types, e.g., epithelial cells, stem cells, and photoreceptors, which reflects by itself the broad tissue distribution of prominin-1, has emphasized divers' phenomena including the organization, remodeling, and dynamics of the plasma membrane as important factors regulating specific properties of the cells. Remarkably, these mechanisms appear conserved despite the considerable difference in cellular function of the cells in question (e.g., stem cell versus photoreceptor). The budding of membrane vesicles containing prominin-1 from the tip of microvilli of polarized epithelial cells and their release during the process of stem cell differentiation, and the organization of photoreceptor cell outer segment are good examples. I believe that the exploration of prominin's function(s) in various model organisms such as mice, zebra fish, axolotls, and flies would bring more insights, not only for cell biological trends, but also in organogenesis and tissue regeneration.

This book is composed of 15 chapters that will describe, on one hand, the molecular and cellular biology of prominin-1 and other members of the prominin family and, on the other hand, the importance of this molecule in the medical field as a valuable marker of stem and cancer stem cells. The opening chapter by my coworkers and myself presents an overview of the identification of prominin-1, its relation to the widely used AC133 epitope, and the general interest of this molecule in regenerative medicine. This chapter is intended as an introduction to the book and provides molecular details of prominin-1 across species including splice variants, tissue distribution, and certain biochemical properties including its specific subcellular localization in plasma membrane protrusions. It connects directly with the topics elaborated in the subsequent chapters of the initial section constructed as a knowledge base essential to the grasp of the physiological function(s) of this glycoprotein with a particular medical interest: CD133. My colleague Christine A. Fargeas presents prominin-2, the prominin-1 paralog, and the evolution of the prominin family of proteins among the animal kingdom. Then, as prominin-1 is not only tightly associated with plasma membrane protrusions, but also released in association with membrane particles into different body fluids, Anne-Marie Marzesco describes such singularity not only from a cell biology aspect, but also clinically, as prominin-1-containing membrane vesicles might be recognized as potential biomarkers in certain diseases. In the fourth chapter, Elisabeth Knust and her colleagues describe the eminent role of prominin in the retina, which demonstrates not only the significance of prominin-1 in the vision but also its evolutionary conserved function in the maintenance of photoreceptive membranes from humans to flies. Kouichi Tabu and his colleagues present in the next chapter the complex gene regulation of *PROM1* in normal and cancerous tissues. In the following 10 chapters, my colleagues describe new aspects of prominin-1-positive cells and the utility of this molecule as a marker of stem cells and cancer stem cells. Numerous tissues and organs are thus virtually dissected with regards to the expression of prominin-1 and the normal and cancerous cells harboring stem cell properties, i.e., self-renewal and multipotential differentiation capacities. In Chap. 6, Wieland B. Huttner, whose laboratory discovered murine prominin-1, presents with his coworker Alex M. Sykes novel features of neuroepithelial cells, precursors of all neurons of the central nervous system, including the asymmetric cell division. Rupert Handgretinger and Selim Kuçi present in great detail the importance of human prominin-1 (CD133) in the hematopoietic system and bring an update on the use of $CD133^+$ cells in autologous and allogeneic hematopoietic stem cell transplantation. In humans, the clinical relevance of $CD133^+$ cells is not limited to blood system; Benedetta Bussolati and Giovanni Camussi describe, for instance, the expression of CD133 in kidney under normal and pathological conditions. Although the possible use of $CD133^+$ progenitors in human studies presents obvious limitations due to immunological barriers, the in vivo experiments reported in mice suggest a potential role of progenitors in renal regeneration. Mariusz Z. Ratajczak and colleagues present the expression of CD133 by very small embryonic/epiblast-like stem cells (VSELs), and its use for their prospective immunoisolation. VSELs could provide a therapeutic alternative to the controversial use of human embryonic stem cells. Afterward, three clinically important organs are reviewed. Alessandro Sgambato and colleagues present the current knowledge on CD133 expression in normal and cancer colon tissues, both in humans and mice, and discuss the apparently conflicting data reported. Moreover, the authors devote great attention to the available information about the functional role of CD133 in colon cancer cells. Likewise, Norman J. Maitland and colleagues dissect the current literature regarding prostate cancer stem cells, with specific reference to the expression of CD133 as a stem cell marker to identify and purify stem cells in normal prostate epithelium and prostate cancer. In Chap. 12, Yuichi Hori reveals new facets of pancreatic progenitor cells and cancer stem cells by studying CD133 expression. Because it is also associated with melanoma stem cells, Aurelio Lorico and colleagues elegantly present the importance of CD133 as a therapeutic target. The two final chapters describe the diverse origins and utilizations of $CD133^{+}$ cells. Peter Donndorf and Gustav Steinhoff give a thorough account of cardiac stem cell therapy, and Yvan Torrente and colleagues discuss potential therapeutic applications of $CD133^{+}$ cells for degenerative diseases including muscular dystrophies.

Taken as a whole, I have attempted to gather almost all topics of significance to the prominin-1 research field and to its medical weight as a cell surface marker of stem and cancer stem cells. I have been greatly encouraged in this project by positive feedback from worldwide-recognized scientists and physicians working in these fields. Finally, it is my privilege to have the opportunity to edit these chapters and, also on behalf of all coauthors, to thank everyone who has helped us to produce this book.

Dresden, Germany Denis Corbeil

Contents

Contributors

Marzia Belicchi Stem Cell Laboratory, Dipartimento di Fisiopatologia Medico-Chirurgica e dei Trapianti, Università degli Studi di Milano, Fondazione IRCCS Cà Granda Ospedale Maggiore Policlinico di Milano, Centro Dino Ferrari, Milan, Italy

Norihisa Bizen Department of Stem cell Regulation, Medical Research Institute, Tokyo Medical and Dental University, Tokyo, Japan

Benedetta Bussolati Department of Internal Medicine, Research Center for Experimental Medicine and Center for Molecular Biotechnology, University of Torino, Torino, Italy

Giovanni Camussi Department of Internal Medicine, Research Center for Experimental Medicine and Center for Molecular Biotechnology, University of Torino, Torino, Italy

Emanuele Caredda Istituto di Patologia Generale, Università Cattolica del Sacro Cuore, Rome, Italy

Achille Cittadini Istituto di Patologia Generale, Università Cattolica del Sacro Cuore, Rome, Italy

Anne T. Collins YCR Cancer Research Unit, Department of Biology, University of York, Wentworth Way, York, UK

Denis Corbeil Tissue Engineering Laboratories (BIOTEC), Technische Universität Dresden, Dresden, Germany

Maddalena Corbi Istituto di Patologia Generale, Università Cattolica del Sacro Cuore, Rome, Italy

Peter Donndorf Department for Cardiac Surgery, University of Rostock, Rostock, Germany

Christine A. Fargeas Tissue Engineering Laboratories (BIOTEC), Technische Universität Dresden, Dresden, Germany

Andrea Farini Stem Cell Laboratory, Dipartimento di Fisiopatologia Medico-Chirurgica e dei Trapianti, Università degli Studi di Milano, Fondazione IRCCS Cà Granda Ospedale Maggiore Policlinico di Milano, Centro Dino Ferrari, Milan, Italy

Mareike Florek Tissue Engineering Laboratories (BIOTEC), Technische Universität Dresden, Dresden, Germany

Blood and Marrow Transplantation, Medical School Stanford University, Stanford, CA, USA

Nagananda Gurudev Max Planck Institute of Molecular Cell Biology and Genetics, Dresden, Germany

Rupert Handgretinger Department of Hematology/Oncology, University Children's Hospital, Tübingen, Germany

Yuichi Hori Division of Medical Chemistry, Department of Biophysics, Kobe University Graduate School of Health Science, Kobe, Japan

Wieland B. Huttner Max Planck Institute of Molecular Cell Biology and Genetics, Dresden, Germany

József Jászai Tissue Engineering Laboratories (BIOTEC), Technische Universität Dresden, Dresden, GermanyInstitute of Anatomy, Technische Universität Dresden, Dresden, Germany

Jana Karbanová Tissue Engineering Laboratories (BIOTEC), Technische Universität Dresden, Dresden, Germany

Elisabeth Knust Max Planck Institute of Molecular Cell Biology and Genetics, Dresden, Germany

Magda Kucia Stem Cell Biology Program at the James Graham Brown Cancer Center, University of Louisville, Louisville, KY, USA

Department of Physiology, Pomeranian Medical University, Szczecin, Poland

Selim Kuçi Department of Hematology/Oncology, University Children's Hospital III, Frankfurt am Main, Germany

Aurelio Lorico Cancer Research Program and College of Pharmacy, Roseman University of Health Sciences, Henderson, NV, USA

Norman J. Maitland YCR Cancer Research Unit, Department of Biology, University of York, Wentworth Way, York, UK

Anne-Marie Marzesco Hertie Institute for Clinical Brain Research, Tübingen, Germany

Javier Mercapide College of Pharmacy, Roseman University of Health Sciences, Henderson, NV, USA

Mirella Meregalli Stem Cell Laboratory, Dipartimento di Fisiopatologia Medico-Chirurgica e dei Trapianti, Università degli Studi di Milano, Fondazione IRCCS Cà Granda Ospedale Maggiore Policlinico di Milano, Centro Dino Ferrari, Milan, Italy

Kasia Mierzejewska Department of Physiology, Pomeranian Medical University, Szczecin, Poland

Emma E. Oldridge YCR Cancer Research Unit, Department of Biology, University of York, Wentworth Way, York, UK

Davide Pellacani YCR Cancer Research Unit, Department of Biology, University of York, Wentworth Way, York, UK

Germana Rappa Cancer Research Program and College of Pharmacy, Roseman University of Health Sciences, Henderson, NV, USA

Janina Ratajczak Stem Cell Biology Program at the James Graham Brown Cancer Center, University of Louisville, Louisville, KY, USA

Department of Physiology, Pomeranian Medical University, Szczecin, Poland

Mariusz Z. Ratajczak Stem Cell Biology Program at the James Graham Brown Cancer Center, University of Louisville, Louisville, KY, USA

Department of Physiology, Pomeranian Medical University, Szczecin, Poland

Alessandro Sgambato Istituto di Patologia Generale, Università Cattolica del Sacro Cuore, Rome, Italy

Gustav Steinhoff Department for Cardiac Surgery, University of Rostock, Rostock, Germany

Maria Svelto Istituto di Patologia Generale, Università Cattolica del Sacro Cuore, Rome, Italy

Alex M. Sykes Max Planck Institute of Molecular Cell Biology and Genetics, Dresden, Germany

Kouichi Tabu Department of Stem cell Regulation, Medical Research Institute, Tokyo Medical and Dental University, Tokyo, Japan

Tetsuya Taga Department of Stem cell Regulation, Medical Research Institute, Tokyo Medical and Dental University, Tokyo, Japan

Shinya Tanaka Department of Pathology, Laboratory of Cancer Research, Hokkaido University Graduate School of Medicine, Sapporo, Japan

Yvan Torrente Stem Cell Laboratory, Dipartimento di Fisiopatologia Medico-Chirurgica e dei Trapianti, Università degli Studi di Milano, Fondazione IRCCS Cà Granda Ospedale Maggiore Policlinico di Milano, Centro Dino Ferrari, Milan, Italy

Part I
Biology of Prominins

Chapter 1
Prominin-1 (CD133): Molecular and Cellular Features Across Species

Denis Corbeil, Jana Karbanová, Christine A. Fargeas, and József Jászai

Abstract Our knowledge of the first member of the prominin family is growing rapidly as the clinical value of prominin-1 (CD133) increases with its ever-wider use as a stem cell marker in normal and cancer tissues. Although the physiological function of this evolutionally conserved pentaspan membrane glycoprotein remains elusive, several studies have revealed new biological features regarding stem cells, cancer stem cells, and photoreceptors. The wide expression of CD133 in terminally differentiated epithelial cells, long overlooked by many authors, has attracted significant interest through the extensive investigation of human PROMININ-1 as a potential target for cancer therapies in various organs. Biochemically, this cholesterol-binding protein is selectively concentrated in plasma membrane protrusions, where it is associated with cholesterol-driven membrane microdomains. Clinically, mutations in the *PROM1* gene are associated with various forms of retinal degeneration, which are mimicked in genetically modified mice carrying either a null allele or mutated form of PROMININ-1. In this introductory chapter, we attempted to review 15 years of prominin-1 study, focusing on its unique protein characteristics across species and the recent developments regarding its cell biology that may shed new light on its intriguing involvement in defining cancer-initiating cells.

Keywords Cancer • CD133 • Eye • Prominin • Splice variant • Stem cell

D. Corbeil, Ph.D. (✉) • J. Karbanová • C.A. Fargeas • J. Jászai
Tissue Engineering Laboratories (BIOTEC), Technische Universität Dresden,
Tatzberg 47-49, 01307 Dresden, Germany
e-mail: corbeil@biotec.tu-dresden.de

D. Corbeil (ed.), *Prominin-1 (CD133): New Insights on Stem & Cancer Stem Cell Biology*, Advances in Experimental Medicine and Biology 777, DOI 10.1007/978-1-4614-5894-4_1,

1.1 Introduction

In modern medicine, stem and cancer stem cells are often described as the good and the bad players, respectively, with primitive properties; the former are essential cellular components in the development and maintenance of tissue homeostasis, and the latter appear to cause the initiation and progression of cancer. Therefore, both stem and cancer stem cells are viewed as potential targets for either regenerative therapy or a therapeutic avenue in cancer treatment [1, 2]. To better understand their cell biology, several groups, including ours, are dissecting the molecular and cellular characteristics of a common cell surface marker of these cells referred to as prominin-1 (CD133). Prominin-1 is expressed in a wide range of somatic stem and progenitor cells, including those in the central nervous and hematopoietic system, marks the cancer stem cells of solid tumors and hematopoietic malignancies, and is often associated with chemoresistant subpopulations. In this chapter, we will describe the molecular characteristics of prominin-1 and highlight the novel biological facets of stem cells, cancer stem cells, and photoreceptor cells that have been revealed by the study of prominin-1.

1.2 Identification of Prominin-1

The first member of the prominin family, prominin-1 (alias CD133, PROML1), was identified in 1997 through the characterization of novel monoclonal antibodies (mAbs) and through molecular cloning by two independent groups in Heidelberg (Germany) and Sunnyvale (United States of America) who were searching for novel markers of mouse neural and human hematopoietic stem and progenitor cells, respectively [3, 4]. The 13A4 mAb revealed mouse prominin-1 as a novel protein of plasma membrane protrusions that are present within the apical domain of polarized epithelial cells found in the embryonic neuroepithelium, adult ependymal layer, and proximal tubules of the kidney [3]. The protein was initially referred to as prominin. The AC133 mAb recognized an epitope on the human glycoprotein initially referred to as AC133 antigen, the expression of which appeared to be restricted to CD34-positive hematopoietic stem and progenitor cells [4]. However, Northern blot analysis of human PROMININ-1 provided data discordant with the immunodetection of the AC133 antigen. Notably, the strong mRNA signal on kidneys was incompatible with the lack of immunodetection in this tissue [4, 5].

The question of the relationship between the mouse and human molecules was rapidly raised and sustained by the detection of mouse splice variants and the related transcripts in both species [6, 7]. In a key publication in 2000, the groups that originally studied prominin joined to carefully demonstrate that human AC133 antigen is, like mouse prominin, expressed in epithelial cells and targeted to plasma membrane protrusions [8]. Most importantly, using the colon carcinoma cell line Caco-2 – a widely used model of enterocytic epithelial differentiation – it was demonstrated that AC133 immunoreactivity diminishes upon differentiation, while mRNA levels increase. This observation led to the hypothesis that AC133 immunoreactivity may be somehow affected by the differential glycosylation state associated with enterocytic differentiation, i.e., the absence of AC133 immunoreactivity detection may not necessarily reflect the absence of protein per se [8]. Further evidence concerning the presence of the human protein in terminally differentiated epithelial cells (e.g., kidneys, mammary glands) was provided by mapping its tissue distribution using an antiserum raised against the PROMININ-1 polypeptide [9]. Along the same line, Karbanová and colleagues reported that, as was demonstrated earlier for its murine counterpart through 13A4 mAb immunohistochemistry [3, 10–12], general expression of PROMININ-1 is widespread in adult human tissues [13]. The analysis of transgenic reporter knock-in mice of the *Prom1* locus led to the same conclusion [14, 15].

In the meantime, the emergence of the prominin family had occurred with the identification and molecular cloning of prominin-2 [16] (for more details concerning prominin-2, see Chap. 2 of this volume by Fargeas) and the characterization of a splice variant in humans [17]. Therefore, in anticipation that human PROMININ-1 may share the characteristics of the murine protein, a unifying nomenclature was proposed to take into account the different paralogs, splice variants, and glycoforms of the prominin family [18]. Moreover, it was stressed that AC133 immunoreactivity may be more restricted than the general expression of human PROMININ-1 (CD133) [18]. The corresponding epitope (AC133 epitope; hereafter AC133) was thought to be dependent on conformation and/or sensitive to changes in glycosylation [5] (reviewed in Ref. [19]).

Here, we will present the molecular and biological characteristics of prominin-1 across species and briefly describe its expression in various tissues and organs, with a special emphasis on those where a medical relevance has been reported.

1.3 Protein Features

The membrane topology of prominin-1 (as of other members of the prominin family) after the cleavage of the signal peptide consists of an N-terminal extracellular domain (EC1), five transmembrane (TM) segments, alternating short intracellular and large extracellular loops (≈20 and >250 amino acid residues, respectively), and an intracellular C-terminal domain (IC3) (Fig. 1.1). A typical prominin-1 molecule comprises approximately 850 amino acid residues depending on the splice variant (see below). The two extracellular loops (EC2 and EC3 domains) contain potential N-glycosylation sites that vary in position according to the animal species from which the prominin-1 sequence is derived [3, 5, 16, 20–23].

The amino acid sequence is poorly conserved among prominin-1 gene products across species. For instance, only 60% identity between primates and rodents was observed [20]. Prominin-1 relatives in other species (fish, amphibian, bird; ≈45%), particularly in invertebrates (worm, fly; <25%), are poorly conserved with mammalian sequences [16, 22–24].

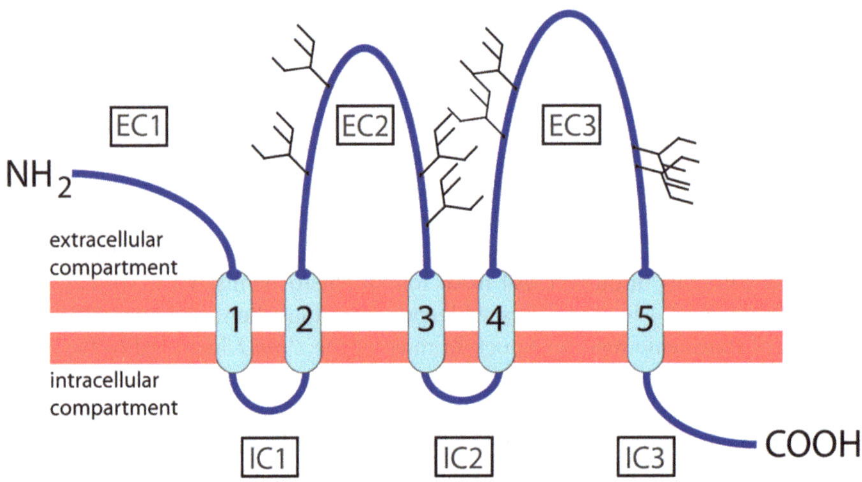

Fig. 1.1 Membrane topology of prominin-1. Prominin-1 is predicted to consist of an extracellular N-terminal domain (*EC1*), five transmembrane domains (1–5) separating two large glycosylated extracellular loops (*EC2* and *EC3*) and two small intracellular loops (*IC1* and *IC2*), and an intracellular C-terminal domain (*IC3*). The locations of the *EC2*, *EC3*, and *IC3* domains have been confirmed by antibody accessibility and epitope insertion analyses [3, 10, 16]. The *EC2* and *EC3* domains contain all of the potential N-glycosylation sites (forks), which vary in position depending on the animal species. Human prominin-1.s2 is illustrated here

However, multiple sequence analysis reveals three common characteristics of prominin proteins: (1) six cysteine residues in the EC2 and EC3 domains are conserved between various prominins, which are likely to form disulfide bridges; (2) a cysteine-rich domain of as yet unknown function is found at the transition of the TM1 and IC1 domains; and (3) a consensus core sequence, i.e., C*X*P*X*(12,13)C*X*(5)[P/S]*X*(4)W*X*(2)h*X*(4)hh*X*h (where *X* stands for any residue; the number of *X* is indicated in parentheses; residues in brackets indicate alternatives for a given position, and h stands for any hydrophobic residue), is observed at the end of the EC3 domain and the beginning of TM5. One of the *Drosophila* prominin sequences constitutes the exception to this rule, as it conforms only partially to the core consensus [24]. Nevertheless, in the absence of known catalytic activity, these sequence characteristics together with the pentaspan membrane topology may help to define the prominin membership of a novel molecule from a particular species. Finally, it is worth mentioning that several prominin-1 molecules exhibit a leucine zipper-like motif in one of the extracellular domains, which may be an indication of an in vivo dimerization process [25], and that prominin-1 (tyrosine residues 828 and 852 within IC3) is a target of Src and Fyn tyrosine kinases in human medulloblastoma cells [26]. The biological relevance of the latter modification remains to be determined.

In mammals, prominin-1 displays only a modest (≈30%) amino acid identity with prominin-2 [16, 27]. Nevertheless, the hydrophobic cluster analysis [28]

reveals several conserved hydrophobic clusters between both proteins, particularly inside the EC2 and EC3 domains, suggesting that the secondary and tertiary structures are highly related (Fargeas and Corbeil, unpublished data). Neither the protein crystal structure of these two loops nor the modeling of the alternative C-terminal domain(s) by nuclear magnetic resonance spectroscopy has been reported.

1.4 Genes, Splice Variants, and Glycoforms

1.4.1 Genomic Organization

The genomic structure of both mammalian prominin paralogs is remarkably similar (introns being concordant in position and phase) and remarkably conserved across species [16, 29]. However, the exonic structure does not correlate with the protein structural domains, as the coding regions of prominin-1 and prominin-2 span 28 and 23 exons, respectively (Fig. 1.2a; note that the numbering of the exons begins with the exon bearing the initial start codon) [16]. Shmelkov and colleagues have described that human PROMININ-1 is under the control of five alternative promoters (named P1–P5), three of which are located in a CpG island and are at least partially regulated by methylation [30, 31]. In certain cancers, aberrant DNA methylation may be an important determinant for the overexpression of PROMININ-1 [32–34]. Up to ten alternative exons that may be included in the 5′-untranslated region have been identified, highlighting complex gene regulation [30]. A potential binding sequence for estradiol receptor has been identified in the *PROM1* gene that was found to be transiently upregulated in uterine luminal epithelium during the implantation process [35]. Readers are invited to look at Chap. 5 by Tabu and colleagues for further details concerning the regulation of the *PROM1* gene.

1.4.2 Prominin-1 Splice Variants

Prominin-1 is subjected to alternative splicing [7, 17, 18, 22, 23, 30, 36, 37]. The functional significance of this variability is not yet understood, but the appearance of certain variants appears to be ontogenetically regulated [37] and/or tissue specific [11, 17]. Currently, 12 alternative splicing variants affecting the open reading frame of mammalian (primates and rodents) prominin-1 have been described in the three extracellular domains or within the IC3 domain (Fig. 1.2b) [11, 29]. As discussed above, a rational nomenclature of the prominin gene products was therefore proposed [18] and updated [29]. In principle, the name of the molecule, i.e., prominin-1 (or prominin-2, prominin-3, etc.), should be followed by a suffix (s) indicating the splice variant, numbered according to the chronology of publication (including database submission in which the sequence is quoted as a distinct splice variant),

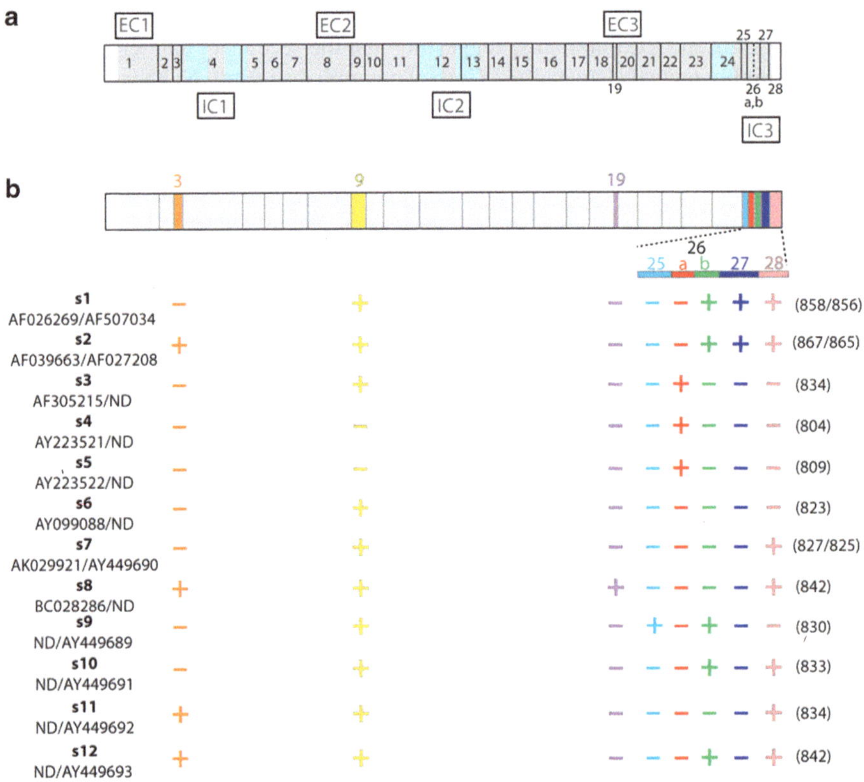

Fig. 1.2 Genomic organization of prominin-1. (**a**) Coding regions appears in *gray*, while the 5′- and 3′-untranslated regions appear in *white*. *Vertical lines* indicate the exon boundaries, and the *blue zones* indicate the predicted transmembrane regions. The numbering of the exons begins with the exon bearing the initial start codon [10, 16, 29]. The various and alternative 5′-untranslated regions were ignored for the sake of simplicity. (**b**) Distinct splice variants of mammalian (mouse and/or human) prominin-1. Facultative exons included in the coding sequences are colored. Splice variants are named according to the prominin nomenclature [18, 29], the GenBank accession number (mouse/human), and the presence (+) or absence (–) of an exon, and the corresponding protein sequence length (number of amino acid) is indicated (ND; not described)

and irrespective of the species. For instance, prominin-1.s1 and prominin-1.s2 refer to the first prominin sequences reported in mouse (prominin) [3] and human (AC133 antigen) [5], respectively, because they differ by the absence or presence of a facultative exon (numbered 3) in the EC1 domain (Fig. 1.2a, b). The use of alternative nomenclature might simply confuse the field; the relationship of the AC133 epitope with the prominin-1 (CD133) protein was already puzzling (see above).

The recent molecular cloning of prominin-1 in nonmammalian vertebrates such as zebrafish (*Danio rerio*), axolotl (*Ambystoma mexicanum*), frogs (*Xenopus laevis*), and birds (*Gallus gallus*) has significantly increased the number of prominin-1 gene products [21–23]. At least 20 potential splice variants (s1 to s20) might exist

Table 1.1 Alternative prominin-1 COOH-termini[§]

Type	Exons included in the coding sequence	COOH-terminal sequence (species depicted)	Splice variant	Species[reference]
A	24, 26b, 27, 28	(H) RRMDSEDVYDDVETIPMKNMENGNNGYHKDHVYGIHNPVMTSPSQH	s1, s2, s13	A[1], H[2,5], M[3,6], R[4]
B	24, 26a	(M) RRMDSEDVYDDSSVSGMWHFTL	s3, s4, s5	M[7]
C	24	(H) RRMDSEDVYDE	s6	F[8], H[7], M[7]
D	24, 28	(H) RRMDSEDVYDDPSQH	s7, s8, s11, s15, s18, s21	A[1], C[1], H[7,9], F[1,10], M[7], R[9]
E	24, [illegible], 26b	(H) RRMDSEDVYDD[illegible]C[b]	s9	H[9]
F	24, 26b, 28	(H) RRMDSEDVYDDVETIPMKNPSQH	s10, s12, s14	A[1], H[9], Rh[9]
G	24, 27, 28	(C) RRMDTEDVYDDMENGNNGYHKEHLYGIHNPVITSSVEQW[C]	s16, s17	C[1], D[11], M[12]
H	24, 26b, F27′	(F) RRMDTEDVYDDIETFPMKTIPTYDTMTRFPRASAPPRHADW	s19, s20	F[1,13], A[14]
I	24, 26b, 27, F27′	(A) RRMDVYDDVETVPMKNLENGNNGYHNEYLYGIHNPIMTSMSSYDT VNRFPRASAPPRQDD	ND	A[15]
J	24, F27′	(F) RRMDTEDVYDDIPTYDTMTRFPRASAPPRHADW	ND	F[16], A[17]

[§]Alternative splicing might generate ten distinct cytoplasmic COOH-terminal tails of prominin-1. Exons 25, 26a, 26b, 27, and F27′ are given in cyan, red, green, blue, and pink, respectively. Types D–J are predicted from nucleotide (cDNA, expressed sequence tag (EST), genomic) sequences, in the absence of protein data with specific antibody.

cDNA complementary DNA, *A* amphibian, *C* chicken, *D* dog, *F* fish, *H* human, *M* mouse, *R* rat, *Rh* rhesus, *ND* not determined

Alternative exon F27′ is found in fish and amphibian.

Splice variant s21 is related to zebrafish prominin-1b.

*Indicates exon 25 introduces a frameshift on the following exon 26b.

**Indicates that exon 27 in dog harbors a stop codon generating a truncated C-terminus in this species (MENGNIGFHRHHSTQTV).

Ref.: 1. Jászai (2011); 2. Yu (2002); 3. Weigmann (1997); 4. Corbeil (2001); 5. Miraglia (1997); 6. Miraglia (1998); 7. Fargeas (2004); 8. McGrail (2010); 9. Fargeas (2007); 10. Fargeas (2003); 11. *Canis lupus* EST DT541441 and DT541268; 12. *Mus musculus* EST EH097137; 13. *Pimephales promelas* EST DT164241; 14. *Xenopus tropicalis* EST CX838199 and CX912260; 15. *Xenopus laevis* EST BJ061920; 16. *Osmerus mordax* EST EL532935, *Pimephales promelas* EST DT181633, *Cyprinus carpio* EST EX822056, *Salmo salar* EST DW541737; 17. *Xenopus laevis* EST CK79685

This table is modified from Ref. [22]

across species. Moreover, in certain fish, a duplication of the prominin-1 gene (e.g., zebrafish, Fugu (*Takifugu rubripes*); named prominin-1a and prominin-1b), which likely arose from the whole-genome duplication within the teleost lineage that resulted in multiple copies of many genes [38], adds a further twist to the complexity of splice variants. Nevertheless, the fish prominin-1 co-orthologs are not affected by alternative splicing to the same extent. For prominin-1b, only one splice form has been identified and is designated as prominin-1b.s21 [22]. Remarkably, at least ten alternative C-termini (Table 1.1; referred to as type A to J) can be generated by intron retention, exon skipping, or the use of a cryptic acceptor site [11, 22, 29].

Some prominin gene products exhibit distinct potential PSD95/Dlg1/ZO-1 (PDZ)-binding domains [29], which may indicate unidentified cytoplasmic PDZ domain-containing protein-interacting partner(s) [39]. Such interactions might regulate the targeting and/or retention of prominin-1 into specific subdomains of the plasma membrane that are characterized by a membrane curvature (e.g., microvillus, leading edge of lamellipodium) [27, 40]. Only tyrosine 828 found in the tyrosine phosphorylation consensus site (R/K)xxx(D/E)xxY is conserved among these prominin-1 splice variants [29]. Further studies are necessary to elucidate the biological relevance of the use of facultative exons, but it appears that a variety of prominin-1-related polypeptides can be created that indirectly highlight alternative and/or complementary function(s).

1.4.3 Glycoforms

In general, N-linked glycans represent approximately 15–20% of the apparent molecular mass (ca. 112–120 kDa) of prominin-1 [3, 5, 8, 11]. The terminal N-glycan moieties of human PROMININ-1 are sialylated [41], while its polypeptide does not appear to carry any O-linked glycan [42]. Mak and colleagues have demonstrated by the heterologous expression of PROMININ-1 mutants that the glycosylation of each potential N-glycan site is not essential for its intracellular transport to the cell surface, but are collectively necessary [43]. In the murine reproductive tract, such as the epididymis and testis, biochemical analysis of prominin-1 suggested that an individual splice variant could exist in several distinct glycoforms [11]. In the absence of catalytic activity or a signaling pathway directly modulated by prominin-1, it is difficult to estimate the physiological significance of this differential posttranslational modification.

As mentioned above, the lack of AC133 immunoreactivity in Caco-2 cells upon their spontaneous enterocytic differentiation, which is concomitant with an alteration of the general glycosylation pattern of intestinal cells, has suggested a link between the accessibility of AC133 and glycosylation status [8]. Similarly, Hemmoranta and colleagues have observed differences in the glycan profiles of AC133-positive and AC133-negative hematopoietic stem and progenitor cells [44]. Whether and how alternative splicing, such as generation of the exon 3 that distinguishes PROMININ-1.s1 from PROMININ-1.s2 (named AC133-2 and AC133-1, respectively, in [17]), affects the recognition of the glycoprotein by the AC133 mAb is currently unknown. Indeed, the exact location of AC133 within human PROMININ-1 is unknown, but it may reside in the second half of the EC3 domain [45]. The differential glycosylation of PROMININ-1 among various human tissues may explain, at least in part, why its transcript is broadly detected, whereas the AC133 immunoreactivity appears to be limited to cells harboring stem cell properties, with some exceptions such as the differentiated cells found in the pancreas [46, 47] and the cones within the retina (see Chap. 4). Moreover, it is important to keep in mind that AC133 may also be masked under certain circumstances, particularly when prominin-1 is embedded in a specific cholesterol-dependent membrane microdomain where protein-lipid and protein-protein interactions are engaged (see below). Thus, a number of parameters other than differential glycosylation may interfere with AC133 detection, and some caution thus needs to be taken, particularly with the immunohistological preparation of samples [9, 13, 46, 48] (reviewed in Refs. [49, 50]). In addition to immunohistochemistry detection, it might be useful to use alternative techniques such as in situ hybridization to ascertain the expression of PROMININ-1 [12, 22].

1.5 Cellular Features of Prominin-1

A remarkable biological feature of prominin-1 is its specific association with curved plasma membrane subdomains that protrude into the surrounding extracellular milieu. Because of such distinctive localization, Huttner and colleagues based its name on the Latin word *prominere* (jut up, to be prominent) [3].

Prominin-1 is specifically localized in various plasma membrane protrusions emerging from the apical domain of polarized epithelial cells such as microvillus, primary and motile cilia, and the midbody [3, 8, 13, 51, 52]. Often, prominin-1 appears concentrated at the tips of the microvillus or cilium [3, 51]. There are other cellular examples that illustrate this characteristic of concentration: (1) in the uropod, a backward-protruding structure of migrating hematopoietic stem and progenitor cells [53, 54]; (2) in myelin sheaths, developing from the oligodendrocyte cell body [37]; and (3) in the plasma membrane evaginations growing out from the photoreceptor connecting cilium [10]. In Chap. 4 of this volume, Gurudev and colleagues further develop this final instance that underlies the photoreceptor outer segment morphogenesis.

Within the membrane structures that protrude from the planar areas of the plasmalemma, prominin-1 interacts directly with membrane cholesterol, as demonstrated using photoactivatable analogue of cholesterol, and associates with a membrane microdomain ("lipid raft"), the integrity of which is dependent on its cholesterol content [52, 55] (reviewed in Ref. [56]). Therein, prominin-1 co-localizes with GM_1 ganglioside [55, 57, 58]. It is of note that prominin-1 bears a potential ganglioside binding site within its EC1 domain [59]. However, such an interaction remains to be demonstrated in vivo.

In addition to its specific localization in plasma membrane protrusions, prominin-1 is also released in various body fluids (internal and external) in association with small membrane vesicles [60]. Prominin-1-containing membrane vesicles have been detected in murine and human cerebrospinal fluids (CSF), seminal fluids, urine, tears, and saliva [52, 60, 61]. This widespread distribution reflects the general expression of prominin-1 in various epithelia (see below). Irrespective of their function(s), which have not yet been elucidated, the detection of these particles may be highly instructive with regard to certain diseases. Huttner and colleagues demonstrated that human CSF contains membrane particles carrying prominin-1 and that their levels varied between healthy adults and glioblastoma patients, which provided new perspectives for studying central nervous system diseases including cancer [61]. Analysis of CSF prominin-1 may provide information about neural stem cell activity in the adult brain, not only regarding the physiologically aging organ but also in the context of neurological diseases [62]. Such vesicles might also be used for monitoring the functional recovery of particular glands (see below) upon tissue engineering and/or cell replacement. In Chap. 3 of this volume, Marzesco further develops these ideas, particularly with respect to the molecular mechanism of their release and the clinical value of these vesicles as novel biomarkers.

1.6 Differentiated Cells, Stem Cells, and Cancer Stem Cells

1.6.1 Differentiated Cells: Widespread Distribution

In rodents, primates, and other species, prominin-1 appears to be physiologically expressed at all stages of life.

In early murine embryos, prominin-1 is expressed in trophoblast cells, but not in the inner cell mass, which are the founder cells of embryonic stem (ES) cells [63]. Prominin-1 is upregulated in murine ES cells as soon as they are committed to differentiation. However, by the terminal stages of differentiation, prominin-1 is no longer detected [63]. This unique expression pattern makes prominin-1 a potential tool to define ES-derived committed and early progenitor cells. In humans, AC133 was detected in ES cell lines [64–66], while embryoid body-derived cells were found to display prominin-1 transcripts [67, 68].

In developing murine embryos, prominin-1 is observed in epithelia of the three germinal layers [3, 69], and its expression persists in several organs during adulthood. Likewise, human PROMININ-1 (AC133) has been detected in various embryonic epithelia [8], and its expression, as revealed using antibodies directed against its polypeptide, is widely distributed similar to its mRNA profile in adults [5, 8, 9, 16]. In a nutshell, murine and human prominin-1 are expressed in the kidneys (see Chap. 8 by Bussolati and Camussi), colon (Chap. 10 by Sgambato and colleagues), prostate (Chap. 11 by Pellacani and colleagues), pancreas (Chap. 12 by Hori), liver [13, 70], mammary glands (lactiferous ducts) [9, 48], the epithelium along the epididymal duct [11, 71, 72], and major cephalic exocrine glands [12, 13]. In all prominin-1-positive epithelial cells examined, prominin-1 was restricted to the apical plasma membrane domain [13, 19, 27]. In non-epithelial cells, prominin-1 is notably detected in photoreceptors (Chap. 4), glial cells [37], and developing spermatozoa [11, 72]. A similar expression pattern is now emerging in nonmammalian species [21, 22, 25, 73] (J.J. and colleagues, manuscript submitted). Although generally considered as a specific stem cell antigen, prominin-1 is broadly expressed in differentiated tissues; nevertheless, this does not disqualify it as a prospective biomarker of stem cells.

1.6.2 Stem Cells

With the restriction detailed above, prominin-1 has emerged as a cell surface marker for the isolation of somatic stem cells with a medical value. Since its original identification by the AC133 mAb, PROMININ-1 has been largely used as an alternative to CD34 to identify and isolate subpopulations of human hematopoietic stem and progenitor cells (reviewed in Ref. [74]). Several clinical trials with leukemia patients have confirmed the success of PROMININ-1-based immunomagnetic selection [75] (see Chap. 7 by Handgretinger and Kuçi).

In the central nervous system, murine prominin-1 was initially detected in neuroepithelial progenitor cells and ependymal cells (see Chap. 6 by Sykes and Huttner); the latter may represent one source of adult neural stem cells [76–78]. Indeed, prominin-1-positive (or $AC133^+$ in the case of humans) progenitor cells have been successfully isolated from the human-fetal and post-mortem brain [79–81] and the mouse postnatal cerebellum [82].

Prominin-1 is considered a general marker of somatic stem cells. Rare prominin-1 (AC133)-positive cells immunoisolated from adult renal interstitium have been shown to differentiate into either tubular or vascular cells when induced with the proper cues [83]. Similarly, a minute population of AC133-positive progenitors was identified in the prostate basal compartment that possesses high proliferative potential and can reconstitute prostatic-like acini in immunocompromised nude mice [84]. Leong and colleagues impressively demonstrated that a single $Lin^-Sca\text{-}1^+prominin\text{-}1^+CD44^+CD117^+$ prostate stem cell could generate a prostate after transplantation [85]. Prominin-1 potentially labels progenitor cells derived from muscle [86], skin [17, 87], and intestines [15, 88] and identifies a small fraction of human bone marrow-derived multipotent mesenchymal stromal cells [89, 90]. AC133 is also useful as a cell surface antigen for the isolation of very small embryonic-like stem cells [91] (see Chap. 9 by Ratajczak and colleagues).

Finally, AC133 is expressed by endothelial progenitor cells [92], which are regarded as potential therapeutic agents of interest in regenerative medicine because they may contribute to postnatal vasculogenesis in physiological and pathological neovascularization [93]. As AC133 is rapidly lost upon differentiation [92, 94], it is used as an attractive cell surface antigen to isolate immature endothelial cells that could contribute to neovascularization in ischemic conditions [95] (reviewed in Ref. [19]). Circulating endothelial progenitors are also regarded as a potential prognostic marker in cancer and cardiac diseases. Several studies have reported elevated prominin-1 mRNA levels, which might reflect the contribution of vasculogenesis to tumor vascularization in patients with breast carcinoma [96], recurrent colon cancer [97], and lung cancer [98, 99]. Additionally, Mehra and colleagues have observed that patients with metastases to the bone had an increased level of prominin-1 mRNA in their peripheral blood mononuclear fraction [100]. The existence of various splice variants implies that one should be cautious about the design and interpretation of assays based on the detection of prominin-1 mRNA.

In addition, AC133-positive cells were proposed as potential biomarkers of vascular disease [101, 102]. Whether AC133 marks the actual endothelial cell subpopulation responsible for postnatal generation of blood vessels has been challenged [103, 104]. Therefore, AC133-positive stem and progenitor cells isolated from various tissues might represent new biological tools for the treatment of tissues after injuries or degenerative diseases. Readers are invited to look at Chap. 14 by Donndorf and Steinhoff and Chap. 15 by Meregalli and colleagues for further details concerning the sources and therapeutic applications of prominin-1 (AC133)-positive cells.

Given the widespread expression of prominin-1, it might be important to consider additional markers that, in combination with prominin-1, will refine the selection procedure. The same might hold true for cancer stem cells.

1.6.3 Cancer Stem Cells

Many reports have highlighted the expression of PROMININ-1 (AC133) in cancerous human tissues. Indeed, the literature concerning this molecule and cancer has grown exponentially since the description of AC133-positive cancer stem cells in human brain tumors by Dirks and colleagues (reviewed in Ref. [105]). As a cell surface antigen, AC133 is used extensively for the prospective isolation and characterization of subpopulations of leukemia cells [106–109] and solid tumors. Although different evaluations of PROMININ-1 in epithelial cancers such as colorectal cancer have generated contradicting data [14, 15, 110–113], rare subpopulations of PROMININ-1-positive cells with self-renewal capacity were found in various tissues and organs such as the brain [114–118], kidney [119], liver [120–122], lung [123, 124], pancreas [125, 126], ovary [127], prostate [128, 129], and skin [130–133]. Various cell lines originating from particular tumors, e.g., anaplastic thyroid carcinoma [134], osteosarcoma [135, 136], hepatocarcinomas [137, 138], and metastatic melanoma [139] (see Chap. 13 by Lorico and Rappa), express PROMININ-1. Such "universal expression" tends to confirm it as an important cell surface antigen of cancer stem cells that is shared by different organ systems and is often associated with a chemoresistant subpopulation [118, 122, 140]. Nevertheless, it should be considered that, given the functional traits of cancer stem cells, not all of them might express PROMININ-1 [2].

A minute fraction of glioma cells was shown to be AC133 positive, was associated with radioresistance, and could be the source of tumor recurrence after conventional radiotherapy [140]. In line with this, an overexpression of PROMININ-1 (and other neural stem cell markers) is associated with disease classes of poor prognosis in a reclassification of gliomas [141]. Similarly, Zeppernick and colleagues have described the expression of PROMININ-1 to be adversely correlated with patient survival in gliomas; the frequency of AC133-positive cells increases with high-grade tumors. Interestingly, these authors observed a specific clustering of AC133-positive cells and hypothesized that this may reflect unbalanced symmetrical versus asymmetrical cell divisions of AC133-positive cells [142].

PROMININ-1 (AC133)-positive glioma stem-like cells have been reported to potentially promote angiogenesis, which constitutes a crucial step in tumor formation [143]. Calabrese and colleagues have shown that AC133-positive cells isolated from primary medulloblastoma and ependymoma can interact physically with endothelial cells in culture as opposed to AC133-negative cells, which cannot [144]. However, when considering PROMININ-1 as a molecular target for the design of novel therapies, it is important to keep in mind that most studies actually address the AC133 epitope of PROMININ-1, and the expression of its polypeptide may be more widespread, particularly in epithelial tissues. Consequently, we do need to be very cautious about some of these therapies (reviewed in Ref. [145]).

1.7 Function of Prominin-1

After 15 years of intensive research on prominin-1 and other members of this family, the physiological function of these molecules remains unknown. It is true that most of the laboratories are focusing on its use as a marker of stem and cancer stem cells and that few of them are actually investigating its role in these cells. Nevertheless, the accumulation of information regarding the molecular and cellular biology of prominin-1 including mutations in human *PROM1* gene [10, 146–150] and the data obtained from its knockout in a murine model [151] have led to a hypothetical role of this molecule (and other prominins) as a scaffold protein involved in the cellular organization of plasma membrane protrusions. As suggested by its interaction(s) with other proteins (cytoplasmic, membranous, and/or soluble), prominin-1 might organize such membranes with the appropriate sets of adhesion/signaling molecules. The binding of prominin-1 to protocadherin 21 in mammalian photoreceptor cells [147] or to eyes shut/spacemaker, a secreted protein involved in the partitioning of rhabdomeres [24, 152] in the *Drosophila* compound eye, are good examples of such organization (see Chap. 4). Moreover, the interaction of cholesterol-based membrane microdomains containing prominin-1 with cytoskeleton elements may be necessary to maintain the temporal and spatial distribution of these protein complexes. As a regulator of signaling pathways, the symmetric versus asymmetric distribution of prominin-1 in dividing stem cells, such as those found in neural [153] and hematopoietic [19, 154] systems or in cancer stem cells [155], might highlight the partition of key players underlying stem cell properties. Similarly, its extracellular release in association with membrane vesicles concomitant with the differentiation of progenitor cells [60, 156, 157] is consistent with such a role. The characterization of these prominin-1-containing membrane microdomains and vesicles at a proteomic level might be instructive with these issues.

Collectively, the experimental and clinical observations reported thus far strongly support the theory that prominin-1, in association with other proteins and membrane lipids (such as cholesterol), provides a scaffolding mechanism in the maintenance and dynamics of the plasma membrane.

1.8 Concluding Remarks and Perspectives

Although the function of prominin-1 has not yet been elucidated, its study over the years has revealed many novel insights: (1) new facets in the cell division of stem and cancer stem cells, including the segregation of certain determinants like prominin-1; (2) unexpected cell biological events, such as the release of membrane particles, during stem cell differentiation; and (3) new aspects into the architecture of photoreceptor cells. In the coming years, further investigation, for instance, using prominin-1 or prominin-2 knockout mice, might reveal fundamental basic phenomena regarding stem and cancer stem cell properties and the membrane polarity of epithelial and photoreceptor cells. The use of various animal models for regeneration (e.g., zebrafish and axolotl), where prominin-1 was recently characterized, may also provide new clues about the role of prominin molecules in organogenesis. There are still numerous outstanding questions concerning the cell biology of prominin-1:

- What are the physiological functions of the prominins and the membrane vesicles containing them?
- What is the significance of the symmetric/asymmetric distribution of prominin-1 in mitotic stem cells?
- What are the other constituents of the membrane microdomains and membrane vesicles containing prominin-1?
- What are the extracellular and cytoplasmic protein partner(s) that interact with prominins and their splice variants?

Clinically, prominin-1 has been considered a useful stem cell marker and promising target for the development of new cancer therapies. The identification of human retinal degenerations directly associated with mutations in the *PROM1* gene makes it a potential candidate for gene replacement therapy. Prominin-1 and the small prominin-1-containing membrane vesicles that are released in various body fluids are of particular medical interest. In addition to its use as a cell surface antigen for somatic stem and cancer stem cell isolation, some therapeutic perspectives related to prominins are possible:

- Molecular target for cancer therapy
- Biomarker (itself and the membrane particles that contain it) for certain cancers
- Biomarker for monitoring tissue engineering/cell replacement therapy
- Gene-based therapy for certain types of retinal degeneration

In conclusion, prominin-1 has revealed new horizons not only in the field of cell biology but also in the clinical domain. We hope that worldwide efforts will help to increase our molecular and cell biological knowledge of this molecule, and more importantly, its translational research will be beneficial to individuals who are ill.

Acknowledgements We warmly thank all the collaborators for their major contributions to the prominin field during the last 15 years. We are indebted to Wieland B. Huttner, whose laboratory discovered murine prominin-1; David W. Buck, whose laboratory discovered human prominin-1; Marion A. Maw, who described the first mutation in the *PROM1* gene; Anne-Marie Marzesco, who noticed prominin-1-containing membrane vesicles; and Peter Carmeliet, whose laboratory generated prominin-1 knockout mice. We thank Andrea Hellwig and Michaela Wilsch-Bräuninger for the excellent electron microscopy demonstrating the preference of prominin-1 for plasma membrane protrusions. The corresponding author (D.C.) wants to express here his particular gratitude to W.B. Huttner for his constant support. Deutsche Forschungsgemeinschaft (DFG) supported D.C. (SFB655 B3; TRR83 #6 and CO298/5-1). Intramural funds of the Medical Faculty of Technische Universität Dresden and DFG supported J.J. (MeDDrive38 grants and CO298/5-1, respectively).

References

1. Visvader JE, Lindeman GJ (2008) Cancer stem cells in solid tumours: accumulating evidence and unresolved questions. Nat Rev Cancer 8:755–768
2. Baccelli I, Trumpp A (2012) The evolving concept of cancer and metastasis stem cells. J Cell Biol 198:281–293
3. Weigmann A, Corbeil D, Hellwig A, Huttner WB (1997) Prominin, a novel microvilli-specific polytopic membrane protein of the apical surface of epithelial cells, is targeted to plasmalemmal protrusions of non-epithelial cells. Proc Natl Acad Sci USA 94:12425–12430
4. Yin AH, Miraglia S, Zanjani ED, Almeida-Porada G, Ogawa M et al (1997) AC133, a novel marker for human hematopoietic stem and progenitor cells. Blood 90:5002–5012
5. Miraglia S, Godfrey W, Yin AH, Atkins K, Warnke R et al (1997) A novel five-transmembrane hematopoietic stem cell antigen: isolation, characterization, and molecular cloning. Blood 90:5013–5021
6. Corbeil D, Röper K, Weigmann A, Huttner WB (1998) AC133 hematopoietic stem cell antigen – human homologue of mouse kidney prominin or distinct member of a novel protein family. Blood 91:2625–2626
7. Miraglia S, Godfrey W, Buck D (1998) A response to AC133 hematopoietic stem cell antigen: human homologue of mouse kidney prominin or distinct member of a novel protein family? Blood 91:4390–4391
8. Corbeil D, Röper K, Hellwig A, Tavian M, Miraglia S et al (2000) The human AC133 hematopoietic stem cell antigen is also expressed in epithelial cells and targeted to plasma membrane protrusions. J Biol Chem 275:5512–5520
9. Florek M, Haase M, Marzesco A-M, Freund D, Ehninger G et al (2005) Prominin-1/CD133, a neural and hematopoietic stem cell marker, is expressed in adult human differentiated cells and certain types of kidney cancer. Cell Tissue Res 319:15–26
10. Maw MA, Corbeil D, Koch J, Hellwig A, Wilson-Wheeler JC et al (2000) A frameshift mutation in prominin (mouse)-like 1 causes human retinal degeneration. Hum Mol Genet 9:27–34
11. Fargeas CA, Joester A, Missol-Kolka E, Hellwig A, Huttner WB et al (2004) Identification of novel Prominin-1/CD133 splice variants with alternative C termini and their expression in epididymis and testis. J Cell Sci 117:4301–4311
12. Jászai J, Janich P, Farkas LM, Fargeas CA, Huttner WB et al (2007) Differential expression of Prominin-1 (CD133) and Prominin-2 in major cephalic exocrine glands of adult mice. Histochem Cell Biol 128:409–419

13. Karbanová J, Missol-Kolka E, Fonseca AV, Lorra C, Janich P et al (2008) The stem cell marker CD133 (Prominin-1) is expressed in various human glandular epithelia. J Histochem Cytochem 56:977–993
14. Shmelkov SV, Butler JM, Hooper AT, Hormigo A, Kushner J et al (2008) CD133 expression is not restricted to stem cells, and both CD133+ and CD133– metastatic colon cancer cells initiate tumors. J Clin Invest 118:2111–2120
15. Zhu L, Gibson P, Currle DS, Tong Y, Richardson RJ et al (2009) Prominin 1 marks intestinal stem cells that are susceptible to neoplastic transformation. Nature 457:603–607
16. Fargeas CA, Florek M, Huttner WB, Corbeil D (2003) Characterization of prominin-2, a new member of the prominin family of pentaspan membrane glycoproteins. J Biol Chem 278: 8586–8596
17. Yu Y, Flint A, Dvorin EL, Bischoff J (2002) AC133-2, a novel isoform of human AC133 stem cell antigen. J Biol Chem 277:20711–20716
18. Fargeas CA, Corbeil D, Huttner WB (2003) AC133 antigen, CD133, prominin-1, prominin-2, etc.: prominin family gene products in need of a rational nomenclature. Stem Cells 21:506–508
19. Fargeas CA, Fonseca AV, Huttner WB, Corbeil D (2006) Prominin-1 (CD133): from progenitor cells to human diseases. Future Lipidol 1:213–225
20. Corbeil D, Fargeas CA, Huttner WB (2001) Rat prominin, like its mouse and human orthologues, is a pentaspan membrane glycoprotein. Biochem Biophys Res Commun 285:939–944
21. McGrail M, Batz L, Noack K, Pandey S, Huang Y et al (2010) Expression of the zebrafish CD133/prominin1 genes in cellular proliferation zones in the embryonic central nervous system and sensory organs. Dev Dyn 239:1849–1857
22. Jászai J, Fargeas CA, Graupner S, Tanaka EM, Brand M et al (2011) Distinct and conserved prominin-1/CD133–positive retinal cell populations identified across species. PLoS One 6:e17590
23. Han Z, Papermaster DS (2011) Identification of three prominin homologs and characterization of their messenger RNA expression in Xenopus laevis tissues. Mol Vis 17:1381–1396
24. Zelhof AC, Hardy RW, Becker A, Zuker CS (2006) Transforming the architecture of compound eyes. Nature 443:696–699
25. Jászai J, Fargeas CA, Florek M, Huttner WB, Corbeil D (2007) Focus on molecules: prominin-1 (CD133). Exp Eye Res 85:585–586
26. Boivin D, Labbé D, Fontaine N, Lamy S, Beaulieu E et al (2009) The stem cell marker CD133 (prominin-1) is phosphorylated on cytoplasmic tyrosine-828 and tyrosine-852 by Src and Fyn tyrosine kinases. Biochemistry 48:3998–4007
27. Corbeil D, Röper K, Fargeas CA, Joester A, Huttner WB (2001) Prominin: a story of cholesterol, plasma membrane protrusions and human pathology. Traffic 2:82–91
28. Gaboriaud C, Bissery V, Benchetrit T, Mornon JP (1987) Hydrophobic cluster analysis: an efficient new way to compare and analyse amino acid sequences. FEBS Lett 224:149–155
29. Fargeas CA, Huttner WB, Corbeil D (2007) Nomenclature of prominin-1 (CD133) splice variants – an update. Tissue Antigens 69:602–606
30. Shmelkov SV, Jun L, St Clair R, McGarrigle D, Derderian CA et al (2004) Alternative promoters regulate transcription of the gene that encodes stem cell surface protein AC133. Blood 103:2055–2061
31. Pleshkan VV, Vinogradova TV, Sverdlov ED (2008) Methylation of the prominin 1 TATA-less main promoters and tissue specificity of their transcript content. Biochim Biophys Acta 1779:599–605
32. Yi JM, Tsai HC, Glöckner SC, Lin S, Ohm JE et al (2008) Abnormal DNA methylation of CD133 in colorectal and glioblastoma tumors. Cancer Res 68:8094–8103
33. Tabu K, Sasai K, Kimura T, Wang L, Aoyanagi E et al (2008) Promoter hypomethylation regulates CD133 expression in human gliomas. Cell Res 18:1037–1046
34. Baba T, Convery PA, Matsumura N, Whitaker RS, Kondoh E et al (2009) Epigenetic regulation of CD133 and tumorigenicity of CD133+ ovarian cancer cells. Oncogene 28:209–218

35. Campbell EA, O'Hara L, Catalano RD, Sharkey AM, Freeman TC et al (2006) Temporal expression profiling of the uterine luminal epithelium of the pseudo-pregnant mouse suggests receptivity to the fertilized egg is associated with complex transcriptional changes. Hum Reprod 21:2495–2513
36. Husain SM, Shou Y, Sorrentino BP, Handgretinger R (2006) Isolation, molecular cloning and in vitro expression of rhesus monkey (Macaca mulatta) prominin-1.s1 complementary DNA encoding a potential hematopoietic stem cell antigen. Tissue Antigens 68:317–324
37. Corbeil D, Joester A, Fargeas CA, Jászai J, Garwood J et al (2009) Expression of distinct splice variants of the stem cell marker prominin-1 (CD133) in glial cells. Glia 57:860–874
38. Hoegg S, Brinkmann H, Taylor JS, Meyer A (2004) Phylogenetic timing of the fish-specific genome duplication correlates with the diversification of teleost fish. J Mol Evol 59:190–203
39. Sheng M, Sala C (2001) PDZ domains and the organization of supramolecular complexes. Annu Rev Neurosci 24:1–29
40. Huttner WB, Zimmerberg J (2001) Implications of lipid microdomains for membrane curvature, budding and fission. Curr Opin Cell Biol 13:478–484
41. Zhou F, Cui C, Ge Y, Chen H, Li Q et al (2010) Alpha2,3-Sialylation regulates the stability of stem cell marker CD133. J Biochem 148:273–280
42. Sgambato A, Puglisi MA, Errico F, Rafanelli F, Boninsegna A et al (2010) Post-translational modulation of CD133 expression during sodium butyrate-induced differentiation of HT29 human colon cancer cells: implications for its detection. J Cell Physiol 224:234–241
43. Mak AB, Blakely KM, Williams RA, Penttilä PA, Shukalyuk AI et al (2011) CD133 protein N-glycosylation processing contributes to cell surface recognition of the primitive cell marker AC133 epitope. J Biol Chem 286:41046–41056
44. Hemmoranta H, Satomaa T, Blomqvist M, Heiskanen A, Aitio O et al (2007) N-glycan structures and associated gene expression reflect the characteristic N-glycosylation pattern of human hematopoietic stem and progenitor cells. Exp Hematol 35:1279–1292
45. Kemper K, Sprick MR, de Bree M, Scopelliti A, Vermeulen L et al (2010) The AC133 epitope, but not the CD133 protein, is lost upon cancer stem cell differentiation. Cancer Res 70:719–729
46. Lardon J, Corbeil D, Huttner WB, Ling Z, Bouwens L (2008) Stem cell marker prominin- 1/ AC133 is expressed in duct cells of the adult human pancreas. Pancreas 36:e1–e6
47. Immervoll H, Hoem D, Steffensen OJ, Miletic H, Molven A (2011) Visualization of CD44 and CD133 in normal pancreas and pancreatic ductal adenocarcinomas: non-overlapping membrane expression in cell populations positive for both markers. J Histochem Cytochem 59:441–455
48. Immervoll H, Hoem D, Sakariassen PO, Steffensen OJ, Molven A (2008) Expression of the "stem cell marker" CD133 in pancreas and pancreatic ductal adenocarcinomas. BMC Cancer 8:48
49. Bidlingmaier S, Zhu X, Liu B (2008) The utility and limitations of glycosylated human CD133 epitopes in defining cancer stem cells. J Mol Med 86:1025–1032
50. Fargeas CA, Karbanová J, Jászai J, Corbeil D (2011) CD133 and membrane microdomains: old facets for future hypotheses. World J Gastroenterol 17:4149–4152
51. Dubreuil V, Marzesco A-M, Corbeil D, Huttner WB, Wilsch-Bräuninger M (2007) Midbody and primary cilium of neural progenitors release extracellular membrane particles enriched in the stem cell marker prominin-1. J Cell Biol 176:483–495
52. Florek M, Bauer N, Janich P, Wilsch-Bräuninger M, Fargeas CA et al (2007) Prominin-2 is a cholesterol-binding protein associated with apical and basolateral plasmalemmal protrusions in polarized epithelial cells and released into the urine. Cell Tissue Res 328:31–47
53. Giebel B, Corbeil D, Beckmann J, Höhn J, Freund D et al (2004) Segregation of lipid raft markers including CD133 in polarized human hematopoietic stem and progenitor cells. Blood 104:2332–2338
54. Fonseca AV, Freund D, Bornhäuser M, Corbeil D (2010) Polarization and migration of hematopoietic stem and progenitor cells rely on the RhoA/ROCK I pathway and an active reorganization of the microtubule network. J Biol Chem 285:31661–31671

55. Röper K, Corbeil D, Huttner WB (2000) Retention of prominin in microvilli reveals distinct cholesterol-based lipid micro-domains in the apical plasma membrane. Nat Cell Biol 2:582–592
56. Corbeil D, Marzesco A-M, Fargeas CA, Huttner WB (2010) Prominin-1, a distinct cholesterol–binding membrane protein, and the organization of the apical plasma membrane of epithelial cells. Subcell Biochem 51:399–423
57. Janich P, Corbeil D (2007) GM1 and GM3 gangliosides highlight distinct lipid microdomains within the apical domain of epithelial cells. FEBS Lett 581:1783–1787
58. Gillette JM, Larochelle A, Dunbar CE, Lippincott-Schwartz J (2009) Intercellular transfer to signalling endosomes regulates an ex vivo bone marrow niche. Nat Cell Biol 11: 303–305
59. Taïeb N, Maresca M, Guo XJ, Garmy N, Fantini J et al (2009) The first extracellular domain of the tumour stem cell marker CD133 contains an antigenic ganglioside-binding motif. Cancer Lett 278:164–173
60. Marzesco A-M, Janich P, Wilsch-Bräuninger M, Dubreuil V, Langenfeld K et al (2005) Release of extracellular membrane particles carrying the stem cell marker prominin-1 (CD133) from neural progenitors and other epithelial cells. J Cell Sci 118:2849–2858
61. Huttner HB, Janich P, Köhrmann M, Jászai J, Siebzehnrubl F et al (2008) The stem cell marker prominin-1/CD133 on membrane particles in human cerebrospinal fluid offers novel approaches for studying CNS disease. Stem Cells 26:698–705
62. Huttner HB, Corbeil D, Thirmeyer C, Coras R, Köhrmann M et al (2012) Increased membrane shedding–indicated by an elevation of CD133-enriched membrane particles – into the CSF in partial epilepsy. Epilepsy Res 99:101–106
63. Kania G, Corbeil D, Fuchs J, Tarasov KV, Blyszczuk P et al (2005) The somatic stem cell marker prominin-1/CD133 is expressed in embryonic stem cell-derived progenitors. Stem Cells 23:791–804
64. Kaufman DS, Hanson ET, Lewis RL, Auerbach R, Thomson JA (2001) Hematopoietic colony-forming cells derived from human embryonic stem cells. Proc Natl Acad Sci USA 98:10716–10721
65. Carpenter MK, Rosler E, Rao MS (2003) Characterization and differentiation of human embryonic stem cells. Cloning Stem Cells 5:79–88
66. Hoffman LM, Carpenter MK (2005) Characterization and culture of human embryonic stem cells. Nat Biotechnol 23:699–708
67. Shamblott MJ, Axelman J, Littlefield JW, Blumenthal PD, Huggins GR et al (2001) Human embryonic germ cell derivatives express a broad range of developmentally distinct markers and proliferate extensively in vitro. Proc Natl Acad Sci USA 98:113–118
68. Levenberg S, Golub JS, Amit M, Itskovitz-Eldor J, Langer R (2002) Endothelial cells derived from human embryonic stem cells. Proc Natl Acad Sci USA 99:4391–4396
69. Ito Y, Hamazaki TS, Ohnuma K, Tamaki K, Asashima M et al (2007) Isolation of murine hair-inducing cells using the cell surface marker prominin-1/CD133. J Invest Dermatol 127:1052–1060
70. Suzuki A, Sekiya S, Onishi M, Oshima N, Kiyonari H et al (2008) Flow cytometric isolation and clonal identification of self-renewing bipotent hepatic progenitor cells in adult mouse liver. Hepatology 48:1964–1978
71. Gashaw I, Dushaj O, Behr R, Biermann K, Brehm R et al (2007) Novel germ cell markers characterize testicular seminoma and fetal testis. Mol Hum Reprod 13:721–727
72. Jászai J, Fargeas CA, Haase M, Farkas LM, Huttner WB et al (2008) Robust expression of prominin-2 all along the adult male reproductive system and urinary bladder. Histochem Cell Biol 130:749–759
73. Han Z, Anderson DW, Papermaster DS (2012) Prominin-1 localizes to the open rims of outer segment lamellae in Xenopus laevis rod and cone photoreceptors. Invest Ophthalmol Vis Sci 53:361–373
74. Bauer N, Fonseca AV, Florek M, Freund D, Jászai J et al (2008) New insights into the cell biology of hematopoietic progenitors by studying prominin-1 (CD133). Cells Tissues Organs 188:127–138

75. Bornhäuser M, Eger L, Oelschlaegel U, Auffermann-Gretzinger S, Kiani A et al (2005) Rapid reconstitution of dendritic cells after allogeneic transplantation of CD133+ selected hematopoietic stem cells. Leukemia 19:161–165
76. Pfenninger CV, Roschupkina T, Hertwig F, Kottwitz D, Englund E et al (2007) CD133 is not present on neurogenic astrocytes in the adult subventricular zone, but on embryonic neural stem cells, ependymal cells, and glioblastoma cells. Cancer Res 67:5727–5736
77. Coskun V, Wu H, Blanchi B, Tsao S, Kim K et al (2008) CD133+ neural stem cells in the ependyma of mammalian postnatal forebrain. Proc Natl Acad Sci USA 105:1026–1031
78. Mirzadeh Z, Merkle FT, Soriano-Navarro M, Garcia-Verdugo JM, Alvarez-Buylla A (2008) Neural stem cells confer unique pinwheel architecture to the ventricular surface in neurogenic regions of the adult brain. Cell Stem Cell 3:265–278
79. Uchida N, Buck DW, He D, Reitsma MJ, Masek M et al (2000) Direct isolation of human central nervous system stem cells. Proc Natl Acad Sci USA 97:14720–14725
80. Tamaki S, Eckert K, He D, Sutton R, Doshe M et al (2002) Engraftment of sorted/expanded human central nervous system stem cells from fetal brain. J Neurosci Res 69:976–986
81. Schwartz PH, Bryant PJ, Fuja TJ, Su H, O'Dowd DK et al (2003) Isolation and characterization of neural progenitor cells from post-mortem human cortex. J Neurosci Res 74:838–851
82. Lee A, Kessler JD, Read T-A, Kaiser C, Corbeil D et al (2005) Isolation of neural stem cells from postnatal cerebellum. Nat Neurosci 8:723–729
83. Bussolati B, Bruno S, Grange C, Buttiglieri S, Deregibus M et al (2005) Isolation of renal progenitor cells from adult human kidney. Am J Pathol 166:545–555
84. Richardson GD, Robson CN, Lang SH, Neal DE, Maitland NJ et al (2004) CD133, a novel marker for human prostatic epithelial stem cells. J Cell Sci 117:3539–3545
85. Leong KG, Wang BE, Johnson L, Gao WQ (2008) Generation of a prostate from a single adult stem cell. Nature 456:804–808
86. Alessandri G, Pagano S, Bez A, Benetti A, Pozzi S et al (2004) Isolation and culture of human muscle-derived stem cells able to differentiate into myogenic and neurogenic cell lineages. Lancet 364:1872–1883
87. Belicchi M, Pisati F, Lopa R, Porretti L, Fortunato F et al (2004) Human skin-derived stem cells migrate throughout forebrain and differentiate into astrocytes after injection into adult mouse brain. J Neurosci Res 77:475–486
88. Snippert HJ, van Es JH, van den Born M, Begthel H, Stange DE et al (2009) Prominin-1/CD133 marks stem cells and early progenitors in mouse intestine. Gastroenterology 136:2187–2194
89. Gregory CA, Prockop DJ, Spees JL (2005) Non-hematopoietic bone marrow stem cells: molecular control of expansion and differentiation. Exp Cell Res 306:330–335
90. Pozzobon M, Piccoli M, Ditadi A, Bollini S, Destro R et al (2009) Mesenchymal stromal cells can be derived from bone marrow CD133+ cells: implications for therapy. Stem Cells Dev 18:497–510
91. Ratajczak J, Zuba-Surma E, Klich I, Liu R, Wysoczynski M et al (2011) Hematopoietic differentiation of umbilical cord blood-derived very small embryonic/epiblast-like stem cells. Leukemia 25:1278–1285
92. Peichev M, Naiyer AJ, Pereira D, Zhu Z, Lane WJ et al (2000) Expression of VEGFR-2 and AC133 by circulating human CD34(+) cells identifies a population of functional endothelial precursors. Blood 95:952–958
93. Asahara T, Kawamoto A (2004) Endothelial progenitor cells for postnatal vasculogenesis. Am J Physiol Cell Physiol 287:C572–C579
94. Gehling UM, Ergun S, Schumacher U, Wagener C, Pantel K et al (2000) In vitro differentiation of endothelial cells from AC133-positive progenitor cells. Blood 95:3106–3112
95. Stamm C, Westphal B, Kleine HD, Petzsch M, Kittner C et al (2003) Autologous bone-marrow stem-cell transplantation for myocardial regeneration. Lancet 361:45–46
96. Sussman LK, Upalakalin JN, Roberts MJ, Kocher O, Benjamin LE (2003) Blood markers for vasculogenesis increase with tumor progression in patients with breast carcinoma. Cancer Biol Ther 2:255–256

97. Lin EH, Hassan M, Li Y, Zhao H, Nooka A et al (2007) Elevated circulating endothelial progenitor marker CD133 messenger RNA levels predict colon cancer recurrence. Cancer 110:534–542
98. Hilbe W, Dirnhofer S, Oberwasserlechner F, Schmid T, Gunsilius E et al (2004) CD133 positive endothelial progenitor cells contribute to the tumour vasculature in non-small cell lung cancer. J Clin Pathol 57:965–969
99. Dome B, Timar J, Dobos J, Meszaros L, Raso E et al (2006) Identification and clinical significance of circulating endothelial progenitor cells in human non-small cell lung cancer. Cancer Res 66:7341–7347
100. Mehra N, Penning M, Maas J, Beerepoot LV, van Daal N et al (2006) Progenitor marker CD133 mRNA is elevated in peripheral blood of cancer patients with bone metastases. Clin Cancer Res 12:4859–4866
101. Powell TM, Paul JD, Hill JM, Thompson M, Benjamin M et al (2005) Granulocyte colony-stimulating factor mobilizes functional endothelial progenitor cells in patients with coronary artery disease. Arterioscler Thromb Vasc Biol 25:296–301
102. Schatteman G (2005) Are circulating CD133+ cells biomarkers of vascular disease? Arterioscler Thromb Vasc Biol 25:270–271
103. Case J, Mead LE, Bessler WK, Prater D, White HA et al (2007) Human CD34+AC133+VEGFR-2+ cells are not endothelial progenitor cells but distinct, primitive hematopoietic progenitors. Exp Hematol 35:1109–1118
104. Timmermans F, Van Hauwermeiren F, De Smedt M, Raedt R, Plasschaert F et al (2007) Endothelial outgrowth cells are not derived from CD133+ cells or CD45+ hematopoietic precursors. Arterioscler Thromb Vasc Biol 27:1572–1579
105. Singh S, Dirks PB (2007) Brain tumor stem cells: identification and concepts. Neurosurg Clin N Am 18:31–38
106. Baersch G, Baumann M, Ritter J, Jurgens H, Vormoor J (1999) Expression of AC133 and CD117 on candidate normal stem cell populations in childhood B-cell precursor acute lymphoblastic leukaemia. Br J Haematol 107:572–580
107. Buhring HJ, Seiffert M, Marxer A, Weiss B, Faul C et al (1999) AC133 antigen expression is not restricted to acute myeloid leukemia blasts but is also found on acute lymphoid leukemia blasts and on a subset of CD34+ B-cell precursors. Blood 94:832–833
108. Horn PA, Tesch H, Staib P, Kube D, Diehl V et al (1999) Expression of AC133, a novel hematopoietic precursor antigen, on acute myeloid leukemia cells. Blood 93:1435–1437
109. Cox CV, Diamanti P, Evely RS, Kearns PR, Blair A (2009) Expression of CD133 on leukemia initiating cells in childhood ALL. Blood 113:3287–3296
110. Ricci-Vitiani L, Lombardi DG, Pilozzi E, Biffoni M, Todaro M et al (2007) Identification and expansion of human colon-cancer-initiating cells. Nature 445:111–115
111. O'Brien CA, Pollett A, Gallinger S, Dick JE (2007) A human colon cancer cell capable of initiating tumour growth in immunodeficient mice. Nature 445:106–110
112. Dalerba P, Dylla SJ, Park IK, Liu R, Wang X et al (2007) Phenotypic characterization of human colorectal cancer stem cells. Proc Natl Acad Sci USA 104:10158–10163
113. Chu P, Clanton DJ, Snipas TS, Lee J, Mitchell E et al (2009) Characterization of a subpopulation of colon cancer cells with stem cell-like properties. Int J Cancer 124:1312–1321
114. Hemmati HD, Nakano I, Lazareff JA, Masterman-Smith M, Geschwind DH et al (2003) Cancerous stem cells can arise from pediatric brain tumors. Proc Natl Acad Sci USA 100:15178–15183
115. Singh SK, Clarke ID, Terasaki M, Bonn VE, Hawkins C et al (2003) Identification of a cancer stem cell in human brain tumors. Cancer Res 63:5821–5828
116. Singh SK, Hawkins C, Clarke ID, Squire JA, Bayani J et al (2004) Identification of human brain tumour initiating cells. Nature 432:396–401
117. Salmaggi A, Boiardi A, Gelati M, Russo A, Calatozzolo C et al (2006) Glioblastoma-derived tumorospheres identify a population of tumor stem-like cells with angiogenic potential and enhanced multidrug resistance phenotype. Glia 54:850–860

118. Liu G, Yuan X, Zeng Z, Tunici P, Ng H et al (2006) Analysis of gene expression and chemoresistance of CD133+ cancer stem cells in glioblastoma. Mol Cancer 5:67
119. Bruno S, Bussolati B, Grange C, Collino F, Graziano ME et al (2006) CD133+ renal progenitor cells contribute to tumor angiogenesis. Am J Pathol 169:2223–2235
120. Rountree CB, Ding W, He L, Stiles B (2009) Expansion of CD133 expressing liver cancer stem cells in liver specific PTEN deleted mice. Stem Cells 27:290–299
121. Ding W, Mouzaki M, You H, Laird JC, Mato J et al (2009) CD133(+) liver cancer stem cells from methionine adenosyl transferase 1A-deficient mice demonstrate resistance to transforming growth factor (TGF)-beta-induced apoptosis. Hepatology 49:1277–1286
122. Zhu Z, Hao X, Yan M, Yao M, Ge C et al (2009) Cancer stem/progenitor cells are highly enriched in CD133(+)CD44(+) population in hepatocellular carcinoma. Int J Cancer 126:2067–2078
123. Eramo A, Lotti F, Sette G, Pilozzi E, Biffoni M et al (2008) Identification and expansion of the tumorigenic lung cancer stem cell population. Cell Death Differ 15:504–514
124. Chen YC, Hsu HS, Chen YW, Tsai TH, How CK et al (2008) Oct-4 expression maintained cancer stem-like properties in lung cancer-derived CD133-positive cells. PLoS One 3:e2637
125. Hermann PC, Huber SL, Herrler T, Aicher A, Ellwart JW et al (2007) Distinct populations of cancer stem cells determine tumor growth and metastatic activity in human pancreatic cancer. Cell Stem Cell 1:313–323
126. Olempska M, Eisenach PA, Ammerpohl O, Ungefroren H, Fandrich F et al (2007) Detection of tumor stem cell markers in pancreatic carcinoma cell lines. Hepatobiliary Pancreat Dis Int 6:92–97
127. Curley MD, Therrien VA, Cummings CL, Sergent PA, Koulouris CR et al (2009) CD133 expression defines a tumor initiating cell population in primary human ovarian cancer. Stem Cells 27:2875–2883
128. Collins AT, Berry PA, Hyde C, Stower MJ, Maitland NJ (2005) Prospective identification of tumorigenic prostate cancer stem cells. Cancer Res 65:10946–10951
129. Miki J, Furusato B, Li H, Gu Y, Takahashi H et al (2007) Identification of putative stem cell markers, CD133 and CXCR4, in hTERT-immortalized primary nonmalignant and malignant tumor-derived human prostate epithelial cell lines and in prostate cancer specimens. Cancer Res 67:3153–3161
130. Frank NY, Margaryan A, Huang Y, Schatton T, Waaga-Gasser AM et al (2005) ABCB5-mediated doxorubicin transport and chemoresistance in human malignant melanoma. Cancer Res 65:4320–4333
131. Klein WM, Wu BP, Zhao S, Wu H, Klein-Szanto AJ et al (2007) Increased expression of stem cell markers in malignant melanoma. Mod Pathol 20:102–107
132. Bongiorno MR, Doukaki S, Malleo F, Aricò M (2008) Identification of progenitor cancer stem cell in lentigo maligna melanoma. Dermatol Ther 21:S1–S5
133. Gedye C, Quirk J, Browning J, Svobodová S, John T et al (2009) Cancer/testis antigens can be immunological targets in clonogenic CD133(+) melanoma cells. Cancer Immunol Immunother 58:1635–1646
134. Zito G, Richiusa P, Bommarito A, Carissimi E, Russo L et al (2008) In vitro identification and characterization of CD133(pos) cancer stem-like cells in anaplastic thyroid carcinoma cell lines. PLoS One 3:e3544
135. Veselska R, Hermanova M, Loja T, Chlapek P, Zambo I et al (2008) Nestin expression in osteosarcomas and derivation of nestin/CD133 positive osteosarcoma cell lines. BMC Cancer 8:300
136. Tirino V, Desiderio V, d'Aquino R, De Francesco F, Pirozzi G et al (2008) Detection and characterization of CD133+ cancer stem cells in human solid tumours. PLoS One 3:e3469
137. Yin S, Li J, Hu C, Chen X, Yao M et al (2007) CD133 positive hepatocellular carcinoma cells possess high capacity for tumorigenicity. Int J Cancer 120:1444–1450
138. Suetsugu A, Nagaki M, Aoki H, Motohashi T, Kunisada T et al (2006) Characterization of CD133+ hepatocellular carcinoma cells as cancer stem/progenitor cells. Biochem Biophys Res Commun 351:820–824

139. Rappa G, Fodstad O, Lorico A (2008) The stem cell-associated antigen CD133 (Prominin-1) is a molecular therapeutic target for metastatic melanoma. Stem Cells 26:3008–3017
140. Bao S, Wu Q, McLendon RE, Hao Y, Shi Q et al (2006) Glioma stem cells promote radioresistance by preferential activation of the DNA damage response. Nature 444:755–760
141. Phillips HS, Kharbanda S, Chen R, Forrest WF, Soriano RH et al (2006) Molecular subclasses of high-grade glioma predict prognosis, delineate a pattern of disease progression, and resemble stages in neurogenesis. Cancer Cell 9:157–173
142. Zeppernick F, Ahmadi R, Campos B, Dictus C, Helmke BM et al (2008) Stem cell marker CD133 affects clinical outcome in glioma patients. Clin Cancer Res 14:123–129
143. Bao S, Wu Q, Sathornsumetee S, Hao Y, Li Z et al (2006) Stem cell-like glioma cells promote tumor angiogenesis through vascular endothelial growth factor. Cancer Res 66: 7843–7848
144. Calabrese C, Poppleton H, Kocak M, Hogg TL, Fuller C et al (2007) A perivascular niche for brain tumor stem cells. Cancer Cell 11:69–82
145. Grosse-Gehling P, Fargeas CA, Dittfeld C, Garbe Y, Alison MR et al (2012) CD133 as a biomarker for putative cancer stem cells in solid tumours: limitations, problems and challenges. J Pathol (2012)
146. Zhang Q, Zulfiqar F, Xiao X, Riazuddin SA, Ahmad Z et al (2007) Severe retinitis pigmentosa mapped to 4p15 and associated with a novel mutation in the PROM1 gene. Hum Genet 122:293–299
147. Yang Z, Chen Y, Lillo C, Chien J, Yu Z et al (2008) Mutant prominin 1 found in patients with macular degeneration disrupts photoreceptor disk morphogenesis in mice. J Clin Invest 118:2908–2916
148. Pras E, Abu A, Rotenstreich Y, Avni I, Reish O et al (2009) Cone-rod dystrophy and a frameshift mutation in the PROM1 gene. Mol Vis 15:1709–1716
149. Permanyer J, Navarro R, Friedman J, Pomares E, Castro-Navarro J et al (2010) Autosomal recessive retinitis pigmentosa with early macular affectation caused by premature truncation in PROM1. Invest Ophthalmol Vis Sci 51:2656–2663
150. Littink KW, Koenekoop RK, van den Born LI, Collin RW, Moruz L et al (2010) Homozygosity mapping in patients with cone-rod dystrophy: novel mutations and clinical characterizations. Invest Ophthalmol Vis Sci 51:5943–5951
151. Zacchigna S, Oh H, Wilsch-Bräuninger M, Missol-Kolka E, Jászai J et al (2009) Loss of the cholesterol-binding protein prominin-1/CD133 causes disk dysmorphogenesis and photoreceptor degeneration. J Neurosci 29:2297–2308
152. Husain N, Pellikka M, Hong H, Klimentova T, Choe KM et al (2006) The agrin/perlecan-related protein eyes shut is essential for epithelial lumen formation in the Drosophila retina. Dev Cell 11:483–493
153. Kosodo Y, Röper K, Haubensak W, Marzesco A-M, Corbeil D et al (2004) Asymmetric distribution of the apical plasma membrane during neurogenic divisions of mammalian neuroepithelial cells. EMBO J 23:2314–2324
154. Fonseca AV, Bauer N, Corbeil D (2008) The stem cell marker CD133 meets the endosomal compartment-new insights into the cell division of hematopoietic stem cells. Blood Cells Mol Dis 41:194–195
155. Lathia JD, Hitomi M, Gallagher J, Gadani SP, Adkins J et al (2011) Distribution of CD133 reveals glioma stem cells self-renew through symmetric and asymmetric cell divisions. Cell Death Dis 2:e200
156. Bauer N, Wilsch-Bräuninger M, Karbanová J, Fonseca AV, Strauss D et al (2011) Hematopoietic stem cell differentiation promotes the release of prominin-1/CD133–containing membrane vesicles – a role of the endocytic-exocytic pathway. EMBO Mol Med 3:398–409
157. Ettinger AW, Wilsch-Bräuninger M, Marzesco AM, Bickle M, Lohmann A et al (2011) Proliferating versus differentiating stem and cancer cells exhibit distinct midbody-release behaviour. Nat Commun 2:503

Chapter 2
Prominin-2 and Other Relatives of CD133

Christine A. Fargeas

Abstract Several molecules related to prominin-1/CD133, which was first characterized as a marker of mouse neuroepithelial stem cells and human hematopoietic stem cells, have been identified in various species. In mammals, a second *prominin* gene, *prominin-2*, has been identified and characterized, whereas in nonmammalian species, up to three *prominin* genes are potentially expressed. The structural similarities between prominin-1 and prominin-2 are, to some extent, reflected by their biochemical properties; both proteins are selectively concentrated in specific plasma membrane subdomains that protrude into the extracellular space and are released in small extracellular membrane vesicles. In contrast to the apically confined prominin-1, prominin-2 is distributed in a nonpolarized apico-basolateral fashion in polarized epithelial cells and appears to be expressed in separate epithelial cells. Their distinctive localization in plasma membrane protrusions is a hallmark of prominins, validating the naming of the family after its first identified member. Insights into the distinctive and/or complementary roles of the two prominins may be obtained by analyzing the evolutionary history of these proteins and the characteristics of orthologs and paralogs in more distantly related species. In addition, the characterization of prominins may shed light on the still elusive function of CD133.

Keywords CD133 • Cholesterol • Epithelial cell • Microvillus • Prominin • Vesicle

C.A. Fargeas, Ph.D. (✉)
Tissue Engineering Laboratories (BIOTEC), Technische Universität Dresden,
Tatzberg 47-51, D-01307 Dresden, Germany
e-mail: christine.fargeas@biotec.tu-dresden.de

D. Corbeil (ed.), *Prominin-1 (CD133): New Insights on Stem & Cancer Stem Cell Biology*, Advances in Experimental Medicine and Biology 777, DOI 10.1007/978-1-4614-5894-4_2,

2.1 Introduction

In spite of the large body of studies on prominin-1/CD133 as a stem cell marker, its function in physiological processes as well as in cancer still remains elusive. I propose to possibly gain insights from a different perspective by presenting the prominin family.

Prominin-1 was first identified 15 years ago as a novel antigenic marker that is present at the apical surface of mouse neuroepithelial cells [1], while its human counterpart, the AC133 antigenic marker, is expressed in a subset of hematopoietic stem and progenitor cells [2, 3]. As described in the first chapter of this book, the intrinsic preference of the protein for plasma membrane protrusions motivated the choice of "prominin" (from the Latin word *prominere*) as the name for the first characterized member of this pentaspan membrane glycoprotein family, which occurs throughout the animal kingdom [4]. Prominin-1-like sequences were soon identified in other vertebrates such as fish and chicken [5, 6] and in invertebrates including worms and flies [1, 5, 7, 8]. Furthermore, the *PROMININ-1* (*PROM1*) gene was found to be the host of mutations that cause retinal degeneration [8] (see Chap. 4).

A second pentaspan membrane glycoprotein structurally related to prominin-1 but encoded by a distinct gene was characterized a few years later in human and rodent through molecular cloning [5]. The tissue distribution profile of this second prominin molecule, named prominin-2, largely overlapped that of prominin-1, with the notable exception of the eye, in which prominin-2 is absent [5]. The existence of prominin-2 suggested that a potential functional redundancy may explain the lack of pathological signs other than retinal phenotype observed in individuals carrying mutations in the *PROM1* gene [8], in spite of the widespread expression of prominin-1 throughout the organism as revealed by Northern blot analysis [1, 3, 5–7, 9]. Further characterization of prominin-2 has since revealed that prominin-1 and prominin-2 harbor similar and distinct cellular specificities.

2.2 Protein and Gene Structure of Prominin-2

Prominin-2 was originally cloned from human, mouse, and rat kidneys [5]. In spite of its low amino acid sequence identity with prominin-1 (26–30% overall identity depending on the species), this 112-kDa glycoprotein has the same pentaspan membrane topology as prominin-1, with an N-terminal domain that is exposed to the extracellular milieu (EC1) and is followed by four alternating, short cytoplasmic (IC1 and IC2) and large glycosylated extracellular loops (EC2 and EC3) and a cytoplasmic C-terminal domain (IC3) (see Fig. 1.1 in Chap. 1, Fig. 2.1). The two extracellular loops contain approximately 250 residues each, and both display potential N-glycosylation sites [5]. Prominin-2 shows no obvious sequence homology to other known proteins except prominin-1.

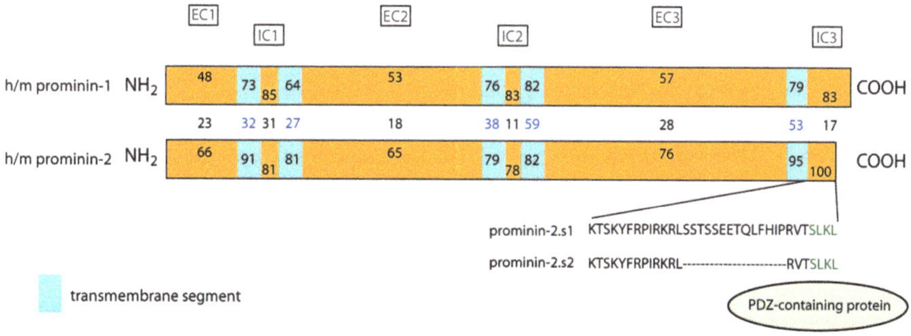

Fig. 2.1 Comparative structure of prominin-1 and prominin-2. Human (*h*) and mouse (*m*) prominin-1 (*top box*) and prominin-2 (*bottom box*) are each comprised of three extracellular domains (EC) and three short intracellular domains (IC) (*orange*) separated by five transmembrane domains (*blue*). Numbers in domains indicate the percentage of amino acid identity between human and mouse orthologs. Numbers between the different domains of prominin-1 and prominin-2 indicate the percentage of identity between the two human paralogs. The sequence of the C-terminal domain is indicated for the two splice variants, prominin-2.s1 and prominin-2.s2, with the putative PDZ-binding site in green (This figure is adapted from Ref. [6])

Prominin-2 shows a higher degree of interspecies conservation than prominin-1 (e.g., 73% and 60%, respectively, between human and mouse) [5]. Strikingly, the cytoplasmic C-terminal tails of human, mouse, and rat prominin-2 are perfectly conserved (100% amino acid identity over 34 residues) and quite distinct from the corresponding domain in prominin-1 (e.g., 17% amino acid identity compared to 26% overall identity between human prominin-1 and prominin-2; see Fig. 2.1) [6, 10]. Moreover this domain is encoded by only two exons, whereas up to four may be involved in the case of prominin-1. A search of predicted genes in the databases at NCBI (National Center for Biotechnology Information) revealed that the last 20 amino acid residues of prominin-2, which are encoded by a single exon (815–834 of human prominin-2), are perfectly conserved among mammals including therians (placentals and marsupials) and monotremes (*Ornithorhynchus anatinus*), with some exceptions (e.g., cat, *Felis catus*; baboon, *Papio anubis*; elephant, *Loxodonta africana*) in which only a single amino acid substitution is present (C.A.F., unpublished data). The conservation rate is approximately 80% in other amniotes (e.g., *Meleagris gallopavo, Gallus gallus breed, Anolis carolinensis, Chrysemys picta*) (not shown) and more divergent in the amphibian *Xenopus laevis* [11]. This sequence conservation suggests that the cytoplasmic tail of prominin-2 is likely important for the function of the molecule.

Interestingly, the C-terminus of mammalian prominin-2 (S-L-K-L) conforms to the pattern of class II PSD95/Dlg1/ZO-1 (PDZ)-binding domains (X-Φ/ψ-X-Φ; X, unspecified amino acids; Φ, hydrophobic residue; ψ, aromatic residue) [12, 13], and this sequence was shown to function as a PDZ-binding site in vitro. By means of a yeast two-hybrid screen using the last 31 residues of the C-terminal domain of prominin-2 as bait, Kathrin Opherk, in the research group of Denis Corbeil (Dresden), has identified a novel splice variant of the glutamate receptor-interacting protein (GRIP1), a multi-PDZ-containing protein involved in anchoring AMPA receptors at the synapse [14], as an interacting partner of prominin-2 [15]. Named GRIP1κ (GenBank accession numbers, AY294283, AY255674), this 4-PDZ-domain splice variant is similar (only differing in its 11 C-terminal residues) to GRIP1τ, a testis-specific isoform with a distinct nuclear distribution in addition to the usual cytoplasmic location of GRIP1 [15, 16]. Further studies are needed to determine the physiological relevance of this putative interaction in the function and intracellular trafficking of prominin-2.

The *PROMININ-2* (*PROM2*) gene is located on chromosome 2 in humans (locus p16.2-p12) and mice (2F1), immediately upstream of the *calsenilin/DREAM/KChIP3* (*KCNIP3*) gene, which overlaps with the 3′ end of *PROM2*. The *PROM2* gene extends over 16 kb with nearly identical genomic structure, i.e., exon/intron boundaries, among mammals [5]. Remarkably, although the amino acid identity between prominin-1 and prominin-2 is very low (see above), their exon/intron organization is highly similar. Most introns are concordant in position and phase, suggesting an early gene duplication event. Nevertheless, this genomic architecture is not correlated with the structural domains of the protein because the coding region spans at least 23 exons [5].

The development of high-throughput sequencing has revealed that most human gene undergo alternative splicing [17]. Prominin-2 is no exception. Like prominin-1, it may be affected by alternative splicing in the N-terminal domain, the first and second extracellular loops or the cytoplasmic C-terminal domain, suggesting the existence of distinct cytoplasmic protein-interacting partners [5]. The prominin-2 splice variants occur through intron retention, exon skipping, or the use of an internal 3′ acceptor site. In particular, exon 23, which encodes the final 20 C-terminal residues, harbors an alternative acceptor site that generates in mouse salivary glands the isoform prominin-2.s2 [4], which has a 14-amino-acid deletion that spares the last 7 residues of the C-terminal domain and thus the potential PDZ-binding site, but removes putative casein kinase II and cAMP- or cGMP-dependent protein kinase phosphorylation sites (Fig. 2.1). Yet, these putative phosphorylation sites occur with high frequency and may therefore be not significant. Several human *PROMININ-2* mRNA variants lacking all or part of exons 5, 6, or 7 (encoding part of EC2) have been amplified by polymerase chain reaction (PCR) (C.A.F., unpublished data). Evidence of the differential splicing of exon 2 (EC1), exons 3 plus 4 (TM1/IC1/TM2), and exon 15 (EC3) and intron retention after exon 16 is present in the databases as expressed sequence tags (EST; GenBank accession number BB617607, CB994746, DA494079, and DA441226). The consequences of these splicing events for the structure of the molecule would vary from short internal deletion to the generation of short proteins with only two transmembrane domains or with an unrelated C-terminus (C.A.F., unpublished data). In human *PROM2*, alternative splicing also affects the 3′-untranslated region (UTR), generating two main mRNA species of 5.0 and 4.2 kb, which are detectable by Northern blot in most tissues expressing prominin-2 [5].

2.3 Cell Biology of Prominin-2

To some extent, the cellular and biochemical properties of prominin-2 reflect its structural similarities with prominin-1.

2.3.1 *Subcellular Localization*

The subcellular localization and cellular trafficking of prominin-2 were investigated by means of a green fluorescent protein (GFP) fusion protein ectopically expressed in Chinese hamster ovary (CHO) and, as a model of polarized epithelial cells, Madin-Darby canine kidney (MDCK) cells [5, 18]. Upon co-transfection with prominin-1, both prominin molecules were found to co-localize in plasma membrane protrusions of CHO cells, indicating that this intrinsic characteristic of prominin-1 is shared with prominin-2 [5]. Like prominin-1 [19, 20], prominin-2 was found in plasma membrane protrusions emerging from the apical domain of MDCK cells [18]. In these protrusions, prominin-2 was observed at the membrane of microvilli and the primary cilium, often concentrated toward their tips, and absent from the neighboring planar areas [18]. GFP-prominin-2 also labeled primary cilia emerging from nearby cells, which engage in long-lasting contacts [18, 21]. Unexpectedly, prominin-2 was also detected at the basal and lateral domains of MDCK cells, in contrast to the exclusive localization of prominin-1 at the apical domain [19]. As revealed by electron microscopy, prominin-2 was concentrated in interdigitated processes growing out from the lateral membrane between adjacent cells and in a few membrane protrusions present at the basal domain. This nonpolarized distribution in polarized epithelial cells may reflect either the absence of apical targeting signal or the presence of two competing signals for each domain [18]. Interestingly, the N-glycan moieties of prominin-2 appear insufficient to mediate an exclusively apical sorting, in contrast to what was observed for other glycoproteins [22, 23].

2.3.2 *Cholesterol Binding and Membrane Microdomain*

In a similar experimental setting that allowed the demonstration of the specific interaction of prominin-1 with membrane cholesterol and membrane microdomains (lipid raft) [24] (for technical details, see Ref. [25]), prominin-2 was shown to interact directly with plasma membrane cholesterol and associate with a cholesterol-dependent membrane microdomain [18]. These biochemical properties may determine the observed preference of prominin-2, like that of prominin-1, for membrane protrusions, irrespective of their subcellular localization.

Table 2.1 Distinctive traits and common features of prominins

Prominin-1	Prominin-2
Strictly apically localized in polarized cells	Apico-basolaterally distributed in polarized cells
Expressed in non-epithelial cells	Singly expressed in the esophagus and the meibomian glands
Singly expressed in the retina	
Used as a stem cell marker	
Common features	
Gene architecture	
Protein structure	
Membrane topology	
Biochemical characteristics	
• Associates with membrane microdomains/interacts with membrane cholesterol	
• Localizes in plasma membrane protrusions	
• Associates with extracellular membrane microvesicles	
Expression in epithelial cells	

2.3.3 Membrane Vesicles

Prominin-2 is associated with small membrane vesicles, as is prominin-1 [26] (see Chap. 3). Upon transfection in MDCK cells, prominin-1 and prominin-2 (as a GFP fusion protein) could be detected in the same extracellular membrane vesicles released into the apical medium, suggesting common intracellular trafficking [18]. Interestingly, in line with its unpolarized (apico-basolateral) distribution and in contrast to prominin-1, prominin-2-containing membrane vesicles were also released into the basolateral medium [18]. In vivo prominin-2-containing membrane vesicles can be isolated from both mouse and human urine [18, 27] and human saliva [28] by high-speed centrifugation.

Collectively, these investigations have demonstrated that prominin-2, like prominin-1, is (i) selectively concentrated in plasma membrane protrusions, (ii) associated with cholesterol-dependent membrane microdomains through direct interaction with membrane cholesterol, and (iii) released into the extracellular milieu in association with small membrane vesicles. Moreover, as a distinctive feature, prominin-2 is distributed in a nonpolarized apico-basolateral manner at the membrane and in extracellular vesicles (Table 2.1). Therefore, prominin-2 could potentially function as an organizer of plasma membrane protrusions similar to prominin-1 (see Chaps. 1 and 4), particularly in the basolateral compartment, in which prominin-1 is absent.

2.4 Prominin-2 Expression

2.4.1 Tissue Distribution

Although prominin-1 and prominin-2 may assume common functions owing to their shared molecular properties, they seem to have their own specificity, as reflected by their differential tissue distribution (Table 2.1). Whereas prominin-1 is expressed in both epithelial and non-epithelial cells, prominin-2 expression seems rather restricted to epithelial cells [5, 27–29]. Northern blot analyses indicate that the highest human *prominin-2* mRNA levels occur in the adult kidney (like prominin-1), while detectable levels occur in the placenta, mammary gland, prostate, trachea, thyroid gland, salivary gland, and all of the tissues of the digestive tract [5]. Notably, *prominin-2* transcripts are strongly and uniquely expressed in the acini of the meibomian gland, the root sheath of the eyelash/cilium [28], and in the esophagus [5]. Conversely, a major site for prominin-1 expression, the retina, is devoid of prominin-2 [5]. Assuming that, under some conditions, the presence of both prominins may be redundant, this latter observation would at least partially explain why the effect of deleterious mutations in *PROM1* is specific to the eye despite the more widespread distribution of prominin-1 expression [8].

Prominin-2 has been detected in different segments of the male genitourinary tract, including the epididymis, the urothelium of the urinary bladder, and the glandular epithelium of the seminal vesicle, where it is more strongly and uniformly expressed than prominin-1 [29]. Prominin-2 is strongly expressed in the rodent and human prostate [5, 29, 30], and in the latter species, it specifically marks basal epithelial cells [29], which are proposed to give rise to secretory luminal cells [31] (see Chap. 11). In organs in which the expression of both prominins appears to overlap based on dot and Northern blots, refinement of their respective expression profiles through immunohistochemistry and in situ hybridization has shown that they mostly occupy distinct tissue compartments [27–29]. Thus, in the human nephron, *prominin-2* transcripts and protein were confined to the basolateral plasma membrane of the epithelial cells of the distal tubules, including the distal convoluted tubule, the connecting duct, and the collecting duct system. In the thick ascending limb of Henle's loop, prominin-2 was also found at the apical domain, whereas prominin-1 expression featured an exclusive apical localization in epithelial cells of the proximal nephron tubules [1, 32] (see Chap. 8). Interestingly, this differential expression may partly explain the apparent quantitatively weaker expression of prominin-2 in urine vesicles compared to prominin-1, although these prominin-containing membrane vesicles may be derived from other epithelia along the urinary tract. Nevertheless, these findings are of note in the context of a recent report on the proteomic analysis of urinary exosomes from normal individuals and IgA nephropathy and thin basement membrane nephropathy (TBMN) patients. Both prominin-1 and prominin-2 were upregulated in pathological samples, and prominin-2 was uniquely associated with TBMN, which is characterized by a general thinning of the glomerular basement membrane [33].

A similar situation prevails in murine salivary glands, where both prominins are expressed following distinctive patterns: prominin-1 is preferentially contained in one segment of the duct system, whereas prominin-2 is localized in acinar cells or duct cells or both compartments, depending on the particular gland [28].

Extraocular neuromuscular junctions constitute a particular type of neuromuscular junction with specific physiological properties and subcellular organization, which were recently shown to be reflected by a unique transcriptome and proteome among skeletal muscles [34]. In this study, rat prominin-2 was preferentially expressed in the subsynaptic compartment over the non-synaptic regions of the extraocular muscle, potentially contributing to its differential postsynaptic morphology.

2.4.2 Regulation of Prominin-2 Expression

The factors that regulate the expression of prominin-2 remain largely unknown. An in silico comparative analysis of the promoter of the human and mouse *Prom2* genes with ConSite [35] has revealed two regions that contain conserved predicted transcription factor binding sites: one within 750 bp upstream from the first initiation codon (ATG) and another in the 5′-flanking region between nucleotides -1750 and -2850 (Denis Corbeil and C.A.F., unpublished data). These binding sites include Snail, Spz1, HFH-2 (HNF-3/Forkhead Homolog), and E74A [36, 37], which is consistent with the tissue-specific and epithelial expression of prominin-2 and would support a relation with the maintenance of the proper architecture of epithelia [38].

Recently, *prom2* was reported as a nonclassical heat-shock gene [39]. Mouse HSF1 (heat-shock factor 1) is required for the expression of classical heat-shock genes in mouse embryonic fibroblasts (MEFs) [40]. After 6 h of recovery from heat shock, *prominin-2* mRNA was upregulated in MEFs, but this upregulation was not observed in HSF1-null MEF cells, which are deficient in the induced expression of the classical heat-shock genes. However, overexpression of a newly characterized HSF relative, mouse HSF3 (or chicken HSF1, its homolog in this species) by transfection restored the inducibility of *prominin-2* mRNA without affecting the expression of the classical genes that mediate heat-shock responses [39]. It would be interesting to investigate this issue further at the protein level.

Rat prominin-2 was first reported to be a testosterone-regulated gene based on its decreased expression in the rat ventral prostate after castration and restored expression upon hormonal replacement [30]. The confinement of prominin-2 expression to the basal compartment of the human prostate epithelium [29] suggests that this hormonal regulation may be indirect, given the dissociated distribution of prom-2 and the androgen receptor (AR) within the prostatic epithelium [41]. Therefore, the hormonal regulation of prominin-2 clearly deserves further investigation. In the same study, Zhang and colleagues attributed pro-apoptotic activity to rat prominin-2 in the prostate based on the expression of a GFP fusion protein construct [30]. However, the evidence provided was questionable, given that this construct would not permit the expression of a prominin-2 fusion protein because of the presence of a stop codon and a single nucleotide insertion in the cDNA, which would introduce a frame shift and the expression of a different amino acid sequence at the C-terminus. These changes resulted in an unusual intracellular localization of the fluorescent signal. Moreover, it is difficult to reconcile this activity with the downregulation of *prominin-2* steady-state mRNA levels upon castration and the concomitant involution of the prostate and upregulation upon androgen replacement that was observed in the same study. In view of the data, the possibility that the transfected constructs had a toxic effect cannot be formally excluded. Such a pro-apoptotic effect has not been observed in other heterologous systems overexpressing mouse or rat prominin-2 [5, 18], although it was not specifically addressed in these studies, and should therefore be considered with caution until confirmation.

2.4.3 Prominin-2 and Cancer

Data regarding the expression of prominin-2 in cancer are scarce and preliminary. Expressed sequence tag clones deposited in the GenBank database indicate that prominin-2 may be expressed in tumors of the breast, lung, tongue, and nervous system [5] (data not shown). A microarray gene profiling and quantitative PCR study suggested *PROM2* as a candidate gene marker for segregating chromophobe renal cell carcinoma from benign oncocytoma [42].

Using a pangenomic oligonucleotide microarray, Winnepenninckx and colleagues observed that the expression of PROMININ-2 was higher in melanomas from patients that did not metastasize to distant sites within 4 years after diagnosis than in those that did, suggesting an anti-metastatic role for this gene [43].

In a recent publication, Duhagon and colleagues proposed that prominin-2 could be a new marker for prostate cancer based on genomic profiling of tumor-initiating prostatospheres (PS) derived from the cancer cell line LNCaP as well as primary prostate cancer stem cells isolated from recurrent prostate tumors through the expression of CD133 [44]. Because prominin-2 is expressed in the basal compartment of the prostate, the authors specifically addressed the expression of prominin-2 and observed its upregulation in LNCaP-derived PS relative to the parental cells. Similar results were obtained with primary tumor-derived PS and confirmed by quantitative PCR in one case of three. Further studies will be needed with a larger panel of samples to determine the potential of prominin-2 as a biomarker for prostate cancer status.

No mutational or pathological relevance that could indicate a physiological function for prominin-2 has been reported, and no phenotypic features of a prominin-2 knockout model have been described.

2.5 Evolution of the Prominin Gene Family

The first phylogenetic analysis of the prominin family performed after the characterization of human and rodent prominin-2 led to the definition of two orthologous groups of mammalian prominin genes [5]. However, none of the complete or partial prominin-related sequences identified at that time in other vertebrate (*Gallus gallus, Danio rerio*) or invertebrate (*Drosophila melanogaster, Caenorhabditis elegans*) phylogenic lineages were clearly related to one or the other group, in particular, *G. gallus* prominin-like (GenBank accession number AF406812) [5]. Advances in whole-genome sequencing revealed that in *G. gallus*, prominin-1 and predicted prominin-2 orthologous genes are located on chromosomes 4 and 22, respectively, while the prominin-like gene is on chromosome 6. Therefore, birds may harbor a third prominin that is more similar to one of the three prominin paralogs identified in zebra fish (*Danio rerio*). In 2008, Wotton and colleagues demonstrated that two of the prominin paralogs identifiable in teleosts clearly belong to the *prom1* group, while the third prominin paralog segregates from the tetrapod *prom2* genes to form the *prom3* orthologous group. They proposed that the *prom3* genes may have evolved from the same ancestral locus as *prom2*, which was generated during the first round of vertebrate genome duplication [45]. Thus, the *Danio rerio* paralogs located on chromosomes 14, 1, and 13 are now referred to as prominin-1a, prominin-1b, and prominin-3, respectively [46, 47]. The recent identification of prominin-1, prominin-2, and prominin-3 in *Xenopus laevis* has confirmed the presence of the three orthologous groups in nonmammalian vertebrates [11]. Moreover, the vast array of whole-genome sequencing projects has enabled the identification or prediction of prominin sequences in many species. The elephant shark (*Callorhinchus milii*) genome, which often exhibits higher homology to human sequences than teleosts, includes a significant number of orthologs to human (mammals) genes that are absent from teleosts [48]. This seems to be the case for *prom2*; partial genomic sequences of homologs of each of the three vertebrate family members can be retrieved through TBLASTN searches against mouse or human prominin-2 sequences from the whole-genome shotgun database (WGS). One genomic contig, AAVX01036519, contains a fragment of a *prominin* gene that is more highly related to human prominin-2 than to tetrapod or teleost prominin-1 or prominin-3 sequences. In addition, a spiny dog shark EST sequence (EE886567) was retrieved from the database based on its high sequence homology with mammalian prominin-1 by a TBLASTN search. This sequence encodes the 219 C-terminal residues of a prominin-1 ortholog (Table 2.2). These sequence relationships confirm that the three distinct orthologous groups emerged very early in the vertebrate lineage. Thus, while prominin-2 appears to have been lost in the fish lineage after the radiation of ray-finned fishes and the concomitant whole-genome duplication [49] that resulted in duplicated *prominin-1* genes, prominin-3 vanishes in the mammalian lineage. However, in the platypus (*O. anatinus*) genome, where *prom1* and *prom2* have been predicted, fragments of a gene with similarity to *prom3* can be identified on chromosome 17 by a TBLASTN search. Whether these sequences mark an intermediate stage in

Table 2.2 Existence of three prominin orthologous groups in cartilaginous fishes – comparison of predicted sequences with bony vertebrate prominin members

Bony Vertebrate Prominin Members (Gene ID)	Spiny dogfish shark*	Elephant shark§		
		A	B	C
Human prominin-1 (8842 PROM1)	50/93 (54)	48/93 (52)	34/86 (39)	41/92 (45)
Chicken prominin-1 (422825 PROM1)	53/93 (57)	59/93 (63)	29/82 (35)	43/92 (47)
Clawed Frog prominin-1 (100036621 prom1)	52/93 (56)	54/93 (58)	29/86 (34)	42/92 (46)
Zebrafish prominin-1a (322857 prom1a)	55/92 (60)	57/92 (62)	33/81 (41)	42/91 (46)
Zebrafish prominin-1b (378834 prom1b)	54/92 (59)	55/92 (60)	31/85 (36)	41/91 (45)
Human prominin-2 (150696 PROM2)	34/79 (43)	35/81 (43)	54/84 (64)	37/78 (47)
Chicken prominin-2 (428200 PROM2)	38/89 (43)	36/90 (40)	46/82 (56)	37/82 (45)
Clawed Frog prominin-2 (100485371 prom2)	27/78 (35)	27/78 (35)	44/84 (52)	28/81 (35)
Chicken prominin-3 (423849 LOC423849)	47/90 (52)	45/90 (50)	37/81 (46)	50/90 (56)
Clawed Frog prominin-3 (100488690 LOC100488690)	45/91 (49)	42/87 (48)	35/85 (41)	47/90 (52)
Zebrafish prominin-3 (556596 prom2)	40/91(44)	42/91 (46)	32/84 (38)	53/91 (58)

*EST sequence EE886567, 657 bp in length, from the spiny dogfish shark (*Squalus acanthias*), encoding the last 216 C-terminal amino acids of the predicted protein and corresponding to a C-type prominin-1 C-terminus [47] (see Chap. 1)

§The three fragments (A, B, and C) of elephant shark (*Callorhinchus milii*) prominin sequences corresponding to exons 20 to 22 of human PROMININ-2 were determined by assembling exonic regions from contig AAVX01091526, the overlapping contigs AAVX01230395 and AAVX01596446, and contig AAVX01036519, respectively, which were previously identified by TBLASTN searches of the GenBank WGS database against human and zebra fish prominin sequences. Introns conform in position and phase to the structure of other prominin genes, and splice sites follow the GT-AG rule

Sequences identity is indicated as the fractional number of exact matches in pairwise alignments with BLAST, with corresponding percentages in parentheses and following the code: red: ≥60, pink: 50–59, green: 40–49, blue: ≤40

mammalian evolution or indicate that monotremes have maintained a functional prominin-3 like many other ancestral characteristics remains to be determined [50]. The presence of three different paralogs in other tetrapods such as birds, reptiles, and amphibians further supports the denomination prominin-3 for the fish paralog that currently appears again as *prom2* gene in the database, after being properly annotated *prom3*.

It is interesting to note that the first orthologous group *prom1*, which is present throughout the different jawed vertebrate lineages, has evolved with longer intronic sequences. These longer intronic sequences differentiate *Prom1* genes from other *Prom* genes by generating an approximately sevenfold lower degree of compactness, depending on the species. This reduction in compactness may be related to a tighter regulation of the expression of the first group [51], in agreement with the commonly observed complex regulation of *prominin-1* gene expression (see Chap. 5).

Table 2.3 Phylogenetic profile of prominins

Human	Platypus	Chicken	Anole	Frog	Zebrafish	Sharks	Drosophila	C. elegans
CD133[a] (8842 PROM1, *chr 4*)	*Predicted prom1* (100084810 PROM1, *chr 4*)	**Prominin-1**[b] (422825 PROM1, *chr 4*)	*AAWZ02006824* (*chr 4*)	**Prominin-1**[b] (100036621 prom1)	**Prominin-1a**[b] (322857 prom1a, *chr 14*) / **Prominin-1b**[b] (378834 prom1b, *chr 1*)	EST 886567 *AAVX01091526*[c]	**Prominin**[b] (246493, *chr 2R*)	**F08B12.1** (181330, *chr X*)
Prominin-2 (150696 PROM2, *chr2*)	*Predicted prom2 (100078032 LOC100078032 unknown)*	*Predicted prominin-2* (428200 PROM2, *chr22*)	Prominin-2 (100564795 LOC100564795, *LGa*)	Prominin-2 (100485371 prom2)		*AAVX01036519*[d]	**Prominin-like** (38372, *chr 3L*)	M28.8 (174641, *chr II*)
	NC_009110 (*chr 17*)	Prominin-3 (423849 LOC423849, *chr 6*)	*AAWZ02021368* (*unknown*)	Prominin-3 (100488690 LOC100488690)	Prominin-3 (556596 prom2, *chr 13*)	*AAVX01230395, AAVX01596446*[e]		M28.9 (174642, *chr II*)
PLACENTALS	MONOTREMES							
MAMMALS		BIRDS	LIZARDS					
AMNIOTES				AMPHIBIANS				
TETRAPODS					RAY-FINNED FISHES			
BONY VERTEBRATES						CARTILAGINOUS FISHES		
VERTEBRATES							ARTHROPODS	NEMATODES

Prominin members are indicated for each species; the GenBank gene ID and chromosomal location are given in parentheses. Bold, as characterized/expressed protein. Regular, as EST or mRNA. Italic, genomic identification. Green, orange, and yellow shadings indicate membership in orthologous groups 1, 2, and 3, respectively. Phylogeny is indicated at the bottom of the table.

*See Chap. 1

§See Chap. 4

[a]AAVX01091526, a 1432-nucleotide (nt) WGS contig from *Callorhinchus milii* displaying four exons of the putative *prom1* gene.

[b]AAVX01036519, a 3032-nt contig from *Callorhinchus milii* displaying three exons of the putative *prom2* gene.

[c]AAVX01230395, a 1892-nt WGS contig from *Callorhinchus milii* displaying two exons of the putative *prom3* gene and AAVX01596446, an overlapping 799-nt WGS contig displaying two exons (see Table 2.2). Inferred from previous publications [5, 45, 46, 55] and personal unpublished results.

Beyond jawed vertebrates, the evolutionary relationship of prominins is more difficult to establish. Evidence in the EST database supports the existence of at least two different prominins in the sea lamprey, both of which are more similar to prominin-1 and prominin-3 than to prominin-2. As for invertebrates, EST and partial genomic sequences demonstrate that more than one prominin gene is present in echinoderms and mollusks. In addition to F08B12.1, the nematode *C. elegans* harbors two prominin-related molecules (M28.8 and M28.9) that are more distantly related sequences [1, 7], whereas two *prominin* genes have been identified in *D. melanogaster* at distinct locations (2R and 3L) and predicted in other insects [8, 52, 53]. Consequently, the evolutionary and functional relationships of these genes are still unclear. Table 2.3 summarizes the phylogenic profile of the prominin family members.

Nevertheless, the propensity to be concentrated in plasma membrane protrusions that is characteristic of mammalian prominin-1 [9, 19, 24] and prominin-2 [5, 18] and that gave its name to these glycoproteins extends to other members of the family across evolution. F08B12.1 was proposed to be a potential ciliary gene through serial analysis of gene expression (SAGE) performed with ciliated and nonciliated *C. elegans* neuronal cells [54]. Prominin may have evolutionarily ancient functions in vision because *Drosophila* prominin is localized to the stalk membrane and the tips of the microvillus in rhabdomeres (see Chap. 4) [52].

2.6 Conclusions

Members of the prominin family are spread throughout the animal kingdom and are still poorly characterized with the exception of human and rodent prominins, for which common features and distinctive traits have been documented. The prominin family has been enlarged by the definition of an additional orthologous group, but comparative studies of these prominin relatives and their thorough biological characterization are needed. Combined evolutionary perspectives may yield interesting insights for future research aimed at deciphering the different functions of this intriguing family of proteins and that of its most well-known member: CD133.

Acknowledgements I would like to thank Denis Corbeil for sharing unpublished data.

References

1. Weigmann A, Corbeil D, Hellwig A, Huttner WB (1997) Prominin, a novel microvilli-specific polytopic membrane protein of the apical surface of epithelial cells, is targeted to plasmalemmal protrusions of non-epithelial cells. Proc Natl Acad Sci USA 94:12425–12430
2. Yin AH, Miraglia S, Zanjani ED, Almeida-Porada G, Ogawa M et al (1997) AC133, a novel marker for human hematopoietic stem and progenitor cells. Blood 90:5002–5012

3. Miraglia S, Godfrey W, Yin AH, Atkins K, Warnke R et al (1997) A novel five-transmembrane hematopoietic stem cell antigen: isolation, characterization, and molecular cloning. Blood 90:5013–5021
4. Fargeas CA, Corbeil D, Huttner WB (2003) AC133 antigen, CD133, prominin-1, prominin-2, etc.: prominin family gene products in need of a rational nomenclature. Stem Cells 21: 506–508
5. Fargeas CA, Florek M, Huttner WB, Corbeil D (2003) Characterization of prominin-2, a new member of the prominin family of pentaspan membrane glycoproteins. J Biol Chem 278: 8586–8596
6. Corbeil D, Röper K, Fargeas CA, Joester A, Huttner WB (2001) Prominin: a story of cholesterol, plasma membrane protrusions and human pathology. Traffic 2:82–91
7. Corbeil D, Röper K, Weigmann A, Huttner WB (1998) AC133 hematopoietic stem cell antigen – human homologue of mouse kidney prominin or distinct member of a novel protein family. Blood 91:2625–2626
8. Maw MA, Corbeil D, Koch J, Hellwig A, Wilson-Wheeler JC et al (2000) A frameshift mutation in prominin (mouse)-like 1 causes human retinal degeneration. Hum Mol Genet 9:27–34
9. Corbeil D, Röper K, Hellwig A, Tavian M, Miraglia S et al (2000) The human AC133 hematopoietic stem cell antigen is also expressed in epithelial cells and targeted to plasma membrane protrusions. J Biol Chem 275:5512–5520
10. Corbeil D, Fargeas CA, Huttner WB (2001) Rat prominin, like its mouse and human orthologues, is a pentaspan membrane glycoprotein. Biochem Biophys Res Commun 285:939–944
11. Han Z, Papermaster DS (2011) Identification of three prominin homologs and characterization of their messenger RNA expression in Xenopus laevis tissues. Mol Vis 17:1381–1396
12. Harris BZ, Lim WA (2001) Mechanism and role of PDZ domains in signaling complex assembly. J Cell Sci 114:3219–3231
13. Sheng M, Sala C (2001) PDZ domains and the organization of supramolecular complexes. Annu Rev Neurosci 24:1–29
14. Dong H, O'Brien RJ, Fung ET, Lanahan AA, Worley PF et al (1997) GRIP: a synaptic PDZ domain-containing protein that interacts with AMPA receptors. Nature 386:279–284
15. Opherk KN (2007) Identification of the novel PDZ domain containing splice variant GRIP1k-a potential interaction partner of Prominin-2. Dissertation, Dresden Medizinische Fakultät Carl Gustav Carus, Technische Universität Dresden, pp 200
16. Nakata A, Ito T, Nagata M, Hori S, Sekimizu K (2004) GRIP1tau, a novel PDZ domain-containing transcriptional activator, cooperates with the testis-specific transcription elongation factor SII-T1. Genes Cells 9:1125–1135
17. Keren H, Lev-Maor G, Ast G (2010) Alternative splicing and evolution: diversification, exon definition and function. Nat Rev Genet 11:345–355
18. Florek M, Bauer N, Janich P, Wilsch-Bräuninger M, Fargeas CA et al (2007) Prominin-2 is a cholesterol-binding protein associated with apical and basolateral plasmalemmal protrusions in polarized epithelial cells and released into the urine. Cell Tissue Res 328:31–47
19. Corbeil D, Röper K, Hannah MJ, Hellwig A, Huttner WB (1999) Selective localization of the polytopic membrane protein prominin in microvilli of epithelial cells – a combination of apical sorting and retention in plasma membrane protrusions. J Cell Sci 112:1023–1033
20. Janich P, Corbeil D (2007) GM1 and GM3 gangliosides highlight distinct lipid microdomains within the apical domain of epithelial cells. FEBS Lett 581:1783–1787
21. Ott C, Elia N, Jeong SY, Insinna C, Sengupta P et al (2012) Primary cilia utilize glycoprotein-dependent adhesion mechanisms to stabilize long-lasting cilia-cilia contacts. Cilia 1:3
22. Scheiffele P, Peranen J, Simons K (1995) N-glycans as apical sorting signals in epithelial cells. Nature 378:96–98
23. Gut A, Kappeler F, Hyka N, Balda MS, Hauri HP et al (1998) Carbohydrate-mediated Golgi to cell surface transport and apical targeting of membrane proteins. EMBO J 17:1919–1929
24. Röper K, Corbeil D, Huttner WB (2000) Retention of prominin in microvilli reveals distinct cholesterol-based lipid micro-domains in the apical plasma membrane. Nat Cell Biol 2:582–592

25. Corbeil D, Marzesco A-M, Fargeas CA, Huttner WB (2010) Prominin-1, a distinct cholesterol–binding membrane protein, and the organization of the apical plasma membrane of epithelial cells. In: Harris JR (ed) Cholesterol binding and cholesterol transport proteins: structure and function in health and disease. Springer, Heidelberg, pp 399–423
26. Marzesco A-M, Janich P, Wilsch-Bräuninger M, Dubreuil V, Langenfeld K et al (2005) Release of extracellular membrane particles carrying the stem cell marker prominin-1 (CD133) from neural progenitors and other epithelial cells. J Cell Sci 118:2849–2858
27. Jászai J, Farkas LM, Fargeas CA, Janich P, Haase M et al (2010) Prominin-2 is a novel marker of distal tubules and collecting ducts of the human and murine kidney. Histochem Cell Biol 133:527–539
28. Jászai J, Janich P, Farkas LM, Fargeas CA, Huttner WB et al (2007) Differential expression of Prominin-1 (CD133) and Prominin-2 in major cephalic exocrine glands of adult mice. Histochem Cell Biol 128:409–419
29. Jászai J, Fargeas CA, Haase M, Farkas LM, Huttner WB et al (2008) Robust expression of prominin-2 all along the adult male reproductive system and urinary bladder. Histochem Cell Biol 130:749–759
30. Zhang Q, Haleem R, Cai X, Wang Z (2002) Identification and characterization of a novel testosterone-regulated prominin-like gene in the rat ventral prostate. Endocrinology 143:4788–4796
31. Signoretti S, Loda M (2006) Defining cell lineages in the prostate epithelium. Cell Cycle 5:138–141
32. Florek M, Haase M, Marzesco A-M, Freund D, Ehninger G et al (2005) Prominin-1/CD133, a neural and hematopoietic stem cell marker, is expressed in adult human differentiated cells and certain types of kidney cancer. Cell Tissue Res 319:15–26
33. Moon PG, Lee JE, You S, Kim TK, Cho JH et al (2011) Proteomic analysis of urinary exosomes from patients of early IgA nephropathy and thin basement membrane nephropathy. Proteomics 11:2459–2475
34. Ketterer C, Zeiger U, Budak MT, Rubinstein NA, Khurana TS (2010) Identification of the neuromuscular junction transcriptome of extraocular muscle by laser capture microdissection. Invest Ophthalmol Vis Sci 51:4589–4599
35. Sandelin A, Wasserman WW, Lenhard B (2004) ConSite: web-based prediction of regulatory elements using cross-species comparison. Nucleic Acids Res 32:W249–W252
36. Batlle E, Sancho E, Francí C, Domínguez D, Monfar M et al (2000) The transcription factor snail is a repressor of E-cadherin gene expression in epithelial tumour cells. Nat Cell Biol 2:84–89
37. Hsu SH, Shyu HW, Hsieh-Li HM, Li H (2001) Spz1, a novel bHLH-Zip protein, is specifically expressed in testis. Mech Dev 100:177–187
38. Perez-Moreno M, Jamora C, Fuchs E (2003) Sticky business: orchestrating cellular signals at adherens junctions. Cell 112:535–548
39. Fujimoto M, Hayashida N, Katoh T, Oshima K, Shinkawa T et al (2010) A novel mouse HSF3 has the potential to activate nonclassical heat-shock genes during heat shock. Mol Biol Cell 21:106–116
40. McMillan DR, Xiao X, Shao L, Graves K, Benjamin IJ (1998) Targeted disruption of heat shock transcription factor 1 abolishes thermotolerance and protection against heat-inducible apoptosis. J Biol Chem 273:7523–7528
41. Cunha GR, Hayward SW, Wang YZ (2002) Role of stroma in carcinogenesis of the prostate. Differentiation 70:473–485
42. Rohan S, Tu JJ, Kao J, Mukherjee P, Campagne F et al (2006) Gene expression profiling separates chromophobe renal cell carcinoma from oncocytoma and identifies vesicular transport and cell junction proteins as differentially expressed genes. Clin Cancer Res 12:6937–6945
43. Winnepenninckx V, Lazar V, Michiels S, Dessen P, Stas M et al (2006) Gene expression profiling of primary cutaneous melanoma and clinical outcome. J Natl Cancer Inst 98:472–482
44. Duhagon MA, Hurt EM, Sotelo-Silveira JR, Zhang X, Farrar WL (2010) Genomic profiling of tumor initiating prostatospheres. BMC Genomics 11:324

45. Wotton KR, Weierud FK, Dietrich S, Lewis KE (2008) Comparative genomics of Lbx loci reveals conservation of identical Lbx ohnologs in bony vertebrates. BMC Evol Biol 8:171
46. McGrail M, Batz L, Noack K, Pandey S, Huang Y et al (2010) Expression of the zebrafish CD133/prominin1 genes in cellular proliferation zones in the embryonic central nervous system and sensory organs. Dev Dyn 239:1849–1857
47. Jászai J, Fargeas CA, Graupner S, Tanaka EM, Brand M et al (2011) Distinct and conserved prominin-1/CD133–positive retinal cell populations identified across species. PLoS One 6:e17590
48. Venkatesh B, Kirkness EF, Loh YH, Halpern AL, Lee AP et al (2007) Survey sequencing and comparative analysis of the elephant shark (Callorhinchus milii) genome. PLoS Biol 5:e101
49. Meyer A, Van de Peer Y (2005) From 2R to 3R: evidence for a fish-specific genome duplication (FSGD). Bioessays 27:937–945
50. Warren WC, Hillier LW, Marshall Graves JA, Birney E, Ponting CP et al (2008) Genome analysis of the platypus reveals unique signatures of evolution. Nature 453:175–183
51. Rayko E, Jabbari K, Bernardi G (2006) The evolution of introns in human duplicated genes. Gene 365:41–47
52. Zelhof AC, Hardy RW, Becker A, Zuker CS (2006) Transforming the architecture of compound eyes. Nature 443:696–699
53. Demontis F, Dahmann C (2007) Apical and lateral cell protrusions interconnect epithelial cells in live Drosophila wing imaginal discs. Dev Dyn 236:3408–3418
54. Blacque OE, Perens EA, Boroevich KA, Inglis PN, Li C et al (2005) Functional genomics of the cilium, a sensory organelle. Curr Biol 15:935–941
55. Han Z, Anderson DW, Papermaster DS (2012) Prominin-1 localizes to the open rims of outer segment lamellae in Xenopus laevis rod and cone photoreceptors. Invest Ophthalmol Vis Sci 53:361–373

Chapter 3
Prominin-1-Containing Membrane Vesicles: Origins, Formation, and Utility

Anne-Marie Marzesco

Abstract The stem cell antigen prominin-1 (CD133) is associated with two major types (small and large) of extracellular membrane vesicles in addition to its selective concentration in various kinds of plasma membrane protrusion. During development of the mammalian central nervous system, differentiating neuroepithelial stem cells release these vesicles into the embryonic cerebrospinal fluid. In glioblastoma patients, an increase of such vesicles, particularly the smaller ones, have been also observed in cerebrospinal fluid. Similarly, hematopoietic stem and progenitor cells release small ones concomitantly with their differentiation. Although the functional significance of these prominin-1-containing membrane vesicles is poorly understood, a link between differentiation of stem (and cancer stem) cells and their release is emerging. In this chapter, I will summarize our knowledge about prominin-1-containing membrane vesicles including a potential role in cell-cell communication and highlight their prospective value as a new biomarker for tumorigenesis diagnostics.

Keywords Apical membrane • Cilium • Microvillus • Midbody • Prominin-1 • Stem cell • Vesicle

A.-M. Marzesco, Ph.D. (✉)
Department of Cellular Neurology, Hertie Institute for Clinical Brain Research, Otfried-Müller-Str. 27, 72076 Tübingen, Germany
e-mail: anne-marie.marzesco@uni-tuebingen.de

D. Corbeil (ed.), *Prominin-1 (CD133): New Insights on Stem & Cancer Stem Cell Biology*, Advances in Experimental Medicine and Biology 777, DOI 10.1007/978-1-4614-5894-4_3,

3.1 Introduction

Cell-cell communication within a group of cells is crucial to tissue formation and to achieve proper multicellular activities of an organism. Extracellular communication between cells involves secreted molecules, sophisticated vehicles such as lipoprotein particles [1, 2], thin membrane protrusions such as cytonemes [3] and tunneling nanotubes [4], and membrane vesicles such as microvesicles [5] and exosomes [6, 7] to relay important information to neighboring cells, often over long distances. Intracellular membrane trafficking has been studied extensively for decades, whereas extracellular membrane trafficking has so far been underestimated. In recent years, there has been growing evidence that extracellular membrane trafficking is involved in important biological processes like embryogenesis, immune response, spreading of pathogens as well as cancer progression.

There is a considerable amount of literature dealing with exosomes, i.e., small membrane vesicles derived from intracellular multivesicular bodies (MVBs) [6–8], and other types of extracellular membrane particles that are released directly from the plasma membrane (e.g., ectosomes, microvesicles, enveloped viruses) [9–11]. By studying the cell biology of the stem cell antigen prominin-1 (CD133; see Chap. 1) in various cell types (epithelial and non-epithelial) including stem cells [12, 13], we have identified new classes of extracellular membrane vesicles, which have only been poorly characterized until now [14–16]. This chapter will focus on those that are distinctively labeled with prominin-1.

3.2 Prominin-1

Prominin-1 is a 115–120 kDa pentaspan membrane glycoprotein which is expressed by a variety of somatic stem cells including neural stem cells (NSCs) as well as differentiated cells such epithelial and photoreceptor cells [12, 13, 17, 18]. Certain cancer stem cells also express it. Biochemically, prominin-1 directly interacts with plasma membrane cholesterol within a distinct cholesterol-based membrane microdomain, which is experimentally characterized by its insolubility in nonionic detergent Lubrol WX but soluble in Triton X-100 [19]. Morphologically, prominin-1 is specifically concentrated in plasma membrane protrusions as those found at the apical plasma membrane (e.g., microvillus, cilium, midbody) of NSCs and other epithelial cell types [12, 20]. Although the physiological function of prominin-1 is still unclear, mutations in the *PROM1* gene lead to photoreceptor degeneration in humans [21–24]. By characterizing a prominin-1 knockout mouse line in which the main phenotype is photoreceptor (cones and rods) degeneration, the importance of prominin-1 in photoreceptor morphogenesis has been corroborated (see Chap. 4) [25]. These observations together with the specific subcellular localization of prominin-1 in plasma membrane protrusions have led to the hypothesis that this cholesterol-binding membrane glycoprotein is somehow involved in the formation and/or dynamics of plasma membrane outgrowths [17].

3.3 Prominin-1-Containing Membrane Vesicles: What Are They?

Two major classes (based on their dimension) of extracellular prominin-1-containing membrane particles (referred to as P2- and P4-type vesicles) have been observed in murine embryonic cerebrospinal fluid (eCSF) [14]. Those of P2 type are large (0.5–1 μm in diameter) with an electron-dense appearance, whereas the P4-type ones are smaller (≈50–80 μm in diameter) and electron translucent. Technically, both types can be harvested from eCSF. Vesicles of P2 and P4 types are pelleted upon differential centrifugation at 1,200× and 110,000× g, respectively. They are also released into the culture medium of cultured primary cells or certain cell lines [14, 16, 26, 27]. For instance, the human colon carcinoma cell line Caco-2 has been extensively used to characterize prominin-1-containing membrane vesicles of P4 type [14, 26] (see below). Prominin-1-containing membrane vesicles could be directly immuno-isolated with antibodies against mouse or human prominin-1 [16, 28].

Finally, it is interesting to note that prominin-2 (see Chap. 2), like its paralog prominin-1, binds to membrane cholesterol and is associated with small membrane particles released in extracellular milieu such as urine [28] and saliva [29]. Co-transfection experiments have demonstrated that both prominin molecules could be found in the same membrane vesicles highlighting a similar mechanism of targeting [28].

3.3.1 Prominin-1-Containing Membrane Vesicles from Stem Cells

The first characterization of extracellular prominin-1-containing membrane vesicles has been made by analyzing eCSF in mice [14]. This fluid allows the expansion of the brain, promotes NSC survival, proliferation, and eventually their differentiation. During embryonic development of the murine central nervous system (CNS), the NSCs are organized as a pseudostratified neuroepithelium that are facing the lumen of the neural tube. Prominin-1 is strongly expressed at the apical plasma membrane of NSCs, and P2/P4 types of prominin-1-containing membrane vesicles are derived there from (see below). In addition to these vesicles, eCSF contains soluble proteins and lipoproteins, which might be involved in brain growth and potentially promote the NSC differentiation, whereas other membranous particles might play a role in the modulation of certain signal transduction by stimulating specific areas of the neuroepithelium along the neural tube (for more details, see Ref. [2]). In humans, Huttner and colleagues have demonstrated that prominin-1-containing membrane vesicles are also released in CSF of both children and adults [30]. Their biochemical characterization revealed that their physical properties are similar to the vesicles of P4 type identified in murine eCSF.

Neural stem cell lines (e.g., NS-5 and CGR8-NS) and also mouse embryonic stem (ES) cell line R1 have been shown to release P2-type-like membrane particles containing midbody ring markers [27].

Besides NSCs, Bauer and colleagues have recently shown that human primary hematopoietic stem and progenitor cells (HSPCs) as well release prominin-1-containing membrane vesicles of the P4 type, indicating that the release of prominin-1 is not a unique characteristic of neural cells [16].

3.3.2 Prominin-1-Containing Membrane Vesicles from Body Fluids

Interestingly, the P4-type prominin-1-containing membrane vesicles are released not only in murine and human CSF but also in various other body fluids such as seminal fluid, saliva, urine, and lacrimal fluid [14, 28]. Their occurrence in various fluids reflects the widespread tissue distribution of prominin-1 and suggests that they may have a role not only in tissue development but as well in maintenance. The vesicles of P2 type have not been characterized yet in other fluids.

3.4 Biogenesis of Prominin-1-Containing Membrane Vesicles

3.4.1 Subcellular Origins

In epithelial cells, the P2- and P4-type prominin-1-containing membrane vesicles have different subcellular origins although both are derived from subdomains of apical plasma membrane.

The P2-type prominin-1-containing membrane vesicles released by NSCs into eCSF are derived from remnant midbody rings (Fig. 3.1a). At the end of cytokinesis, the central part of the cytoplasmic bridge formed by ingression of the furrow between two nascent daughter cells and referred to as the midbody has been shown to be severed [31, 32]. After abscission, the midbody remnant is normally asymmetrically inherited by one of the daughter cells, where it is degraded by autophagy [33]. Alternatively, a second cleavage might occur leading to the release of midbody central core (often called Flemming body) into the extracellular space as P2-type prominin-1-containing vesicles [15, 27].

The P4-type prominin-1-containing membrane vesicles might originate from released P2-type vesicles or directly from distinct apical plasma membrane protrusions of NSCs: midbody, primary cilium, and microvillus (Fig. 3.1a) [14, 15]. The observation of budding from microvilli of polarized epithelial cells such as Caco-2 cells has shed light on the molecular mechanism of these vesicles formation (see below) [14, 26].

In non-epithelial cells such as human primary HSPCs, a significant pool of prominin-1, in addition to that located at plasma membrane protrusions (microvillus-like structures), is associated with internal vesicles found within MVBs [16].

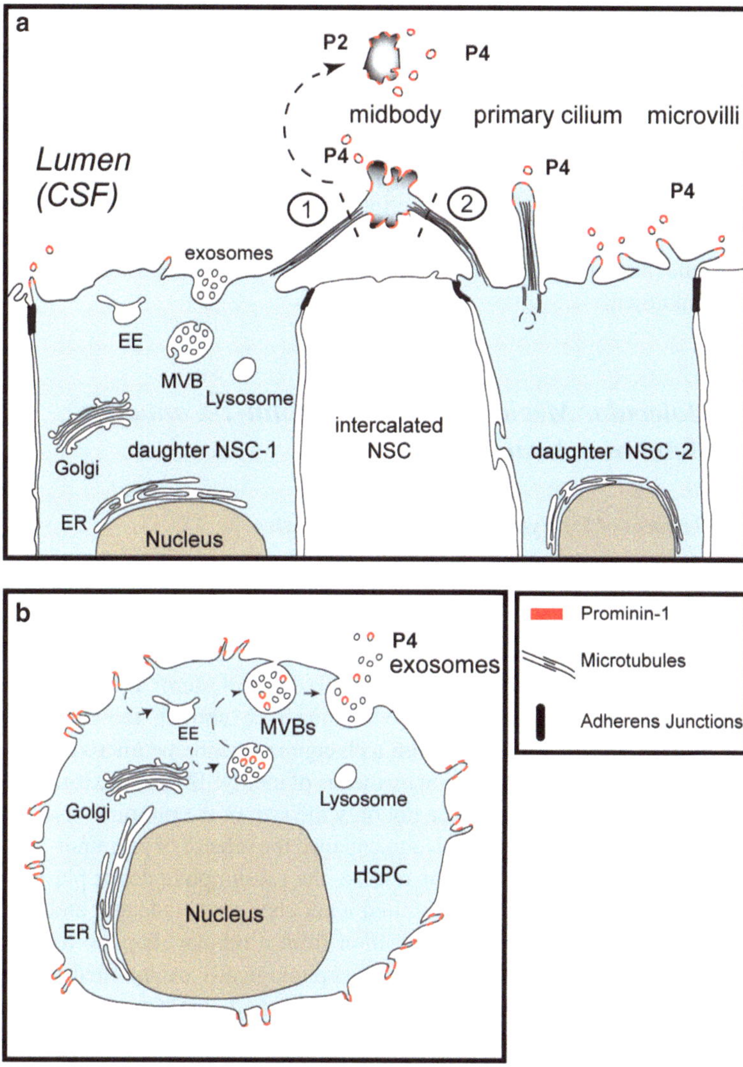

Fig. 3.1 Schematic representation of the biogenesis of prominin-1-containing membrane vesicles. (**a**) In neural stem cells (NSCs), large prominin-1-containing membrane vesicles (P2 type) are derived from membrane fission events (*dashed lines 1 and 2*) of the central portion of the midbody, an intercellular cytoplasmic bridge formed between nascent daughter cells at the end of cytokinesis, whereas smaller ones (P4 type) are budding from three distinct structures found at the apical plasma membrane – midbody, primary cilium, and microvilli – or directly from released P2-type membrane vesicles. (**b**) In non-epithelial cells such as hematopoietic stem and progenitor cells (HSPCs) P4-type membrane vesicles are derived from microvillus-like structures and/or multivesicular bodies (MVBs) as exosomes. Prominin-1 molecules (or prominin-1-containing membrane microdomains) are indicated in *red*. *EE*, early endosome; *ER*, endoplasmic reticulum

Therefore, P4-type prominin-1-containing membrane vesicles have been shown to be associated with a subset of exosomes (Fig. 3.1b) [16]. The association of prominin-1 with exosomes has not been observed in NSCs [14]. The biogenesis of exosome and their intracellular trafficking is much better characterized than those of membrane vesicles derived from the plasma membrane. The formation of internal vesicles within MVBs (late endosomes) occurs by inward budding of the endosomal limiting membrane followed by their release into the extracellular milieu upon fusion of MVBs with the plasma membrane (Fig. 3.1b). The molecular signals and cytoplasmic molecules that trigger the internalization of prominin-1 en route to MVBs are currently unknown.

3.4.2 Molecular Mechanisms of Prominin-1-Containing Membrane Vesicle Release

3.4.2.1 Release of P4 Type of Membrane Vesicles

Insights into the molecular mechanism underlying the release of P4-type prominin-1-containing membrane vesicles have been obtained by manipulating the level of membrane cholesterol in Caco-2 cells [26]. Keeping in mind that, within the membrane of microvilli, prominin-1 binds to membrane cholesterol and associates with cholesterol-dependent membrane microdomains (see above), the release of the P4-type vesicles may be associated with a change in membrane microdomain organization. Indeed, cholesterol depletion by means of methyl-β-cyclodextrin and lovastatin treatments was shown to reduce not only the size of the prominin-1-containing membrane microdomains [19] but also to enhance the release of prominin-1-positive vesicles [26]. Concomitant with their release, the presumptive donor plasma membrane structure (i.e., microvillus) exhibited a net change in its length and morphology, which is characterized by the transition from a tubular shape to a "pearling" state [26]. The pearling of microvilli was morphologically exemplified by periodic constriction sites regularly distributed along its length. A single constriction site was sometimes observed near to the tip of the microvilli, suggesting where pearling may initiate (see Fig. 3 in Ref. [26]). Such pearling of microvilli has already been documented in vivo, and therefore, it might be a physiological intermediate state in the fission of vesicles from microvilli [34]. Moreover, the occurrence of membrane vesicles in the vicinity of microvilli of the intestinal brush border membrane of chicken receiving a low-cholesterol diet is in agreement with a role of membrane cholesterol and membrane microdomain in biogenesis of membrane vesicles [34, 35].

The pearling of tubes has been shown using in vitro systems such as lipid bilayer tubes to be induced by altering two major parameters: exerted tension and curvature [36, 37]. In the case of the microvillus, which is a tubular membrane supported by parallel arrays of actin filaments that are linked to the plasma membrane, the depolymerization of actin might lower the tension exerted and hence lead to pearling [38–40]. Furthermore, a microvillus is characterized by the two-dimensional curvature of its membrane throughout the cylindrical portion of its length. Pearling of the microvillus exhibits a curvature of the membrane in all three dimensions at the constriction site. The intrinsic curvature of the membrane reflects thus the driving forces of its major constituents, integral membrane proteins, and lipids. It has been shown that manipulation of membrane cholesterol levels may lead to changes in lipid-ordered membrane microdomains resulting in changes of curvature [41–47]. Altogether, these observations are in agreement with the concept that the organization of cholesterol-dependent membrane microdomains may control the release of prominin-1-containing membrane vesicles of P4 type from the microvillus, and perhaps also the primary cilium. It might be more than a coincidence that lipid composition of cultured Caco-2 cells changes upon their enterocytic differentiation, a phenomenon occurring about seven days post-confluency just like the beginning of the release of prominin-1-positive vesicles [14, 35].

3.4.2.2 Release of P2 Type of Membrane Vesicles

In contrast to the P4-type vesicles, little is known about the biogenesis of P2-type membrane vesicles, which are derived from midbody structure. The molecular machinery involved in the severing of the midbody is not well understood [48–52]. The remnant midbody ring has been shown to be either retained by one of the daughter cells or released subsequently as a P2-type-like membrane vesicle [27]. Recently, new insights into the release of the remnant midbody were brought in a study using various stem and cancer cell lines and specific knockdown experiments by means of small interfering RNAs combined with automated image analysis [27]. It was demonstrated that main players involved in the midbody abscission (i.e., apoptosis-linked gene-2-interacting protein X (Alix) and the tumor-suppressor gene 101 (Tsg101), or their interacting partner, centrosomal protein 55 (Cep55)) are also engaged in the release of P2-type-like membrane vesicles [27].

Finally, it is interesting to note that prominin-1 itself is dispensable for the release of P2-type vesicles into CSF during embryonic development since such vesicles are still observed using citron rho-interacting kinase as an alternative marker of P2-type vesicles in prominin-1 knockout mice [25] (A-M Marzesco, WB Huttner, unpublished data).

3.5 Physiological Role

Although we start to understand the molecular mechanisms underlying the biogenesis of prominin-1-containing membrane vesicles, their physiological function(s) remains elusive. At the present stage, we could solely propose hypotheses that are based on their occurrence and the physiological phenomena correlated with them. Thus, they might represent new vehicles that either carry stem cell-specific membrane microdomains whose loss could promote cellular differentiation, and/or spread biological information implicated in embryonic development, and eventually tumorigenesis. However, the widespread detection of prominin-1-containing membrane vesicles in various physiological fluids suggests that their physiological roles are not only restricted to stem cells.

3.5.1 Embryonic Development of the Central Nervous System

Until recently, the main attention concerning the brain development was focusing on its general organization [53] and the potential of NSCs to proliferate and differentiate [54]. Prior to neurogenesis, important changes in the organization of the neuroepithelium occur not only at the level of tissue architecture but also at the cellular one. The neuroepithelium becomes thicker and pseudostratified while NSCs proliferate. After closure of the neural tube, the neurogenic progenitors start to give rise to neurons that migrate basally and arrange themselves in distinct horizontal layers. It was mostly overlooked that the neural tube is filled with CSF. As mentioned above, the components (soluble proteins, lipoproteins, and membrane vesicles) present in CSF are involved as extracellular signals in the unique spatiotemporal control of neurogenesis through dynamic interactions with NSCs. Interestingly, the P2 and P4 types of prominin-1-containing membrane vesicles appeared in the CSF at the very onset and during the early phase of neurogenesis (Fig. 3.2a, b) [14, 15]. Despite the relationship between the release of these vesicles and neural differentiation, it should be stressed that the physiological signification of the plasma membrane release by means of vesicles is currently unclear.

Two potential roles for these prominin-1-positive membrane vesicles, which are not mutually exclusive, may be considered. Our biochemical knowledge about prominin-1, i.e., (1) its interaction with membrane cholesterol and (2) its association with particular membrane microdomains, are relevant in these contexts. Firstly, given that membrane microdomains (lipid rafts) are actively involved in signal transduction, those associated with stem cells may carry the key determinants allowing the maintenance of their stem/progenitor properties, and their disposal – release of prominin-1-positive membrane vesicles – may contribute to differentiation. It is of note that prominin-1 within the P4 type of membrane vesicles binds to cholesterol within membrane microdomain, as reported for the donor membranes such as microvilli [19]. Secondly, as proposed for other constituents of CSF, these prominin-1-positive vesicles may participate in intercellular communication. The increase of release of P4-type membrane vesicles during a short window period (embryonic stage E10.5–E12) of development when microvilli are highly abundant in the floor plate, a signaling center, is consistent with our hypothesis. Likewise, the release of midbody as P2-type membrane vesicles could provide information to the surrounding tissues about the history of cell division, i.e., symmetric proliferative cell divisions versus asymmetric neurogenic cell divisions [15].

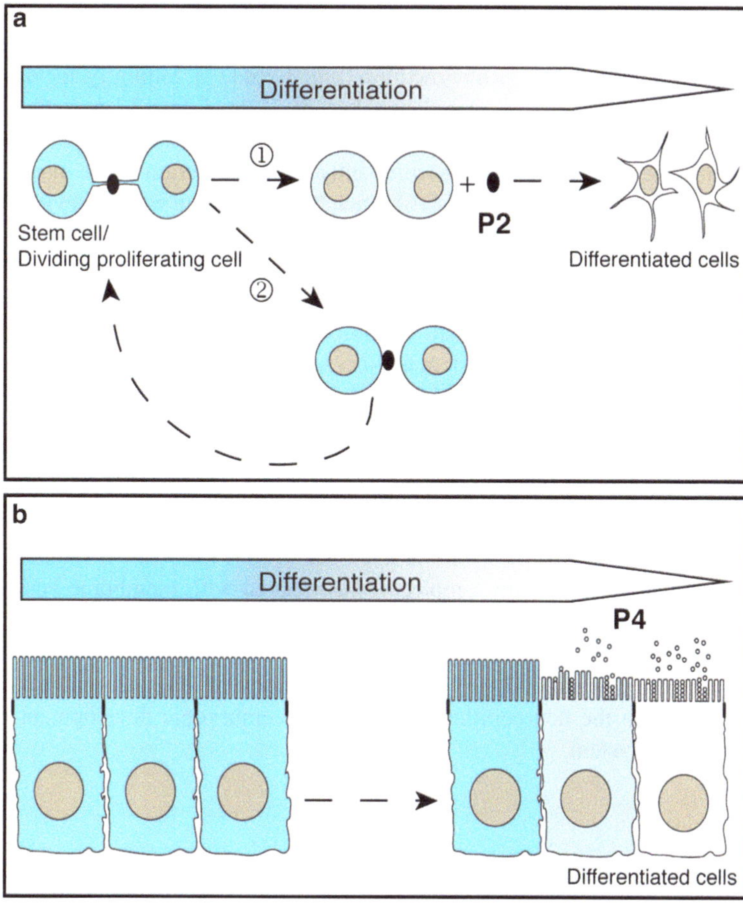

Fig. 3.2 Relationship between the release of prominin-1-containing membrane vesicles and cellular differentiation. (**a**) P2-type prominin-1-containing membrane vesicles released (1) from, or the midbody retention (2) by, the daughter cell will lead to cellular differentiation and proliferation, respectively. (**b**) The prominin-1-containing membrane vesicles of P4 type are released from microvilli of NSCs or the fusion of MVBs with plasma membrane of HSPCs (not depicted) concomitant with cellular differentiation

3.5.2 *Cellular Differentiation*

In addition to the in vivo observation made in the murine neural system (see above), other studies using ex vivo or in vitro cell culture models that are still responsive to differentiating stimuli have contributed to emphasize a link between the release of both types (P2 and P4) of membrane vesicles and cellular differentiation.

For instance, membrane vesicles of P4 type have been described to be released concomitantly with cellular differentiation. In Caco-2 cells, the enterocytic differentiation that occurs spontaneously around seven days after cell confluence [55, 56] correlated with the beginning of the release of P4-type prominin-1-containing membrane vesicles into the culture medium (Fig. 3.2b) [14]. Similarly, the differentiation of primary HSPCs triggered by phorbol 12-myristate 13-acetate (PMA) treatment [57] resulted in a net release of P4-type prominin-1-containing membrane vesicles into the culture medium concomitantly with a depletion of the intracellular pool of prominin-1 found in MVBs [16]. Such data are in line with a previous study demonstrating an increase in exosome release from human Burkitt lymphoma cells upon phorbol ester treatment [58].

P2-type-like membrane vesicles have been shown to be released by various cell lines grown under different growth conditions [27]. Interestingly, the extent of release is significantly greater in stem cell-derived lines (neural, hematopoietic, and embryonic) than cancer-derived and immortalized cells and in cell lines responsive to differentiation stimulus [27]. Furthermore, cells with impaired release of P2-type-like membrane vesicles upon knockdown of Alix, Tsg101, or Cep55 were more susceptible to differentiation factors [27]. Altogether, these data suggest that release of the midbody occurs in differentiation-responsive stem or cancer cells while cell division gives rise to two post-mitotic daughter cells (Fig. 3.2a). It remains to be determined whether prominin-1 molecules associated with membrane vesicles of P2 type bind to membrane cholesterol within a specific membrane microdomain as those of P4 type. Moreover, the physiological evidence of such phenomena (retention versus release of midbody) remains to be documented in healthy and cancerous tissues other than the neuroepithelium during the embryonic development of the central nervous system.

3.5.3 Membrane Vesicles in Tumorigenesis and Other Diseases

Many solid cancer cells are able to modify their environment in order to promote tumor growth and metastasis, and recent information suggest that they do so through extracellular membrane trafficking such as exosomes and other plasma membrane-derived vesicles. Within cancerous tissues, cancer stem cells represent a small subset of cells which constitutes a reservoir of self-sustaining cells with the ability to self-renew and maintain the tumor growth [59]. The expression of prominin-1 by cancer stem cells such as those in gliomas [60] and the release of prominin-1-containing membrane vesicles by neural stem cells [12, 30] may be more than a coincidence. This relation prompted us to investigate whether the status of glioblastoma (e.g., the duration time) will result in an elevation of prominin-1-containing membrane vesicles of P4 type in CSF of patients. In healthy human subjects, the level of P4-type membrane vesicles declines during childhood and remains thereafter constant in a narrow range between individuals [30]. Remarkably, in adult patients suffering from glioblastoma, the level of P4-type membrane vesicles is surprisingly increased, and strikingly, at the final stage of disease (>30 months) a drastic decrease is observed [30]. In this context, it might be envisioned that the altered release of the membrane vesicles may not only contribute either to the differentiation of cancer stem cells as observed for normal stem cells (Fig. 3.2a) or the modification the surrounding tissue microenvironment including the recruitment of novel blood vessels but also participates in tumor progression by the transformation of healthy tissues. It is now urgent to decipher the molecular contents of such vesicles in order to identify certain factors (protein, lipid, RNA) that could provide clues about tumor progression. Interestingly, Bauer and colleagues have elegantly demonstrated that prominin-1-containing membrane vesicles/exosomes released by HSPCs were internalized by primary mesenchymal stromal cells, consistent with a role in intercellular communication [16].

Finally, the analysis of CSF (or external fluids: urine, saliva) for prominin-1-containing membrane vesicles might be of clinical value. Certain human diseases are diagnosed following neuroradiological and histopathological observations by altered amounts of marker proteins in corresponding tissue-derived body fluids. For example, it has been shown that exosomes are detected in CSF of patients with neurodegenerative brain diseases [61–63]. Monitoring prominin-1 as a biomarker, Huttner and colleagues could detect, just like in brain cancers, an alteration of prominin-1-containing membrane vesicles in the CSF of patients suffering from temporal lobe epilepsy [64], thus opening a new avenue for diagnostic purposes.

3.6 Conclusion and Outlook

In recent years, the study of somatic stem cell antigen prominin-1 has brought the spotlight on membrane vesicles that are released by stem cells during the process of differentiation, a phenomenon that could eventually also apply to cancer stem cells. Although the potential functions of such vesicles are still speculative, it is nevertheless worthy of consideration that the disposal of stem cell-specific cholesterol-based membrane microdomains by means of release of prominin-1-positive membrane vesicles might regulate the machinery that controls cell fate decision such as proliferation and differentiation. Likewise, such vesicles might be part of novel membrane-based cell-cell communication pathways responsible for proper embryonic development, organ homeostasis, and cancer progression throughout healthy tissues. The molecular characterization of prominin-1-containing membrane vesicles such as proteome and lipidome may reveal in a near future novel aspects of stem and cancer stem cell biology.

Acknowledgements I thank especially D. Corbeil, as well as F. Baumann and N. Varvel, for critical reading and comments on this manuscript. I would like also to acknowledge the main collaborators involved in this research: W.B. Huttner, D. Corbeil, M. Wilsch-Bräuninger, V. Dubreuil, A.W. Ettinger, N. Bauer, P. Janich, and M. Florek.

References

1. Greco V, Hannus M, Eaton S (2001) Argosomes: a potential vehicle for the spread of morphogens through epithelia. Cell 106:633–645
2. Bachy I, Kozyraki R, Wassef M (2008) The particles of the embryonic cerebrospinal fluid: how could they influence brain development? Brain Res Bull 75:289–294
3. Ramirez-Weber FA, Kornberg TB (1999) Cytonemes: cellular processes that project to the principal signaling center in Drosophila imaginal discs. Cell 97:599–607
4. Gerdes HH, Carvalho RN (2008) Intercellular transfer mediated by tunneling nanotubes. Curr Opin Cell Biol 20:470–475
5. Muralidharan-Chari V, Clancy JW, Sedgwick A, D'Souza-Schorey C (2010) Microvesicles: mediators of extracellular communication during cancer progression. J Cell Sci 123:1603–1611
6. Simons M, Raposo G (2009) Exosomes-vesicular carriers for intercellular communication. Curr Opin Cell Biol 21:575–581
7. Bobrie A, Colombo M, Raposo G, Thery C (2011) Exosome secretion: molecular mechanisms and roles in immune responses. Traffic 12:1659–1668
8. Hurley JH, Odorizzi G (2012) Get on the exosome bus with ALIX. Nat Cell Biol 14:654–655
9. Pap E (2011) The role of microvesicles in malignancies. Adv Exp Med Biol 714:183–199
10. Chen BJ, Lamb RA (2008) Mechanisms for enveloped virus budding: can some viruses do without an ESCRT? Virology 372:221–232
11. Ganser-Pornillos BK, Yeager M, Pornillos O (2012) Assembly and architecture of HIV. Adv Exp Med Biol 726:441–465
12. Weigmann A, Corbeil D, Hellwig A, Huttner WB (1997) Prominin, a novel microvilli-specific polytopic membrane protein of the apical surface of epithelial cells, is targeted to plasmalemmal protrusions of non-epithelial cells. Proc Natl Acad Sci USA 94:12425–12430
13. Yin AH, Miraglia S, Zanjani ED, Almeida-Porada G, Ogawa M et al (1997) AC133, a novel marker for human hematopoietic stem and progenitor cells. Blood 90:5002–5012
14. Marzesco AM, Janich P, Wilsch-Bräuninger M, Dubreuil V, Langenfeld K et al (2005) Release of extracellular membrane particles carrying the stem cell marker prominin-1 (CD133) from neural progenitors and other epithelial cells. J Cell Sci 118:2849–2858
15. Dubreuil V, Marzesco AM, Corbeil D, Huttner WB, Wilsch-Bräuninger M (2007) Midbody and primary cilium of neural progenitors release extracellular membrane particles enriched in the stem cell marker prominin-1. J Cell Biol 176:483–495
16. Bauer N, Wilsch-Bräuninger M, Karbanová J, Fonseca AV, Strauss D et al (2011) Haematopoietic stem cell differentiation promotes the release of prominin-1/CD133-containing membrane vesicles – a role of the endocytic-exocytic pathway. EMBO Mol Med 3:398–409
17. Corbeil D, Röper K, Fargeas CA, Joester A, Huttner WB (2001) Prominin: a story of cholesterol, plasma membrane protrusions and human pathology. Traffic 2:82–91
18. Fargeas CA, Fonseca A-V, Huttner WB, Corbeil D (2006) Prominin-1 (CD133): from progenitor cells to human diseases. Future Lipidol 1:213–225
19. Röper K, Corbeil D, Huttner WB (2000) Retention of prominin in microvilli reveals distinct cholesterol-based lipid microdomains in the apical plasma membrane. Nature Cell Biol 2:582–592
20. Corbeil D, Marzesco AM, Wilsch-Bräuninger M, Huttner WB (2010) The intriguing links between prominin-1 (CD133), cholesterol-based membrane microdomains, remodeling of apical plasma membrane protrusions, extracellular membrane particles, and (neuro)epithelial cell differentiation. FEBS Lett 584:1659–1664
21. Maw MA, Corbeil D, Koch J, Hellwig A, Wilson-Wheeler JC et al (2000) A frameshift mutation in prominin (mouse)-like 1 causes human retinal degeneration. Hum Mol Genet 9:27–34
22. Zhang Q, Zulfiqar F, Xiao X, Riazuddin SA, Ahmad Z et al (2007) Severe retinitis pigmentosa mapped to 4p15 and associated with a novel mutation in the PROM1 gene. Hum Genet 122:293–299
23. Yang Z, Chen Y, Lillo C, Chien J, Yu Z et al (2008) Mutant prominin 1 found in patients with macular degeneration disrupts photoreceptor disk morphogenesis in mice. J Clin Invest 118:2908–2916

24. Permanyer J, Navarro R, Friedman J, Pomares E, Castro-Navarro J et al (2010) Autosomal recessive retinitis pigmentosa with early macular affectation caused by premature truncation in PROM1. Invest Ophthalmol Vis Sci 51:2656–2663
25. Zacchigna S, Oh H, Wilsch-Bräuninger M, Missol-Kolka E, Jászai J et al (2009) Loss of the cholesterol-binding protein prominin-1/CD133 causes disk dysmorphogenesis and photoreceptor degeneration. J Neurosci 29:2297–2308
26. Marzesco AM, Wilsch-Bräuninger M, Dubreuil V, Janich P, Langenfeld K et al (2009) Release of extracellular membrane vesicles from microvilli of epithelial cells is enhanced by depleting membrane cholesterol. FEBS Lett 583:897–902
27. Ettinger AW, Wilsch-Brauninger M, Marzesco AM, Bickle M, Lohmann A et al (2011) Proliferating versus differentiating stem and cancer cells exhibit distinct midbody-release behaviour. Nature Commun 2:503
28. Florek M, Bauer N, Janich P, Wilsch-Bräuninger M, Fargeas CA et al (2007) Prominin-2 is a cholesterol-binding protein associated with apical and basolateral plasmalemmal protrusions in polarized epithelial cells and released into urine. Cell Tissue Res 328:31–47
29. Jászai J, Janich P, Farkas LM, Fargeas CA, Huttner WB et al (2007) Differential expression of Prominin-1 (CD133) and Prominin-2 in major cephalic exocrine glands of adult mice. Histochem Cell Biol 128:409–419
30. Huttner HB, Janich P, Kohrmann M, Jászai J, Siebzehnrubl F et al (2008) The stem cell marker prominin-1/CD133 on membrane particles in human cerebrospinal fluid offers novel approaches for studying central nervous system disease. Stem Cells 26:698–705
31. Mullins JM, McIntosh JR (1982) Isolation and initial characterization of the mammalian midbody. J Cell Biol 94:654–661
32. Pohl C, Jentsch S (2008) Final stages of cytokinesis and midbody ring formation are controlled by BRUCE. Cell 132:832–845
33. Pohl C, Jentsch S (2009) Midbody ring disposal by autophagy is a post-abscission event of cytokinesis. Nat Cell Biol 11:65–70
34. Hobbs DG (1980) The origin and distribution of membrane-bound vesicles associated with the brush border of chick intestinal mucosa. J Anat 131:635–642
35. Jindrichova S, Novakova O, Bryndova J, Tvrzicka E, Lisa V et al (2003) Corticosteroid effect on Caco-2 cell lipids depends on cell differentiation. J Steroid Biochem Mol Biol 87:157–165
36. Bar-Ziv R, Moses E (1994) Instability and "pearling" states produced in tubular membranes by competition of curvature and tension. Phys Rev Lett 73:1392–1395
37. Roux A, Cuvelier D, Nassoy P, Prost J, Bassereau P et al (2005) Role of curvature and phase transition in lipid sorting and fission of membrane tubules. EMBO J 24:1537–1545
38. Bar-Ziv R, Tlusty T, Moses E, Safran SA, Bershadsky A (1999) Pearling in cells: a clue to understanding cell shape. Proc Natl Acad Sci USA 96:10140–10145
39. McConnell RE, Tyska MJ (2007) Myosin-1a powers the sliding of apical membrane along microvillar actin bundles. J Cell Biol 177:671–681
40. Nambiar R, McConnell RE, Tyska MJ (2009) Control of cell membrane tension by myosin-I. Proc Natl Acad Sci USA 106:11972–11977
41. Evans E, Rawicz W (1990) Entropy-driven tension and bending elasticity in condensed-fluid membranes. Phys Rev Lett 64:2094–2097
42. Needham D, Nunn RS (1990) Elastic deformation and failure of lipid bilayer membranes containing cholesterol. Biophys J 58:997–1009
43. Chen Z, Rand RP (1997) The influence of cholesterol on phospholipid membrane curvature and bending elasticity. Biophys J 73:267–276
44. Huttner WB, Zimmerberg J (2001) Implications of lipid microdomains for membrane curvature, budding and fission. Curr Opin Cell Biol 13:478–484
45. Wang W, Yang L, Huang HW (2007) Evidence of cholesterol accumulated in high curvature regions: implication to the curvature elastic energy for lipid mixtures. Biophys J 92:2819–2830
46. Baumgart T, Hess ST, Webb WW (2003) Imaging coexisting fluid domains in biomembrane models coupling curvature and line tension. Nature 425:821–824
47. Yanagisawa M, Imai M, Taniguchi T (2008) Shape deformation of ternary vesicles coupled with phase separation. Phys Rev Lett 100:148102

48. Barr FA, Gruneberg U (2007) Cytokinesis: placing and making the final cut. Cell 131:847–860
49. Steigemann P, Gerlich DW (2009) Cytokinetic abscission: cellular dynamics at the midbody. Trends Cell Biol 19:606–616
50. Burton K, Taylor DL (1997) Traction forces of cytokinesis measured with optically modified elastic substrata. Nature 385:450–454
51. Darenfed H, Mandato CA (2005) Wound-induced contractile ring: a model for cytokinesis. Biochem Cell Biol 83:711–720
52. Guizetti J, Schermelleh L, Mantler J, Maar S, Poser I et al (2011) Cortical constriction during abscission involves helices of ESCRT-III-dependent filaments. Science 331:1616–1620
53. Nakai J, Fujita S (1994) Early events in the histo- and cytogenesis of the vertebrate CNS. Int J Dev Biol 38:175–183
54. Götz M, Huttner WB (2005) The cell biology of neurogenesis. Nat Rev Mol Cell Biol 6:777–788
55. Pinto M, Appay M-D, Simon-Assmann P, Chevalier G, Dracopoli J et al (1982) Enterocyte differentiation of cultured human colon cancer cells by replacement of glucose by galactose in the medium. Biol Cell 44:193–196
56. Louvard D, Kedinger M, Hauri HP (1992) The differentiating intestinal epithelial cell: establishment and maintenance of functions through interactions between cellular structures. Annu Rev Cell Biol 8:157–195
57. Rossi F, McNagny M, Smith G, Frampton J, Graf T (1996) Lineage commitment of transformed haematopoietic progenitors is determined by the level of PKC activity. EMBO J 15:1894–1901
58. Clayton A, Court J, Navabi H, Adams M, Mason MD et al (2001) Analysis of antigen presenting cell derived exosomes, based on immuno-magnetic isolation and flow cytometry. J Immunol Methods 247:163–174
59. Clarke MF, Dick JE, Dirks PB, Eaves CJ, Jamieson CH et al (2006) Cancer stem cells-perspectives on current status and future directions: AACR workshop on cancer stem cells. Cancer Res 66:9339–9344
60. Singh SK, Hawkins C, Clarke ID, Squire JA, Bayani J et al (2004) Identification of human brain tumor initiating cells. Nature 432:396–401
61. Fevrier B, Vilette D, Archer F, Loew D, Faigle W et al (2004) Cells release prions in association with exosomes. Proc Natl Acad Sci USA 101:9683–9688
62. Rajendran L, Honsho M, Zahn TR, Keller P, Geiger KD et al (2006) Alzheimer's disease beta-amyloid peptides are released in association with exosomes. Proc Natl Acad Sci USA 103:11172–11177
63. Aguzzi A, Rajendran L (2009) The transcellular spread of cytosolic amyloids, prions, and prionoids. Neuron 64:783–790
64. Huttner HB, Corbeil D, Thirmeyer C, Coras R, Kohrmann M et al (2012) Increased membrane shedding – indicated by an elevation of CD133-enriched membrane particles-into the CSF in partial epilepsy. Epilepsy Res 99:101–106

Chapter 4
Prominent Role of Prominin in the Retina

Nagananda Gurudev, Mareike Florek, Denis Corbeil, and Elisabeth Knust

Abstract Prominin molecules represent a new family of pentaspan membrane glycoproteins expressed throughout the animal kingdom. The name originates from its localization on membrane protrusion, such as microvilli, filopodia, lamellipodia, and microspikes. Following the original description in mouse and human, representative prominin members were found in fish (e.g., *Danio rerio*), amphibian (*Ambystoma mexicanum*, *Xenopus laevis*), worm (*Caenorhabditis elegans*), and flies (*Drosophila melanogaster*). Mammalian prominin-1 was identified as a marker of somatic and cancer stem cells and plays an essential role in the visual system, which contributed to increased interest of the medical field in this molecule. Here we summarize recent data from various fields, including *Drosophila*, which will aid to our understanding of its still elusive function.

Keywords CD133 • Cone • Photoreceptor • Retina • Rhabdomere • Rod

4.1 Mammalian Prominin

The first member of this family, prominin-1 (in human also known as CD133, PROML1, and AC133 antigen), has become widely known as a cell-surface antigen used for the isolation of stem and progenitor cells originating from various tissues/organs, notably from the neural and hematopoietic system [1]. Identification of

N. Gurudev • E. Knust, Ph.D. (✉)
Max Planck Institute of Molecular Cell Biology and Genetics,
Pfotenhauerstr. 108, 01307 Dresden, Germany
e-mail: knust@mpi-cbg.de

M. Florek • D. Corbeil
Tissue Engineering Laboratories (BIOTEC), Technische Universität Dresden,
Tatzberg 47-49, 01307 Dresden, Germany

D. Corbeil (ed.), *Prominin-1 (CD133): New Insights on Stem & Cancer Stem Cell Biology*, Advances in Experimental Medicine and Biology 777, DOI 10.1007/978-1-4614-5894-4_4,

cancer stem cells based on human PROMININ-1 and its particular AC133 epitope has also contributed to increasing interest in PROMININ-1 within the medical field [2–4].

Prominin-1 is expressed in various stem and progenitor cells, but its expression is not limited to these cell types. For instance, prominin-1 is detected in photoreceptor cells (see below), epithelial cells of developing and adult organs, and in glial cells [5–10]. The primary transcript of mammalian prominin-1 is subjected to differential splicing, mostly occurring in the 3′ region, resulting in various protein isoforms with different C-terminal tails, which exhibit tissue-specific expression during development (for membrane topology see Fig. 4.1a) [10–12].

The role of prominins remains elusive. Several lines of cell biological observations have suggested a potential role in the organization and/or maintenance of plasma membrane protrusions, i.e., those being marked by a significant membrane curvature. Such observations include preferential concentration of prominins at the membrane of microvilli and primary cilia and their direct interaction with membrane cholesterol within a particular cholesterol-dependent membrane microdomain [13–15]. Although hints regarding prominin's physiological function(s) are scarce, most of them arose from studies of prominin-1 in the context of the retina. These include diseases that are linked to mutations in the *PROM1* gene, animal models, and the identification of interacting proteins.

4.2 Prominin-1 and the Mammalian Visual System

Retinal dystrophies, such as macular degeneration or retinitis pigmentosa (RP), account for millions of blindness cases worldwide. Therefore, understanding the mechanisms leading to blindness and establishing regenerative therapies have a high socioeconomic impact. Interestingly, although prominin-1 is expressed in a variety of organs (see above), the most striking effects elicited upon loss of function of prominin-1 are observed in the visual system.

4.2.1 Retinal Dystrophies: Mutations in PROM1

Since its discovery in 1997, five mutations in the human *PROM1* gene (Online Mendelian Inheritance in Man (OMIM) #604365; locus 4p15.32) have been reported. Maw and colleagues described the first mutation more than a decade ago [16]. They observed a single-nucleotide deletion (ΔG_{1878}), which is predicted to shift the reading frame from codon 614 onward (with reference to the PROMININ-1.s2 splice variant; for nomenclature of splice variants see Ref. [17]). This mutation, which causes a premature termination of translation after addition of 12 amino acids unrelated to PROMININ-1 (Fig. 4.1a, G614EfsX12), is associated with autosomal recessive retinal degeneration in a consanguineous Indian pedigree [16]. Clinically, the affected individuals have reported night blindness, loss of peripheral vision from childhood

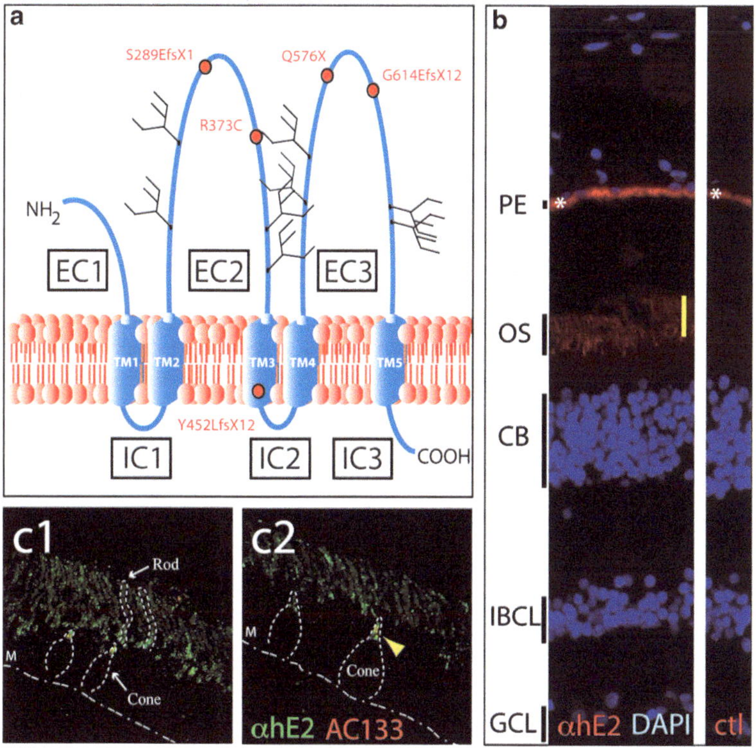

Fig. 4.1 Mutations in PROMININ-1 and its expression in human photoreceptor cells. (**a**) Membrane topology and mutations affecting the open reading frame of PROMININ-1 and causing retinal degeneration are indicated (*red circles*). Prominin-1 consists after the cleavage of the signal peptide of an N-terminal extracellular domain (*EC1*) five transmembrane (TM) domains, alternating short intracellular (IC) and large (>250 amino acid residues) extracellular (EC) loops, and an intracellular C-terminal domain (*IC3*). The *EC2* and *EC3* domains contain all potential N-glycosylation sites (*forks*), which vary in position according to the animal species. For details of the mutations, see main text and Table 4.1. (**b**) Expression of PROMININ-1 in human photoreceptor cells. Human samples [Department of Pathology, Technical University of Dresden; archival materials that had not be used for genetic analysis] were processed as described [6]. Samples were incubated in blocking solution containing 0.2% saponin and then immunolabeled with rabbit antiserum αhE2 (*red*) directed against the PROMININ-1 polypeptide [6]. Nuclei were visualized with 4,6-diamidino-2-phenylindole (DAPI). As a negative control (ctl), pre-immune serum was used. Note the specific PROMININ-1 staining (*yellow bar*) in the outer segment (*OS*) of photoreceptor cell layer. *Asterisks* indicate the autofluorescent signal observed in the pigmented epithelial (*PE*) cell layer, which is detached during histological preparation. *CB* cell bodies of photoreceptors, *IBCL* integrating bipolar cell layer. *GCL* ganglion cell layer. (**c1** and **c2**) Human samples were double immunolabeled with antiserum αhE2 (*green*) and mouse monoclonal antibody CD133/1 (AC133 epitope; Miltenyi Biotec, Germany; *red*). The OS of two rods (c1) and OS/inner segment of two cones (c1 and c2) are indicated by *dashed lines*. A *yellow arrowhead* shows double immunolabeling of the cone OS (c2). *M* indicates the outer limiting membrane

Table 4.1 Mutations in *PROM1* gene that affect its open reading frame

Mutation	Molecular defect	Disease	Reference
ΔG_{1878}	Shift of reading frame from codon 614[a] onward, premature termination of translation after additional 12 amino acids unrelated to PROMININ-1	Autosomal recessive RP	[16]
$C_{1726}\rightarrow T$	Premature stop codon (Gln576X)	Severe forms of rod-cone degeneration consistent with RP	[18]
Insertion of T at position 1349	Frameshift starting at codon 452 and a putative stop codon 12 amino acids downstream in the translated protein	Autosomal recessive cone-rod dystrophy	[19]
ΔG_{869}	Frameshift from codon 289 onward and caused a premature STOP codon after the addition of one amino acid	RP and macular degeneration	[20]
C_{1117}GC>*T*GC	Single amino acid substitution (Arg373Cys)	Autosomal dominant macular dystrophy	[21, 22]

[a]With reference to the PROMININ-1.s2 splice variant (for nomenclature of splice variants, see Ref. [17])
RP retinitis pigmentosa

with progression to profound visual impairment, and extinguished electroretinograms (ERG) by their third decade. Fundus examination revealed narrowed arteries, optic disc pallor, pigment deposits, and macular degeneration. Zhang and colleagues have reported a second mutation (nonsense mutation $C_{1726}\rightarrow T$) also creating a premature stop (Gln576X) in a consanguineous Pakistani pedigree (Fig. 4.1a, Q576X) [18]. The symptoms of five out of six affected individuals were night blindness and decreased visual acuity that appeared in early childhood. Nystagmus (i.e., uncontrollable movements of the eyes) was also observed. Funduscopic examination revealed waxy-pale discs, attenuation of the retinal arteries, macular degeneration, and typical bone-spicule pigmentation in the mid-peripheral retina. A typical bull's-eye macular degeneration was observed in three patients. ERG recordings revealed extensive loss of rod and cone function in two affected individuals indicating severe forms of rod-cone degeneration consistent with RP [18]. Pras and colleagues have described a third mutation (homozygous insertion of T at position 1349) creating a frameshift starting at codon 452 and a putative stop codon 12 amino acids downstream in the translated protein (Fig. 4.1a, Y452LfsX12) in a consanguineous Arab family [19]. This autosomal recessive cone-rod dystrophy with a high myopia was characterized by severe visual impairment evident in the first decade of life, bull's-eye macular appearance, peripheral retinal pigment clumps, and cone-rod type ERG changes [19]. Permanyer and colleagues have reported more recently another autosomal recessive mutation (ΔG_{869}; Fig. 4.1a, S289EfsX1) in the *PROM1* gene that is responsible for severe RP with macular degeneration and myopia in a consanguineous pedigree from Spain [20]. The three severely impaired members displayed night blindness in early childhood and bilateral progressive decline in visual acuity. Upon ophthalmic survey, they were diagnosed with retinal dystrophy, in which not only rods but also cones were affected. Again, funduscopic examination revealed waxy-pale discs and discrete attenuation of retinal arterioles. Alteration in the retinal pigment epithelium in the macula area was notable in all cases [20].

The first two mutations disrupt the extracellular (EC) 3 domain by creating a truncated version of PROMININ-1 lacking half of the EC3 domain and the entire transmembrane domain (TM) 5 and intracellular domain (IC) 3 domains whereas the third mutation generates a shorter PROMININ-1 mutant lacking IC2 domain onward (Fig. 4.1a). The last mutation removes about two-thirds of the protein including the two extracellular loops. These truncated PROMININ-1 versions might not be transported to the cell surface as we have previously demonstrated for a similar mutant (G614EfsX12) of murine prominin-1 [16].

Finally, a fifth mutation (C_{1117}GC>*T*GC) has been identified, which results in a single amino acid substitution (R373C) and generates three forms of autosomal-dominant macular dystrophies (i.e., Stargardt-like macular dystrophy (Stargardt disease 4 (STGD4); OMIM 603786), bull's-eye macular dystrophy (MCDR2; OMIM 608051), and cone-rod dystrophy) [21, 22] (Fig. 4.1a, R373C). STGD4 was characterized by bilateral, symmetric, atrophic lesions in the macula and by the presence of yellow fundus flecks, whereas MCDR2 is characterized by bilateral annular retinal pigment epithelium atrophy at the macula [23, 24]. The third pedigree was characterized by both cone and rod photoreceptor degeneration. Comparison of disease haplotypes revealed that affected members in the STGD4, MCDR2, and cone-rod dystrophy families had distinct haplotypes, indicating that the *PROM1* mutation in each family arose independently [21]. The introduction of an additional cysteine residue in the EC2 domain may promote an irreversible disulfide bridging with a second prominin-1 molecule or another protein. The dominant nature of this mutation supports the hypothesis that prominin-1 undergoes a dimerization process (see below) [25].

Thus, *PROM1* is the host of several mutations that are responsible for various retinal phenotypes and might thus become an ideal target for gene therapy.

4.2.2 Photoreceptor Morphogenesis

To understand the importance of PROMININ-1 in the retina in the context of human diseases, it is essential to know its precise cellular and subcellular localizations.

In vertebrates, the retina has an "orderly" laminated structure, where nuclei and cellular processes are segregated into alternate, anatomically distinct layers. Photoreceptor cells that lie in the outer domain of the retina in close contact with the retinal pigment epithelium are responsible for the generation of the neuronal signal in response to light. Two types of photoreceptor cells are found in the vertebrate retina, rods, and cones. The latter mediate vision in bright light including color perception, while rods mediate vision in dim light. Rod outer segments (ROS) house several discs that contain all the elements of the photo-transduction machinery required for photon collection and amplification. The continuous formation of new discs at the proximal bases and the shedding of older ones from their distal ends renew the outer segments. Nascent discs develop from the connecting cilium as flattened, cholesterol-enriched plasma membrane evaginations [26, 27]. Further plasma membrane outgrowths between the rims of two adjacent membrane evaginations lead to their merging and eventually their sealing, which results in the release of newly formed discs into the cytoplasm of ROS. In cones, the discs retain the topology of lamellar evaginations in continuity with the plasma membrane [26] (Note that the vesicular targeting model has been proposed as an alternative scenario to explain the biogenesis of the outer segment.) ([28, 29]; for additional information, see Ref. [30]).

In mice, prominin-1 is specifically located in the outer segment of photoreceptors [16, 21, 31], where it appears to be concentrated in newly synthesized plasma membrane evaginations at the base of ROS [16] (Fig. 4.2e). In humans, a similar localization is observed using an antiserum (αhE2) directed against PROMININ-1 polypeptide [6] (Fig. 4.1b; outer segment, OS; yellow bar). It is highly interesting to note that under the same immunohistochemistry procedures, the immunoreactivity for the anti-PROMININ-1 monoclonal antibody AC133, which is commonly used to detect human stem cells, is observed mainly in cone outer segment (COS), suggesting that its epitope is conformational dependent [32] and that the accessibility of the AC133 epitope is different between rods and cones, (Fig. 4.1c1, c2; arrowhead). In agreement with this possibility, the use of sensitive techniques for the retrieval of a particular epitope allows detection with AC133 in rods as well [33]. This differential accessibility may be a consequence of the morphological distinction between rods and cones, i.e., close or open disc structures, respectively. Prominin-1 expression in cones indirectly suggests that it is most likely involved in a particular biogenesis step common to both rods and cones and not specific for the release of the disc membrane within the ROS cytoplasm.

Direct evidence for the structural and mechanistic involvement of prominin-1 in photoreceptor outer segment morphogenesis was provided by studies using two distinct murine models. First, together with our colleague Peter Carmeliet (Belgium), we observed, using animals lacking prominin-1, an age-dependent deterioration of the photoreceptor structure. In 6-month-old *Prom-1*$^{-/-}$ mice, outer segments were disorganized and mostly degraded, while in older animals (e.g., 1-year-old), the entire photoreceptor layer was missing, resulting in blindness [31]. The photoreceptor defect was specific, since the morphology of two other protruding structures, microvilli and basal infolding of retinal pigment epithelial cells, appeared unaffected in *Prom-1*$^{-/-}$ eyes. Similarly, other retinal cell types were not affected [31].

The second study used transgenic knock-in mice carrying the human R373C *PROM1* mutation (see above). By 16 days after birth, these animals displayed outer segments with greatly overgrown and abnormally oriented disc membranes.

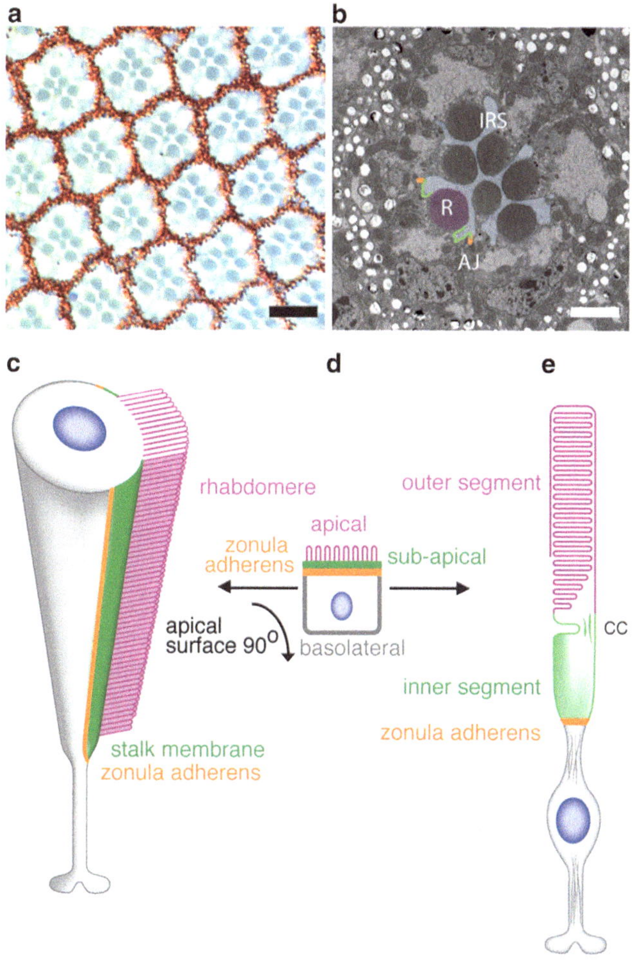

Fig. 4.2 The *Drosophila* and vertebrate photoreceptor neurons have conserved cell polarity and topology. (**a**) Horizontal section through a wild-type *Drosophila* eye. Ommatidia are separated from each other by pigment cells (*red pigment grana*). In each ommatidium within a single section, seven photoreceptors can be recognized by their dark rhabdomeres, which point toward each other and are stereotypically arranged in a trapezoid-like pattern. (**b**) Electron micrograph of a wild-type *Drosophila* ommatidium. *R* rhabdomere (*magenta*), *IRS* interrhabdomeral space (*light blue*), *AJ* adherens junction between photoreceptor cells (*brown*). *Green*: stalk membranes between the rhabdomere and the adherens junction. Scale bars: (**a**) 5 μm; (**b**) 1.7 μm. (**c**–**e**) Cartoons of a *Drosophila* photoreceptor cell (**c**), the polarized neuroepithelial cell (**d**), and the vertebrate rod photoreceptor cell (**e**). The *Drosophila* and the vertebrate rod photoreceptor cell develop from polarized epithelial cells (**d**). These are characterized by a highly specialized apical membrane, separated into the most apical region (*magenta*) and the subapical region (*green*). Adherens junctions (*brown*) separate the apical from the basolateral membrane and connect adjacent cells. In adult photoreceptor cells, the apical domain expands by forming either stacked microvilli to build the rhabdomere in *Drosophila* (**c**) or by stacked membrane discs to build the vertebrate photoreceptor cell outer segment (**e**). The subapical membrane domain (*green*) differentiates into the stalk membrane in *Drosophila* (**c**) and the inner segment in vertebrates (**e**). In vertebrate photoreceptor cells, the connecting cilium (*CC*) connects the inner and outer segment (**c–e** Adapted and modified Ref. [74])

By 4 months of age, the remaining photoreceptors contained either vesiculated outer segment membranes or stacks of disc membranes that were of excessive size and aligned perpendicular to normal disc orientation, pointing to a defect in disc morphogenesis [21]. Both mutant (human R373C) and endogenous (mouse) prominin-1 were distributed throughout the photoreceptor layer (cone and rod), which is consistent with the dominant nature of this mutation. Remarkably, a similar phenotype has been documented in eyecup cultures and in retinas *in vivo* following cytochalasin-D treatment, which inhibits actin polymerization [34, 35]. These investigations provided evidence that the overgrowth was caused by continued membrane evaginations and a failure in the initiation of new disc membrane, i.e., in the initial protrusion of the ciliary plasma membrane. In agreement with these results was the observation that actin could be co-immunoprecipitated with prominin-1, together with protocadherin 21 (PCDH21). PCDH21 is a photoreceptor-specific cadherin, which plays an essential role for the structural integrity of the outer segment and for photoreceptor survival [36]. Its proteolytic cleavage is a prerequisite of outer segment assembly [37]. In the R373C *PROM1* transgenic mouse lines, PCDH21 revealed impaired proteolytic cleavage and was mislocalized in a pattern similar to that of prominin-1 [21]. Because prominin-1 and protocadherin 21 are both present on the nascent photoreceptor discs [16, 36], their interaction is likely to occur within the same membrane and/or between adjacent membranes of two nascent discs. There they may play a role in maintaining the appropriate spacing between neighboring disc membranes and hence inhibit the fusion step that releases the newly synthesized discs within the ROS cytoplasm. The interaction of a particular, prominin-1–containing cholesterol-dependent membrane microdomain [13, 38] with cytoskeleton elements such as actin as well as unidentified cytoplasmic protein(s) may be necessary to maintain the temporal and spatial distribution of the prominin-1/PCDH21 complex [21]. Moreover, such protein-lipid complexes might be active at the base of ROS, i.e., on membrane evaginations, and within the entire COS, explaining somehow the morphological difference between rod and cone outer segments.

Together, the clinical and experimental observations support the hypothesis that prominin-1 in association with other proteins (secreted, membranous, and/or cytoplasmic) and eventually membrane lipids may provide a new scaffolding mechanism in mammalian disc membrane morphogenesis.

4.3 Prominin-1 in Non-mammalian Vertebrate Eyes

Prominin-1 is also expressed in non-mammalian retinas such as zebrafish (*Danio rerio*), axolotls (*Ambystoma mexicanum*) and frogs (*Xenopus laevis*) [25, 39–41]. It is interesting to note that the gene is duplicated in zebrafish, and both prominin-1 molecules (called prominin-1a and prominin-1b) are simultaneously expressed in photoreceptor cells [40].

Although the retina of *Xenopus laevis* has been studied for more than half a century [42], experimental evidence for prominin family members in frogs was reported only recently by David Papermaster's group. The *X. laevis* prominin family consists of three members: xlProminin-1, the closest ortholog of human PROMININ-1; xlProminin-2; and xlProminin-3 [41]. For a phylogenic analysis of prominin homologs from various species, readers are referred to the literature [41, 43] (see also Chap. 2). As reported for human (see above), xlProminin-1 was localized at the outer rims of open disc lamellae of COS, and, in a light cycle-dependent way, at the open lamellae at the base of ROS [44]. In addition, xlProminin-1 was also found diffusely distributed in the ROS, as described in murine ROS [16], which may reflect trapped xlProminin-1 in closed discs moving toward the distal end of the ROS [44]. Again, the sub-cellular localization of prominin-1 supports the model of disc morphogenesis in rods as proposed by Steinberg and colleagues [26], in which the basal discs are open to the exterior environment and are structurally distinct from the closed distal ones in organization of their membrane proteins and lipids [45].

Importantly, a differential distribution of xlProminin-1 and the retinal degeneration slow (rds) protein, also known as peripherin-2, was noted in COS and ROS. Peripherin-2/rds is an integral membrane glycoprotein of the tetraspanin family. Mutations in peripherin-2/rds are associated with various late-onset retinal dystrophies in human and dysmorphic photoreceptors in *X. laevis* (reviewed in Refs. [46, 47]). In contrast to xlProminin-1, peripherin-2/rds distributes to the inner rim on the same side as the connecting cilium in COS and to the entire rim of closed mature discs in ROS [44, 48]. This suggests that xlProminin-1 may play an anti-fusogenic function to preserve the open lamellae structure of the outer segments by counteracting the membrane fusogenic activity of peripherin-2/rds [49].

The large size of the *X. laevis* photoreceptor cells makes them ideally suited to directly test the role of xlProminin-1 in membrane turnover and directed intracellular transport, as well as to dissect its role in the different cone subtypes of the frog retina.

4.4 Prominin in Fly Eyes

Drosophila melanogaster contains two gene products related to mammalian prominin-1 and prominin-2, named Prominin (prom) and Prominin-like [43, 50, 51]. Strikingly, despite an entirely different organization of insect photoreceptor cells compared to those of vertebrates, *prom* plays also a crucial role in their morphogenesis. It is tempting to speculate that the cell biological function of *Drosophila* Prom in photoreceptors is comparable to that reported for vertebrate prominin-1. Given the power of *Drosophila* genetics and the reiterative nature of the fly eyes [52, 53], the fly retina provides an attractive system to study the role of this evolutionarily conserved molecule in retinal development and pathology [54].

4.4.1 The Organization of the Fly Retina

Drosophila melanogaster has a compound eye, composed of ≈750 individual units, called ommatidia, which are arranged in a highly stereotypic hexagonal pattern (reviewed in Ref. [55]). Each ommatidium contains eight photoreceptor cells, which form a cylindrical structure, with the apical surfaces facing a central lumen [56] (Fig. 4.2a). As in vertebrates, insect photoreceptors develop an extremely elaborated apical surface, called the rhabdomere, which is formed by about 50,000 densely packed microvilli, each about 1.5-μm in length and 50-nm in width [57]. The rhabdomere, which forms the light-sensing organelle and harbors the visual pigment rhodopsin, is the functional equivalent of the outer segment of photoreceptors in vertebrates. Both rhabdomeric microvilli and the outer segment membrane discs are aligned orthogonal to the eye's optical axis. The rhabdomere is separated from the zonula adherens (ZA) by the stalk membrane, a supporting structure, which corresponds to the inner segment of vertebrate photoreceptors. The ZA connects neighboring photoreceptor cells with each other (Fig. 4.2b).

4.4.2 Fly Photoreceptor Morphogenesis

Similar to vertebrates, *Drosophila* photoreceptors develop from a polarized neuroepithelium, which is formed by the eye imaginal disc, a monolayer of polarized epithelial cells, tightly held together by adherens junctions [58] (Fig. 4.3d). During the third instar larval stage, photoreceptors are specified and groups of eight form the ommatidial cluster [59]. From 37% pupal development (p.d.) onward, the apical poles adopt a lateral position. Thereby, the apical membranes of the eight photoreceptors of an ommatidium, which are characterized by disorganized membrane projections, are in contact with each other [60, 61]. By 55% p.d., the apical membrane domain becomes separated into two distinct regions, the central incipient rhabdomere, distinguished by forming microvilli, and the adjacent smooth stalk membrane. This is followed by an elongation of photoreceptor cells, which results in an increase in the depth of the retina from about 20 to 100 μm. Concomitantly, a central lumen is formed, called the interrhabdomeral space (IRS), which separates the apical surfaces from each other [61]. Formation of the lumen depends on *eyes shut* (*eys*) (also known as *spacemaker*, *spam*), which encodes a secreted protein related to the proteoglycans agrin and perlecan, produced by photoreceptor cells and released into the IRS, probably via the stalk membrane [62]. Eys is necessary and sufficient for the formation of the IRS [50, 63].

Segregation of apical proteins to form the rhabdomere and the stalk membrane is a key event observed during photoreceptor cell morphogenesis. While F-actin highlights the forming microvillar domain, the developing stalk membranes concentrate members of the Crumbs protein complex [64–66]. The Crumbs complex is an evolutionarily conserved, membrane-associated protein complex, composed of the

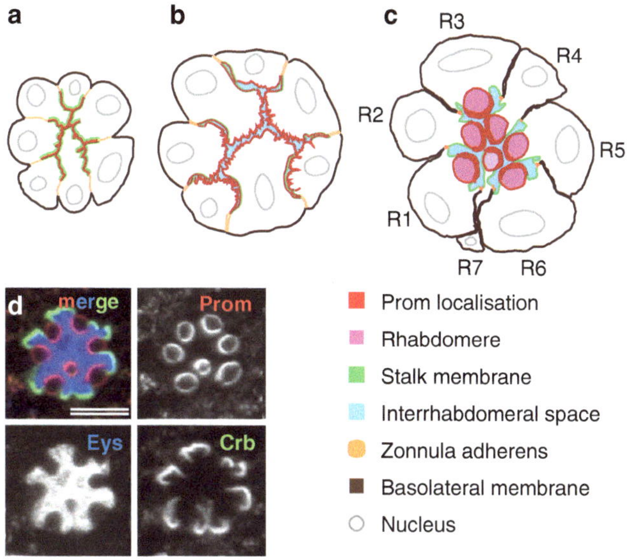

Fig. 4.3 Prom localization during morphogenesis of *Drosophila* photoreceptor cells. (**a–c**) Schematic cross sections of ommatidia at 37% pupal development (p.d.) (**a**), 55% p.d. (**b**), and adult eyes (**c**) to demonstrate localization of prom (*red*). (**a**) At 37% p.d., prom is localized on the entire apical membrane. At this stage, the apical surfaces of the photoreceptor cells are in contact with each other. (**b**) At 55% p.d., the apical membranes differentiate into a central, microvilli containing region (*red*) and the stalk membrane (*green*). Formation of the interrhabdomeral space (IRS; *blue*) is initiated. Prom remains localized to the entire apical membrane. (**c**) Adult ommatidium, displaying the stereotypical trapezoid arrangement of completely developed rhabdomeres (*pink*). Prom is enriched at the tip and the base of the rhabdomeres of the outer photoreceptor cells *R1–R7*. The apical compartment consists of two specialized membrane domains, the stalk membrane (*green*) and the rhabdomere (central-apical compartment) (*magenta*). An IRS separates individual rhabdomeres from each other. (**d**) Confocal image of a tangential section of an adult ommatidium stained for prom, eys, and Crb. Scale bar: 5 μm

transmembrane protein Crumbs and the scaffolding proteins Stardust, *Drosophila* PATJ and *Drosophila* Lin-7. It is involved in photoreceptor morphogenesis and prevents light-dependent retinal degeneration (reviewed in Ref. [67]).

4.4.3 Role of Prominin in Photoreceptor Morphogenesis

Immunolabeling studies in the adult flies revealed that Prom is localized to the rhabdomere of photoreceptors, showing the highest concentration at the tip and a second enrichment at their base [50] (Fig. 4.3d). The exact role of Prom at these sites, which are characterized by curvature of the membrane, is not well understood, but it is consistent with the preference of prominin-1 to highly curved membranes (see above). Eys can be found in the IRS, but mostly marks its periphery, while the Crumbs complex is restricted to the stalk membrane of photoreceptor cells (Fig. 4.3d).

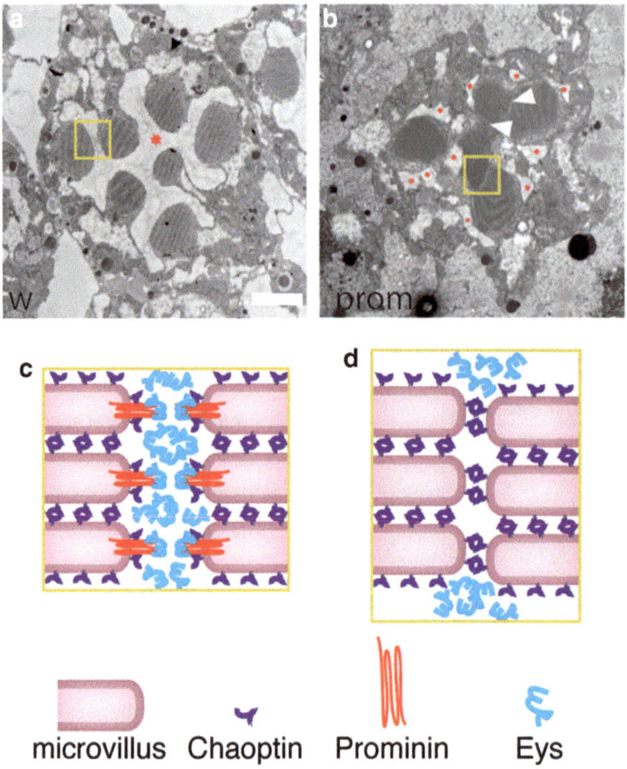

Fig. 4.4 Prom is essential for separation of rhabdomeres in *Drosophila* ommatidia. (**a, b**) Electron micrographs showing cross sections through a wild-type (*w*; **a**) and *prom* mutant (**b**) *Drosophila* ommatidium. *White arrowheads* in (**b**) mark interrhabdomeral contacts in *prom* mutants. The interrhabdomeral space (*red asterisks*) is not continuous in *prom* mutants due to fused rhabdomeres. *Yellow squares* indicate the regions shown in (**c**) and (**d**). Scale bar: 1.7 μm. (**c, d**) Model for inter-rhabdomere membrane separation in wild-type (**c**) and *prom* mutant (**d**). It is hypothesized that in wild-type the GPI-anchored membrane protein chaoptin maintains rhabdomere integrity by mediating intra-rhabdomeral microvillar adhesion. In addition, it mediates inter-rhabdomeral fusion at early developmental stages (not shown). The cooperative activity of Prom and the secreted protein, Eys, antagonizes inter-rhabdomeral adhesion, resulting in the formation of individual rhabdomeres (**c**). Absence of Prom prevents inter-rhabdomeral separation, since Eys alone is not sufficient to counteract Chaoptin-mediated adhesion between microvilli of apposing rhabdomeres (**d**)

Rhabdomeres of wild-type ommatidia are well separated from each other by the central IRS (Figs. 4.3c, 4.4a). Remarkably, loss of *prom* prevents proper separation of the rhabdomeres, manifested by fused rhabdomeres and a fragmentation of the IRS [50] (Fig. 4.4b). This phenotype is similar, though weaker than that observed in *eys* mutant flies, where all rhabdomeres are fused and the IRS is completely missing [50]. This led to the speculation that Prom and Eys act together to antagonize the adhesion that initially connects the apical membranes of photoreceptor cells with each other. Adhesion between the microvilli of rhabdomeres is mediated by Chaoptin, a glycosylphosphatidylinositol-anchored membrane protein enriched on rhabdomeral microvilli [68]. Photoreceptor cells lacking Chaoptin fail to properly align their microvilli at early stages of development, which results in massive microvillar disorganization or, in extreme cases, in the complete absence of rhabdomeres in adult photoreceptor cells [69]. From these results and biochemical studies, it was concluded that Chaoptin mediates adhesion between microvilli through homophilic interactions [70]. The phenotype of *prom* and *eys* mutant ommatidia suggests that in wild-type flies, adhesion between microvilli of different rhabdomeres – also mediated by Chaoptin – is neutralized by Prom after binding to Eys. This results in the separation of the microvilli of adjacent rhabdomeres and thus the formation of the IRS (Fig. 4.4a, c). Without Prom, separation does not occur (Fig. 4.4b, d). Indeed, removing of one gene copy of *chaoptin* strongly suppresses the *prom* mutant phenotype [50]. In the absence of Prom, Eys is normally secreted so that an IRS can be formed.

Together, the available data have led to the following model: Initially, apical membranes are in contact with each other, held together by Chaoptin. Although Prom is localized apically at this stage, it is unable to work against Chaoptin-mediated adhesion. Separation of the apical membranes during photoreceptor cell morphogenesis coincides with the secretion of Eys, suggesting that only interaction of Eys with Prom is able to counteract chaoptin function. In fact, cell culture experiments support a direct interaction between Prom and Eys [50]. It is worth noting that Prom/Eys do not act on Chaoptin-mediated adhesion between microvilli within the rhabdomere.

Thus, Prom, in concert with other protein partner(s), seems to organize the photoreceptive membranes in diverse organisms. It will be of interest to find out whether an analogous interaction exists between vertebrate Prominin-1 and EYS and, if so, whether this interaction is also required to antagonize an as yet unidentified adhesion molecule in the vertebrate retina.

Interestingly, *prom*, *eys*, and *chaoptin* are evolutionarily conserved in arthropods. While *Drosophila* and other insects, e.g., the housefly *Musca domestica*, develop an open rhabdom, characterized by an IRS, others, like the honey bee *Apis mellifera* or the midge *Anopheles gambiae*, form a closed rhabdom, with all rhabdomeres fused to each other, very similar to the *eys* mutant rhabdomeres of *Drosophila*. Strikingly, *eys* is not expressed in the eyes of insects with a closed rhabdom [50, 63], suggesting that expression of *eys* was instrumental for the evolutionary transformation from a closed to an open rhabdom in arthropods.

The second member of the *Drosophila* Prominin family, Prominin-like, was detected in the proteome of *Drosophila* sperm [71]. A Prominin-like – green fluorescent protein fusion protein – when ectopically expressed in *Drosophila* wing imaginal discs is enriched at apical protrusions and weakly detected at lateral membrane protrusions [51]. These observations are in agreement with those published previously on vertebrate prominin-2 overexpression [43]. However, due to lack of mutations in *prominin-like*, data on its physiological relevance are still missing.

Many questions regarding Prom in *Drosophila* are yet to be answered. For example: How does Prom become apically localized in fly photoreceptor cells? Does it also define a membrane microdomain as described for the mammalian ortholog? Does its loss result in retinal degeneration, as described for mouse and human, and if yes, what is the molecular mechanism? What are the other molecules (proteins and lipids) that might interact with Prom during development and homeostasis of *Drosophila* photoreceptor cells?

4.5 Outlook

So far we know that prominin-1 is conserved from humans to insects. Furthermore, in most cases prominin-1 is enriched at apical membrane protrusions, independently of whether it is expressed endogenously or ectopically. Nonetheless, we are still lacking a detailed picture on its physiological role in the diverse cell types. It is still unclear whether prominin-1 just plays a role in supporting protrusive structures, like lamellae in vertebrate outer segments, or whether it additionally is part of a signal transduction pathway. Given the observation that the human ortholog of *Drosophila eys*, EYS, also leads to recessive RP when mutated [72, 73], it might be interesting to find out whether human eys also interacts with human PROMININ-1 in a similar way as the corresponding proteins in *Drosophila*.

Using both mice and flies as genetically tractable organisms, we are now in a very good position to further dissect the role of Prominin in order to get a deeper understanding of its role in retinal development and degeneration as observed in human diseases. In addition, studying the retinas of fish, frogs, or axolotls, which, in contrast to mice, contain nearly equal numbers of cones and rods, may help to analyze a possible function of prominin-1 for the particular differentiation of these two cell types.

Acknowledgements We thank Shirin Meher Pocha for critical comments on the manuscript. D.C. and E.K. are supported by the Deutsche Forschungsgemeinschaft SFB655.

References

1. Fargeas CA, Fonseca AV, Huttner WB, Corbeil D (2006) Prominin-1 (CD133): from progenitor cells to human diseases. Future Lipid 1:213–225
2. Singh SK, Clarke ID, Terasaki M, Bonn VE, Hawkins C et al (2003) Identification of a cancer stem cell in human brain tumors. Cancer Res 63:5821–5828
3. Singh SK, Hawkins C, Clarke ID, Squire JA, Bayani J et al (2004) Identification of human brain tumour initiating cells. Nature 432:396–401
4. O'Brien CA, Pollett A, Gallinger S, Dick JE (2007) A human colon cancer cell capable of initiating tumour growth in immunodeficient mice. Nature 445:106–110
5. Weigmann A, Corbeil D, Hellwig A, Huttner WB (1997) Prominin, a novel microvilli-specific polytopic membrane protein of the apical surface of epithelial cells, is targeted to plasmalemmal protrusions of non-epithelial cells. Proc Natl Acad Sci USA 94:12425–12430

6. Florek M, Haase M, Marzesco A-M, Freund D, Ehninger G et al (2005) Prominin-1/CD133, a neural and hematopoietic stem cell marker, is expressed in adult human differentiated cells and certain types of kidney cancer. Cell Tissue Res 319:15–26
7. Jászai J, Janich P, Farkas LM, Fargeas CA, Huttner WB et al (2007) Differential expression of prominin-1 (CD133) and prominin-2 in major cephalic exocrine glands of adult mice. Histochem Cell Biol 128:409–419
8. Karbanová J, Missol-Kolka E, Fonseca AV, Lorra C, Janich P et al (2008) The stem cell marker CD133 (prominin-1) is expressed in various human glandular epithelia. J Histochem Cytochem 56:977–993
9. Shmelkov SV, Butler JM, Hooper AT, Hormigo A, Kushner J et al (2008) CD133 expression is not restricted to stem cells, and both CD133+ and CD133- metastatic colon cancer cells initiate tumors. J Clin Invest 118:2111–2120
10. Corbeil D, Joester A, Fargeas CA, Jászai J, Garwood J et al (2009) Expression of distinct splice variants of the stem cell marker prominin-1 (CD133) in glial cells. Glia 57:860–874
11. Fargeas CA, Joester A, Missol-Kolka E, Hellwig A, Huttner WB et al (2004) Identification of novel prominin-1/CD133 splice variants with alternative C-termini and their expression in epididymis and testis. J Cell Sci 117:4301–4311
12. Kemper K, Tol MJ, Medema JP (2010) Mouse tissues express multiple splice variants of prominin-1. PLoS One 5:e12325
13. Röper K, Corbeil D, Huttner WB (2000) Retention of prominin in microvilli reveals distinct cholesterol-based lipid micro-domains in the apical plasma membrane. Nat Cell Biol 2:582–592
14. Corbeil D, Röper K, Fargeas CA, Joester A, Huttner WB (2001) Prominin: a story of cholesterol, plasma membrane protrusions and human pathology. Traffic 2:82–91
15. Florek M, Bauer N, Janich P, Wilsch-Bräuninger M, Fargeas CA et al (2007) Prominin-2 is a cholesterol-binding protein associated with apical and basolateral plasmalemmal protrusions in polarized epithelial cells and released into the urine. Cell Tissue Res 328:31–47
16. Maw MA, Corbeil D, Koch J, Hellwig A, Wilson-Wheeler JC et al (2000) A frameshift mutation in prominin (mouse)-like 1 causes human retinal degeneration. Hum Mol Genet 9:27–34
17. Fargeas CA, Huttner WB, Corbeil D (2007) Nomenclature of prominin-1 (CD133) splice variants – an update. Tissue Antigens 69:602–606
18. Zhang Q, Zulfiqar F, Xiao X, Riazuddin SA, Ahmad Z et al (2007) Severe retinitis pigmentosa mapped to 4p15 and associated with a novel mutation in the PROM1 gene. Hum Genet 122:293–299
19. Pras E, Abu A, Rotenstreich Y, Avni I, Reish O et al (2009) Cone-rod dystrophy and a frameshift mutation in the PROM1 gene. Mol Vis 15:1709–1716
20. Permanyer J, Navarro R, Friedman J, Pomares E, Castro-Navarro J et al (2010) Autosomal recessive retinitis pigmentosa with early macular affectation caused by premature truncation in PROM1. Invest Ophthalmol Vis Sci 51:2656–2663
21. Yang Z, Chen Y, Lillo C, Chien J, Yu Z et al (2008) Mutant prominin 1 found in patients with macular degeneration disrupts photoreceptor disk morphogenesis in mice. J Clin Invest 118:2908–2916
22. Michaelides M, Gaillard MC, Escher P, Tiab L, Bedell M et al (2010) The PROM1 mutation p.R373C causes an autosomal dominant bull's eye maculopathy associated with rod, rod-cone, and macular dystrophy. Invest Ophthalmol Vis Sci 51:4771–4780
23. Michaelides M, Johnson S, Poulson A, Bradshaw K, Bellmann C et al (2003) An autosomal dominant bull's-eye macular dystrophy (MCDR2) that maps to the short arm of chromosome 4. Invest Ophthalmol Vis Sci 44:1657–1662
24. Arrigoni FI, Matarin M, Thompson PJ, Michaelides M, McClements ME et al (2011) Extended extraocular phenotype of PROM1 mutation in kindreds with known autosomal dominant macular dystrophy. Eur J Hum Genet 19:131–137
25. Jászai J, Fargeas CA, Florek M, Huttner WB, Corbeil D (2007) Focus on molecules: prominin-1 (CD133). Exp Eye Res 85:585–586

26. Steinberg RH, Fisher SK, Anderson DH (1980) Disc morphogenesis in vertebrate photoreceptors. J Comp Neurol 190:501–508
27. Boesze-Battaglia K, Hennessey T, Albert AD (1989) Cholesterol heterogeneity in bovine rod outer segment disk membranes. J Biol Chem 264:8151–8155
28. Chuang JZ, Zhao Y, Sung CH (2007) SARA-regulated vesicular targeting underlies formation of the light-sensing organelle in mammalian rods. Cell 130:535–547
29. Obata S, Usukura J (1992) Morphogenesis of the photoreceptor outer segment during postnatal development in the mouse (BALB/c) retina. Cell Tissue Res 269:39–48
30. Sung CH, Chuang JZ (2010) The cell biology of vision. J Cell Biol 190:953–963
31. Zacchigna S, Oh H, Wilsch-Bräuninger M, Missol-Kolka E, Jászai J et al (2009) Loss of the cholesterol-binding protein prominin-1/CD133 causes disk dysmorphogenesis and photoreceptor degeneration. J Neurosci 29:2297–2308
32. Bidlingmaier S, Zhu X, Liu B (2008) The utility and limitations of glycosylated human CD133 epitopes in defining cancer stem cells. J Mol Med 86:1025–1032
33. Immervoll H, Hoem D, Sakariassen PO, Steffensen OJ, Molven A (2008) Expression of the "stem cell marker" CD133 in pancreas and pancreatic ductal adenocarcinomas. BMC Cancer 8:48
34. Williams DS, Linberg KA, Vaughan DK, Fariss RN, Fisher SK (1988) Disruption of microfilament organization and deregulation of disk membrane morphogenesis by cytochalasin D in rod and cone photoreceptors. J Comp Neurol 272:161–176
35. Vaughan DK, Fisher SK (1989) Cytochalasin D disrupts outer segment disc morphogenesis in situ in rabbit retina. Invest Ophthalmol Vis Sci 30:339–342
36. Rattner A, Smallwood PM, Williams J, Cooke C, Savchenko A et al (2001) A photoreceptor-specific cadherin is essential for the structural integrity of the outer segment and for photoreceptor survival. Neuron 32:775–786
37. Rattner A, Chen J, Nathans J (2004) Proteolytic shedding of the extracellular domain of photoreceptor cadherin. Implications for outer segment assembly. J Biol Chem 279:42202–42210
38. Janich P, Corbeil D (2007) GM1 and GM3 gangliosides highlight distinct lipid microdomains within the apical domain of epithelial cells. FEBS Lett 581:1783–1787
39. McGrail M, Batz L, Noack K, Pandey S, Huang Y et al (2010) Expression of the zebrafish CD133/prominin1 genes in cellular proliferation zones in the embryonic central nervous system and sensory organs. Dev Dyn 239:1849–1857
40. Jászai J, Fargeas CA, Graupner S, Tanaka EM, Brand M et al (2011) Distinct and conserved prominin-1/CD133–positive retinal cell populations identified across species. PLoS One 6:e17590
41. Han Z, Papermaster DS (2011) Identification of three prominin homologs and characterization of their messenger RNA expression in Xenopus laevis tissues. Mol Vis 17:1381–1396
42. Röhlich P, Szél A (2000) Photoreceptor cells in the Xenopus retina. Microsc Res Tech 50:327–337
43. Fargeas CA, Florek M, Huttner WB, Corbeil D (2003) Characterization of prominin-2, a new member of the prominin family of pentaspan membrane glycoproteins. J Biol Chem 278:8586–8596
44. Han Z, Anderson DW, Papermaster DS (2012) Prominin-1 localizes to the open rims of outer segment lamellae in Xenopus laevis rod and cone photoreceptors. Invest Ophthalmol Vis Sci 53:361–373
45. Albert AD, Boesze-Battaglia K (2005) The role of cholesterol in rod outer segment membranes. Prog Lipid Res 44:99–124
46. Goldberg AF (2006) Role of peripherin/rds in vertebrate photoreceptor architecture and inherited retinal degenerations. Int Rev Cytol 253:131–175
47. Farjo R, Naash MI (2006) The role of Rds in outer segment morphogenesis and human retinal disease. Ophthalmic Genet 27:117–122
48. Arikawa K, Molday LL, Molday RS, Williams DS (1992) Localization of peripherin/rds in the disk membranes of cone and rod photoreceptors: relationship to disk membrane morphogenesis and retinal degeneration. J Cell Biol 116:659–667
49. Damek-Poprawa M, Krouse J, Gretzula C, Boesze-Battaglia K (2005) A novel tetraspanin fusion protein, peripherin-2, requires a region upstream of the fusion domain for activity. J Biol Chem 280:9217–9224

50. Zelhof AC, Hardy RW, Becker A, Zuker CS (2006) Transforming the architecture of compound eyes. Nature 443:696–699
51. Demontis F, Dahmann C (2007) Apical and lateral cell protrusions interconnect epithelial cells in live Drosophila wing imaginal discs. Dev Dyn 236:3408–3418
52. Hardie R (2001) Phototransduction in Drosophila melanogaster. J Exp Biol 204:3403–3409
53. Knust E (2007) Photoreceptor morphogenesis and retinal degeneration: lessons from Drosophila. Curr Opin Neurobiol 17:541–547
54. Cook T, Zelhof A, Mishra M, Nie J (2011) 800 facets of retinal degeneration. Prog Mol Biol Transl Sci 100:331–368
55. Dickson B, Hafen E (1993) Genetic dissection of eye development in Drosophila. In: Bate M, Arias AM (eds) The development of Drosophila melanogaster. CSHL Press, NewYork, pp 1327–1362
56. Wolken JJ, Mellon AD, Contis G (1957) Photoreceptor structures. II. Drosophila melanogaster. J Exp Zool 134:383–409
57. Arikawa K, Hicks JL, Williams DS (1990) Identification of actin filaments in the rhabdomeral microvilli of Drosophila photoreceptors. J Cell Biol 110:1993–1998
58. Cagan RL, Ready DF (1989) The emergence of order in the Drosophila pupal retina. Dev Biol 136:346–362
59. Cagan R (2009) Principles of Drosophila eye differentiation. Curr Top Dev Biol 89:115–135
60. Waddington CH, Perry MM (1960) The ultrastructure of the developing eye of Drosophila. Proc Roy Soc B 153:155–178
61. Longley RL, Ready DF (1995) Integrins and the development of three-dimensional structure in the Drosophila compound eye. Dev Biol 171:415–433
62. Cook B, Zelhof AC (2008) Photoreceptors in evolution and disease. Nat Genet 40:1275–1276
63. Husain N, Pellikka M, Hong H, Klimentova T, Choe KM et al (2006) The agrin/perlecan-related protein eyes shut is essential for epithelial lumen formation in the Drosophila retina. Dev Cell 11:483–493
64. Richard M, Roepman R, Aartsen WM, van Rossum AG, den Hollander AI et al (2006) Towards understanding CRUMBS function in retinal dystrophies. Hum Mol Genet 15:R235–R243
65. Berger S, Bulgakova NA, Grawe F, Johnson K, Knust E (2007) Unraveling the genetic complexity of Drosophila stardust during photoreceptor morphogenesis and prevention of light-induced degeneration. Genetics 176:2189–2200
66. Izaddoost S, Nam SC, Bhat MA, Bellen HJ, Choi KW (2002) Drosophila Crumbs is a positional cue in photoreceptor adherens junctions and rhabdomeres. Nature 416:178–183
67. Bulgakova NA, Knust E (2009) The Crumbs complex: from epithelial-cell polarity to retinal degeneration. J Cell Sci 122:2587–2596
68. Reinke R, Zipursky SL (1988) Cell-cell interaction in the Drosophila retina: the bride of sevenless gene is required in photoreceptor cell R8 for R7 cell development. Cell 55:321–330
69. Van Vactor D Jr, Krantz DE, Reinke R, Zipursky SL (1988) Analysis of mutants in chaoptin, a photoreceptor cell-specific glycoprotein in Drosophila, reveals its role in cellular morphogenesis. Cell 52:281–290
70. Krantz DE, Zipursky SL (1990) Drosophila chaoptin, a member of the leucine-rich repeat family, is a photoreceptor cell-specific adhesion molecule. EMBO J 9:1969–1977
71. Dorus S, Busby SA, Gerike U, Shabanowitz J, Hunt DF et al (2006) Genomic and functional evolution of the Drosophila melanogaster sperm proteome. Nat Genet 38:1440–1445
72. Collin RW, Littink KW, Klevering BJ, van den Born LI, Koenekoop RK et al (2008) Identification of a 2 Mb human ortholog of Drosophila eyes shut/spacemaker that is mutated in patients with retinitis pigmentosa. Am J Hum Genet 83:594–603
73. Abd El-Aziz MM, Barragan I, O'Driscoll CA, Goodstadt L, Prigmore E et al (2008) EYS, encoding an ortholog of Drosophila spacemaker, is mutated in autosomal recessive retinitis pigmentosa. Nat Genet 40:1285–1287
74. Xia H, Ready DF (2011) Ectoplasm, ghost in the R cell machine? Dev Neurobiol 71:1246–1257

Chapter 5
Gene Regulation of Prominin-1 (CD133) in Normal and Cancerous Tissues

Kouichi Tabu, Norihisa Bizen, Tetsuya Taga, and Shinya Tanaka

Abstract A pentaspan membrane glycoprotein prominin-1 (frequently called CD133 in human) is widely used as a surface marker to identify and isolate normal stem/progenitor cells from various organs, although it is also expressed in some types of differentiated cells. Since CD133 was identified as a universal marker to isolate cancer stem cells (CSCs) in tumors derived from multiple tissues, much attention has been directed toward the relationship between its gene regulation and identity of CSCs (i.e., cancer stemness). *Prominin-1* (*PROM1*) gene possesses five alternative promoters yielding multiple first exons within the 5′-untranslated region (UTR) and also splicing variants affecting the open reading frame (ORF) sequence, implicating the complicated gene regulation in a context-dependent manner. This chapter aims to organize the accumulated findings on prominin-1 with a focus on its altered expression and regulation in normal and cancerous cells and to discuss potential regulatory networks underlying cancer stemness.

Keywords Cancer stem cell • CD133 • Epigenetics • Hypoxia • Metabolism • Prominin-1 • Transcription

---

K. Tabu, Ph.D. (✉) • N. Bizen • T. Taga
Department of Stem cell Regulation, Medical Research Institute, Tokyo Medical and Dental University, 1-5-45, Yushima, Bunkyo-ku, 113-8510 Tokyo, Japan
e-mail: k-tabu.scr@mri.tmd.ac.jp

S. Tanaka
Department of Pathology, Laboratory of Cancer Research, Hokkaido University Graduate School of Medicine, N15 W7, Kita-ku, 060-8638 Sapporo, Japan

D. Corbeil (ed.), *Prominin-1 (CD133): New Insights on Stem & Cancer Stem Cell Biology*, Advances in Experimental Medicine and Biology 777, DOI 10.1007/978-1-4614-5894-4_5,

5.1 Introduction

Prominin-1 (PROM1, also known as CD133 or AC133 in human) is a cell surface glycoprotein containing five transmembrane regions and two large glycosylated extracellular loops, typically comprising 842 (for mouse) and 865 (for human) amino acid residues with variable molecular weights depending on the length of splice variants and their glycosylation levels [1]. Prominin-1 was described independently by three groups in 1997 [2–4]. Huttner and colleagues reported it as a novel marker selectively localized at the apical plasma membrane, especially in microvilli and primary cilia of murine neuroepithelial stem cells [2], while Buck and colleagues described it on a subpopulation of human CD34(+) hematopoietic stem cells (HSCs) derived from fetal liver, bone marrow, and cord blood as the AC133 antigen [3]. In 2004, CD133, the human ortholog of mouse prominin-1, was identified as a surface marker to isolate brain tumor stem cells from human specimens [5], and hereafter, considerable attention has been devoted to understanding this molecule and its regulation. Actually, in a variety of human tumor tissues, a CD133(+) subpopulation exhibits a greater ability to recapitulate original tumors in immunodeficient mice [6–9]. Based on these observations, CD133 was initially expected as a possible target for cancer eradication, but it is still controversial whether CD133 directly leads to tumorigenesis or whether it is just a concomitant marker, which appears in a tumorigenic process.

The physiological function of prominin-1 remains unknown. *PROM1* null mice are born and normally age with no major phenotype despite its expression in many tissues [1, 10], except that a progressive degeneration of photoreceptors develops with complete loss of vision in adults [11, 12], in keeping with its critical expression in photoreceptor cells [13]. It is evident that deciphering the regulatory mechanisms upstream of *PROM1* gene are important for a better understanding of the nature of normal and cancerous stem cells. Here, we will review recent progress on *PROM1* gene regulation within the framework of stem cells as well as basic information of its gene structure.

5.2 Prominin-1 Gene Structure

The *PROM1* gene is located on chromosome 4p15.32 in the human and 5B3 in the mouse with a conserved genomic structure between animal species [14]. They are composed of at least 37 and 34 exons in the human and mouse, respectively. In mammals, 12 distinct splice variants affecting the prominin-1 open reading frame were described [1, 15]. From them, six alternative cytoplasmic C-terminal tails could be generated highlighting the complexity of prominin-1 molecules. Moreover, at least ten alternative exons from exon 1 (Exon 1A, B, C, D1, D2, D3, E1, E2, E3, and E4) were originally found to be transcribed within the 5′-untranslated region (UTR) [16]. The analyses of different human tissues and cell lines revealed that at least seven *prominin-1* transcripts with different exon 1s are driven by five alternative promoters (P1, P2, P3, P4, and P5). In addition, all these promoters are TATA-less, and three of them (P1, P2, and P3) are located within a cytosine-guanine-rich DNA region (CpG island). The expression of 5′-UTR isoforms exhibits tissue-dependent patterns. Liver, pancreas, placenta, lung, spleen, colon, kidney, and brain express exon 1A- and exon 1B-containing transcripts, while prostate and small intestine express only the exon 1A-containing transcript and ovary only the exon 1B-containing transcript. None of these isoforms have been detected in the heart, skeletal muscle, thymus, or bone marrow. Exon 1D is detected solely in the testis, and Exon 1C and 1E isoforms were not detected in any normal tissues until now. Our colleagues and we have previously reported that CD133 expression levels correlate with human glioma WHO grading and that malignant gliomas predominantly express exon 1A, 1B, and 1C isoforms [17]. In addition, at least three novel exon 1s, designated as exons 1C2, 1D4, and 1E5, and at least three elongated forms of known exon 1C, 1D1, and 1E2 were identified in human kidney, testis, brain, and glioma tissues, all of which conform to the GT-AG exon-intron consensus rule. The complete structure of *PROM1* gene promoter containing all identified exon 1s and their expression in normal and glioma tissues are digested in Fig. 5.1a, b, respectively. It should be noted however that additional combinations of exon 1 isoforms might be found in other normal and cancerous cells.

5.3 Prominin-1 in Normal Stem Cells

Since prominin-1 expression was initially observed in murine neuroepithelial stem cells, it has been well investigated in mouse cerebrum. Its expression is spatiotemporally regulated along cerebrum development and neural stem cell (NSC) differentiation. Following the formation of neural tube, the apical surface of the polarized neuroepithelial cells (NECs) is facing toward the lumen of the ventricular zone (VZ; Fig. 5.2a) [18]. Prominin-1 is selectively concentrated therein particularly in plasma membrane protrusions. Corbeil and colleagues have proposed that prominin-1 as a cholesterol-binding protein is somehow involved in remodeling of plasma membrane protrusions [19, 20].

At a later stage of cerebrum development (e.g., E10.5), NECs undergo asymmetric division and start to produce radial glial cells (RGCs) concomitantly with the disappearance of tight junctions (Fig. 5.2b) [21]. RGCs can respond to extracellular signals from their surrounding microenvironment that determine the specific cell lineages, such as neurons and glial cells. In E14.5, prominin-1 colocalizes with the RGC markers GLAST and Nestin in dorsal VZ lining the cortex but is absent at the ventral side of VZ lining the striatum [22]. This heterogeneous expression of prominin-1 can be explained by regional differences in development of RGCs. Prominin-1 is therefore expressed in a specific type of RGCs at a later stage of neural development.

In thc carly postnatal stage, RGCs disappear from the brain during the first 14 postnatal days (P) and give rise not only to mature glial cells but also to ependymal cells (ECs) and neurogenic astrocytes called "type B cells" in the adult subventricular zone (SVZ) (Fig. 5.2c) [23, 24]. Contrary to its location in E14.5, prominin-1 was reported to be located at the apical surface of GLAST-positive cells present at the

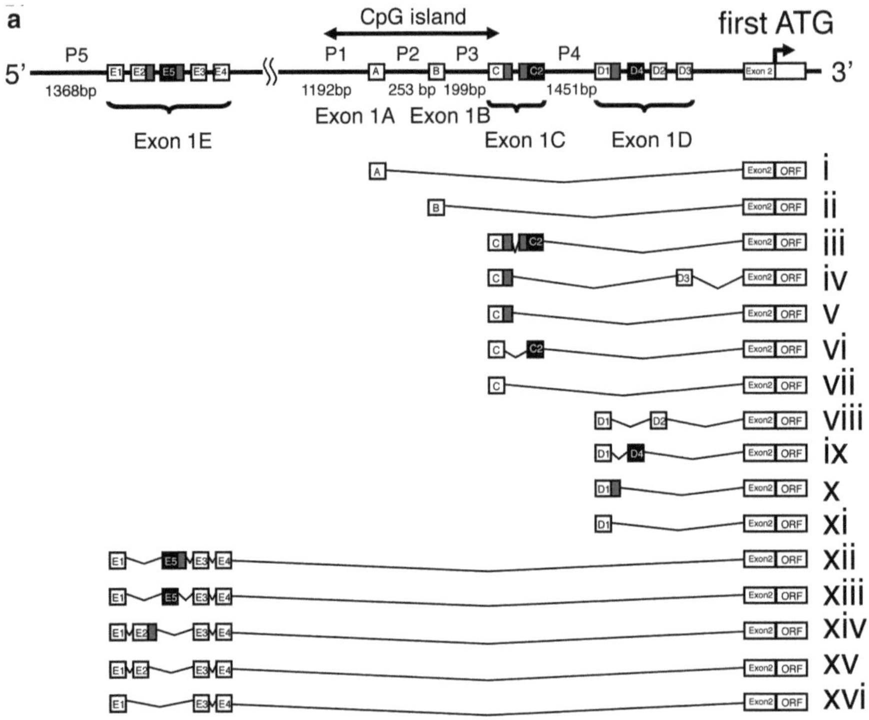

b

5′-UTR isoforms	liver	pancreas	placenta	lung	spleen	colon	prostate	small intestine	ovary	heart	skeletal muscle	thymus	bone marrow	kidney	testis	brain	glioma tissues
i	+	+	+	+	+	+	+	+	-	-	-	-	-	+	-	+	+
ii	+	+	+	+	+	+	-	-	+	-	-	-	-	+	-	+	+
iii	-	-	-	-	-	-	-	-	-	-	-	-	-	-	-	-	+
iv	-	-	-	-	-	-	-	-	-	-	-	-	-	-	-	+	-
v	-	-	-	-	-	-	-	-	-	-	-	-	-	+	-	+	+
vi	-	-	-	-	-	-	-	-	-	-	-	-	-	-	-	+	-
vii	-	-	-	-	-	-	-	-	-	-	-	-	-	+	-	-	+
viii	-	-	-	-	-	-	-	-	-	-	-	-	-	-	+	-	-
ix	-	-	-	-	-	-	-	-	-	-	-	-	-	-	+	-	-
x	-	-	-	-	-	-	-	-	-	-	-	-	-	-	+	-	-
xi	-	-	-	-	-	-	-	-	-	-	-	-	-	-	+	-	-
xii	-	-	-	-	-	-	-	-	-	-	-	-	-	-	+	-	-
xiii	-	-	-	-	-	-	-	-	-	-	-	-	-	-	+	-	-
xiv	-	-	-	-	-	-	-	-	-	-	-	-	-	-	+	-	-
xv	-	-	-	-	-	-	-	-	-	-	-	-	-	-	+	-	-
xvi	-	-	-	-	-	-	-	-	-	-	-	-	-	-	+	-	-
ref.	(16)													(17)			

Fig. 5.1 Structures of the human *PROM1* promoter region and expression of the spliced 5′-UTR isoforms in various tissues. (**a**) The *PROM1* gene possesses five promoters (P5, P1 through P4, in this order), yielding different 5′-UTR isoforms (i–xvi) with five alternated first exons (Exon 1A to 1E). A CpG island resides in the P1 through P3 promoter region. *Empty boxes* represent the exon 1s identified by Shmelkov and colleagues [16]. *Closed and gray boxes* represent the novel exon 1s and elongation of exon 1s, respectively, all of which were identified by Tabu and colleagues [17]. (**b**) Expression of PROM1 5′-UTR isoforms (i–xvi) in human normal and cancerous (glioma) tissues. Semiquantitative RT-PCR data in Refs. [16, 17] are summarized by "+" (detected) and "–" (not detected)

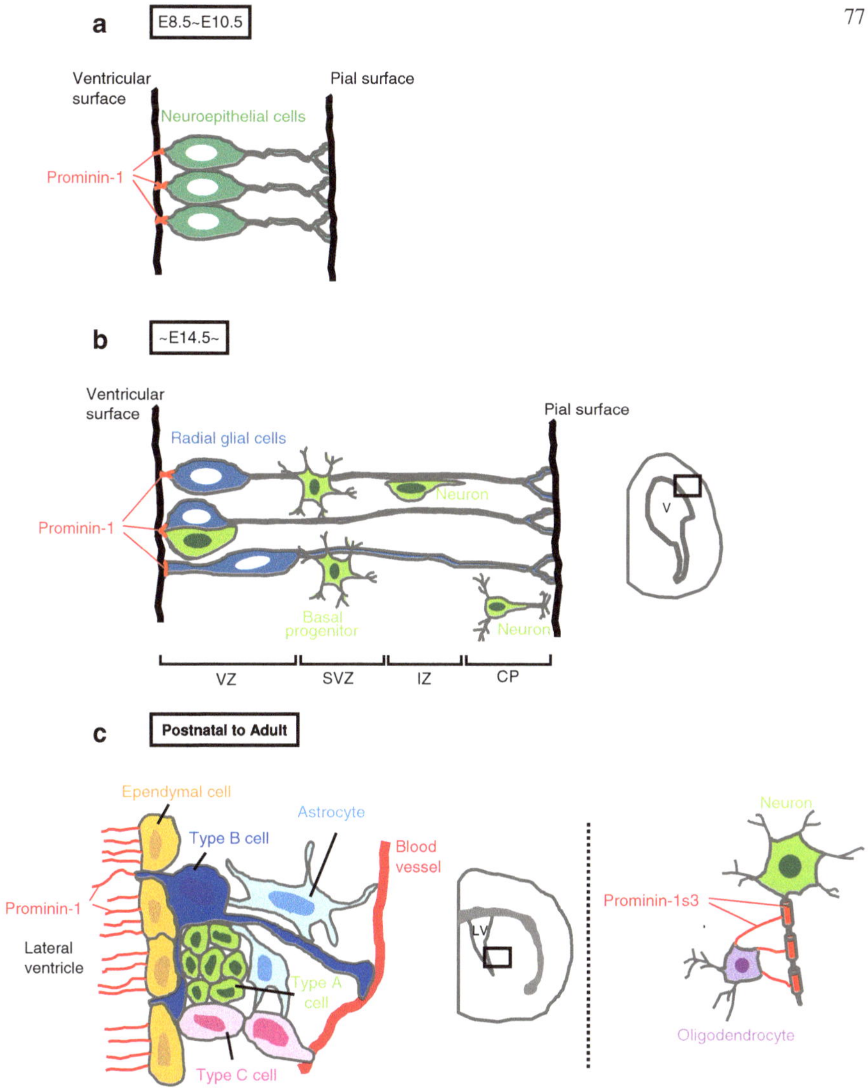

Fig. 5.2 Cerebrum development and transition of prominin-1 expression. (**a**) After neural tube closure, neuroepithelial cells (NECs) divide symmetrically to expand their own pool. Prominin-1 is already expressed in NECs at E8.5 and localized at their apical plasma membrane (*red*). (**b**) NECs give rise to RGCs from mid-gestational stage (around E10.5). RGCs proliferate asymmetrically and generate post-mitotic neurons via basal progenitors. Post-mitotic neurons migrate radially to pial surface and are located in precise positions of cortical plate (CP). Prominin-1 expression is localized at the apical membrane of dorsal RGCs but not of ventral RGCs. *VZ* ventricular zone, *SVZ* subventricular zone, *IZ* intermediate zone, *CP* cortical plate. (**c**) Adult neural stem cells (NSCs) in the subventricular zone (SVZ) (Modified from Ref. [61]). Adult NSCs called type B cells (*blue*), which are defined as GFAP-expressing radial cells, are located at the SVZ of lateral ventricle (LV) and lie down on ependymal cells (ECs). The cells penetrate into LV via the apical cilia and contact blood vessel via the basal end. Type B cells generate transit-amplifying cells (type C cells; *peach*), actively dividing progenitors, and subsequently provide neuroblasts (type A cells; *green*). Prominin-1 expression is detected in cilia of ECs and type B cells. In addition, one splicing variant prominin-1.s3 is identified as a component of myelin sheath in adult mature oligodendrocytes (*purple*)

ventral side of the ventricular wall at P4 but to be absent from those at the dorsal side [22]. Those GLAST-positive cells co-express β-IV-tubulin, a marker for ciliated cells, i.e., they are ECs. At the dorsal side, GLAST-positive cells are negative for both of prominin-1 and β-IV-tubulin but are characterized as SOX2-positive RGCs, suggesting again spatiotemporal regulation of prominin-1 gene along differentiation of RGCs into ECs. Prominin-1 is therefore expressed in mature ECs at the ventral side of VZ at a postnatal stage. These postnatal prominin-1-positive ECs are known to have the ability to continuously produce neurons as more dormant NSCs [25].

In adult SVZ, prominin-1 is still present on ECs and GFAP-positive type B cells having single cilia [22]. The identity of NSCs in adult brain has been controversial, whether it is ECs or type B cells, but the latter are strongly accepted as adult NSCs [24]. In addition, one splice variant prominin-1.s3, which has a different cytoplasmic C-terminal domain, is found to be a component of myelin sheath in adult mature oligodendrocytes [26]. Intriguingly, one sophisticated study has recently proposed oligodendrocyte precursor cells (OPCs) as the cellular origin of gliomas [27]. Given that prominin-1 appears to be expressed in both oligodendrocytes and glioma stem cells (GSCs), there might be common mechanisms for the differentiation of OPCs and the development of GSCs. Taken together, prominin-1 is constantly expressed in certain types of stem and progenitor cells during tissue development, and its expression seems to be regulated in region- and developmental stage-dependent manner.

5.4 Prominin-1 in Cancer Stem Cells

The presence of prominin-1 (CD133)-expressing CSCs has been reported in various types of solid tumors. Like normal NSCs, CSCs can establish tumor hierarchy through differentiation, and their differentiation lineages determine tumor subtypes. The CSC concept has clinically important implications, because some features of CSCs make them particularly difficult to be eliminated. At first, CSCs reside in the relatively quiescent state in the cell cycle, allowing them to escape from conventional chemotherapies that generally target proliferating cells [28]. In addition, CSCs over express some membrane transporters associated with detoxification and consequently drug resistant by expelling chemotherapeutic reagents [29]. Moreover, CSCs are resistant to radiotherapy by virtue of their elevated abilities of scavenging reactive oxygen species (ROS) [30] and repairing DNA damages [31]. Therefore, in addition to simple transcriptional mechanisms, a comprehensive understanding of regulatory networks of prominin-1 expression is important for elucidation of cancer stemness and development of more effective cancer therapies.

5.5 PROM1 Gene Regulation

In this section, we will give an overview of important regulators of *PROM1* gene mainly identified in cancerous cells (Fig. 5.3).

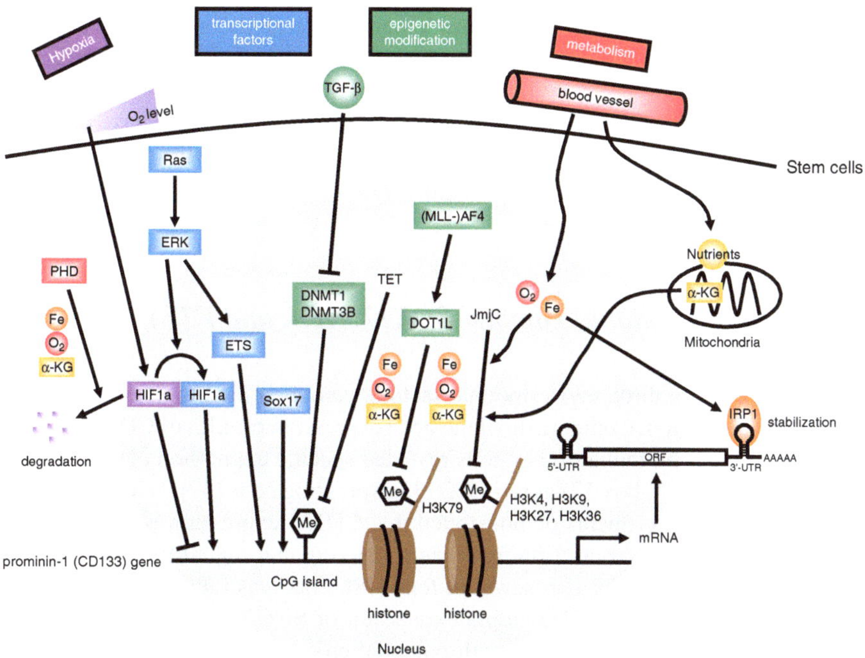

Fig. 5.3 The regulatory networks of *prominin-1* (*PROM1*) gene. Human *PROM1* gene transcription is eventually determined by cell-intrinsic program, such as transcriptional factors (*Sox17, ETS,* and *HIF1α*), and epigenetic regulators of DNA (*DNMTs* and *TET*) and histone (*JmjC* proteins including *DOTL1*). But the extracellular oxygen level and cellular metabolism are also mutually interconnected; oxygen, iron, and *α-KG* are indispensable for the enzymatic activities of oxygenase (*TET* and *JmjC* proteins). Therefore, chromatin status across *PROM1* gene is altered in accordance with surrounding microenvironments of cells. Theoretically, the distance from blood vessel seems to be one of critical determinants

5.5.1 DNA Hypomethylation

DNA methylation is an important component to regulate gene expression in numerous cellular processes. For example, genomic hypomethylation contributes to tumorigenesis by inducing oncogene activation and genomic instability [32]. Human *PROM1* promoters P1, P2, and P3 have 54 CpG sites at high percentage (50%, 68%, and 69%, respectively), and abnormal DNA hypomethylation status of the CpG islands in the P1–P3 promoters has been reported to be positively correlated with increased CD133 expression in multiple types of CSCs [17, 33, 34]. Interestingly, transforming growth factor (TGF)-β induces DNA demethylation of the *PROM1* P1 promoter by inhibiting DNA (cytosine-5-)-methyltransferase 1 (DNMT1) and DNMT3B expression [35]. In addition, the TGF-β receptor I inhibitor SB431542 significantly decreases CD133+/Nestin+ population sizes and decreases sphere-forming activity, self-renewal, and in vivo tumorigenic potential [36]. Importantly, demethylating drugs such as decitabine (5-aza-2′deoxycytidine) and 5-azacytidine

(5-AC) have been utilized as anticancer reagents. Certainly, some differentiated cells undergo malignant transformation by silencing of tumor suppressor genes through promoter hypermethylation. However, treatment with 5-AC restores *PROM1* gene expression [17], implying that DNA methylation might have opposite effects for CSCs and relatively differentiated cancerous cells. Further studies with a focus on the mechanisms underlying gene hypomethylation could improve our understanding of cancer stemness.

5.5.2 Transcriptional Factors (SOX17, AF4, and ETS)

To our knowledge, three transcriptional factors were identified to date to regulate *PROM1* transcription. Comparison of the gene expression profiles of CD133(+) and CD133(−) gastric epithelial tumor cells revealed specific expression of Sox17 (sex determining region Y-box 17) gene in CD133 expressing cells [37]. Overexpression and knockdown experiments demonstrated that CD133 expression is regulated by Sox17 in vitro. Sox17 transcriptional factor was originally identified as a critical regulator of survival and self-renewal of fetal HSCs but was found dispensable for adult HSC maintenance [38]. Ectopic expression of Sox17 can confer fetal HSC properties on both adult HSCs and multipotential progenitors and result in transformation into leukemia in the long run [39]. Interestingly, Sox17 is highly expressed in OPCs in adult brain and plays important roles in controlling both cell cycle exit and differentiation [40], which is largely consistent with an above-mentioned observation that prominin-1 is expressed in adult oligodendrocytes and that OPCs are cells of origin transformed into glioma stem cells.

Another approach of a large-scale shRNA screening identified AF4 (ALL1-fused gene from chromosome 4 protein) gene as a regulator of CD133 in the colon adenocarcinoma Caco-2 cells [41]. AF4 gene is involved in reciprocal translocation to the MLL (mixed lineage leukemia) gene, which is correlated with the high risk of ALL (acute lymphoblastic leukemia) in early childhood [42]. AF4 is a transcriptional activator having roles in transcriptional elongation and also in chromatin modification through the recruitment and activation of DOT1L (Dot1-like) histone H3K79 methyltransferase [43]. Both AF4 and MLL-AF4 fusion proteins can promote CD133 transcription [41], and previous studies have shown that histone H3 associated with the *PROM1* promoter is dimethylated at K79 [44], suggesting an epigenetic regulation of *PROM1* gene through DNA methylation and histone modification.

In addition to these two transcriptional factors, our group has revealed that the Ras/ERK/ETS pathway regulates gene expression via two E26 transformation-specific (ETS) binding sites within a *PROM1* promoter [45]. The ETS family of transcriptional factors is the nuclear target of the receptor tyrosine kinase (RTKs)/Ras/ERK signaling. Our study also confirmed that transduction of the oncogenic Ras gene (RasV12) into immortalized normal human astrocytes increased the ability of sphere and tumor formation, as well as *PROM1* gene expression [45]. In addition, inhibition of ERK pathway by U0126 resulted in downregulation of CD133 and decrease of side population size. Importantly, expression of *ABCG2* and *MDR1* genes, both of which are cell surface transporters conferring multiple drug resistance to tumor cells, is known to be regulated by ERK signaling [46, 47]. In addition, Ras controls radioresistance in some normal and cancerous cells, and its inhibition results in radiosensitization of tumor cells [48]. Therefore, activated Ras plays extensive roles for multiple CSC features, such as tumorigenicity, chemo-/radioresistance, and *PROM1* gene expression.

5.5.3 Hypoxia

The external microenvironment in which stem cells exist is called niche. Low-oxygen areas away from tumor vessels are recognized as a niche that plays a role in stem cell regulation [49]. When primary glioma spheres are incubated in physiological concentration of oxygen (7%) less than atmospheric one (21%), CD133 expression increases together with other stem cell genes, Sox2, Oct4, and Nestin [50]. Particularly, hypoxia-inducible factor (HIF)-1 transcriptional factor is known as a major factor for adaptation to changes in oxygen levels. Interestingly, activated Ras phosphorylates and stabilizes HIF-1α via ERK activation [51]. However, in gastrointestinal cancer cell lines, HIF-1α downregulates *PROM1* gene expression downstream of Akt/mTOR signaling [52]. Considering that our data of *PROM1* gene upregulation by Ras/ERK were obtained under normoxia, Ras might have additional effects that convert HIF-1α function to positively regulate *PROM1* gene.

5.5.4 Iron

Iron (Fe) is essential for life, because many Fe-containing proteins catalyze a wide variety of biological processes involving energy metabolism, respiration, and DNA synthesis. IRP1 (iron-responsive element-binding protein 1) senses intracellular iron levels and has dual functions; mitochondrial IRP1 catalyzes the conversion of citrate to isocitrate in TCA cycle as aconitase, and cytosolic IRP1 posttranscriptionally regulates mRNA through binding to IREs (iron-responsive elements) in the UTR of transcripts [53]. When cytosolic iron level is low, binding of IRP1 to IREs in the 5′-UTR of mRNA represses its translation, whereas binding to IREs in the 3′-UTR of mRNA generally stabilizes its transcripts. When cytosolic iron levels are high, IRPs do not bind to IREs. Recently, human CD133 mRNA has been reported not to have any classical stem loop sequence CAGUGU within either the 5′-UTR or the 3′-UTR but to contain the noncanonical CAGAGU sequence in the 3′-UTR [54]. In fact, the treatment of Caco-2 cells with deferoxamine (an iron chelator) significantly downregulates the CD133 protein level, which is sufficiently convincing because some CD133(+) CSCs are located in the perivascular region [55] that cells are accessible to serum-derived iron. In addition, iron and oxygen are necessary for prolyl hydroxylase domain (PHD) enzyme family to hydroxylate on proryl reside resulting in degradation of HIF-α [56]. Interestingly, ten-eleven translocation (TET) proteins having an enzymatic activity for the conversion from 5-methylcytosine to 5-hydroxymethylcytosine (5-hmC), which may be an intermediate of DNA demethylation, are iron- and oxygen-dependent oxygenases [57]. In addition, JmjC domain-containing proteins also require iron and oxygen to catalyze the demethylation of mono-, di-, and trimethyl lysine residues via an oxidative reaction [58], implying potential difference in chromatin status of *PROM1* genes in cells located between perivascular and hypoxic region.

5.5.5 Low Mitochondrial Activity

The importance of mitochondria in tumor biology has initially been proposed by Warburg [59]. He postulated that cancer cells had enhanced glycolysis as a consequence of mitochondrial dysfunctions, which are called Warburg effect. While mitochondria produce ATP, which drives the cellular processes needed for life, it produces ROS leading to cell death and aging. Therefore, low mitochondrial activity in cancerous cells could be considered as one of survival strategies to adapt hypoxic stress or to avoid an excess of ROS production. One study has shown that the treatment of U251 glioma cells with Rotenone, a mitochondrial complex I (NADH dehydrogenase) blocker to inhibit electron transport chain, enriched CD133(+) population in U251 glioma cells [60]. They further confirmed that depletion of mitochondrial DNA (mtDNA) by long-term treatment with very low doses of ethidium bromide posttranscriptionally induced the increase of CD133 expression and the acquisition of CSC properties. Moreover, retransfer of wild type parental mtDNA into mtDNA-depleted cells can reverse their effects, demonstrating that mitochondrial status is closely associated with CSC properties including CD133 expression. Although the detailed mechanisms by which depletion of mtDNA increases CD133 expression remain to be examined, it might be possibly explained by the decreased production of α-ketoglutarate (α-KG). α-KG is a key intermediate in the mitochondrial TCA cycle, coming after isocitrate and before succinyl CoA. Importantly, as well as iron and oxygen, α-KG is also required for the enzymatic activity of a HIF regulator PHD hydroxygenase and epigenetic regulators TET and JmjC oxygenases, suggesting strong links between cellular metabolism and epigenetic dynamics in CD133 gene regulation.

5.6 Conclusion

The 15-year studies since the identification and molecular cloning of prominin-1 (CD133) have demonstrated that this molecule is of immense importance as an indicator to study stem cell behavior. At present, those precious findings provide additional clues to deepen our knowledge on tissue development and to enhance the prospects for cancer therapies. For instance, the region-dependent expression of prominin-1 in RGCs suggests the involvement of external stimuli under the specific microenvironments and implies the heterogeneity within stem cells. In addition, the identification of manifold intrinsic and extrinsic factors upstream of *PROM1* gene reveals one fascinating link between cellular metabolism and chromatin modifications. Beyond simple regulations of genes by transcriptional and epigenetic mechanisms, establishment of molecular underpinnings to regulate cellular metabolism in specific locations could facilitate not only the understanding of stem cell identity but also the development of effective therapeutic strategies against human neoplasms.

References

1. Fargeas CA, Joester A, Missol-Kolka E et al (2004) Identification of novel prominin-1/CD133 splice variants with alternative C-termini and their expression in epididymis and testis. J Cell Sci 117:4301–4311
2. Weigmann A, Corbeil D, Hellwig A, Huttner WB (1997) Prominin, a novel microvilli-specific polytopic membrane protein of the apical surface of epithelial cells, is targeted to plasmalemmal protrusions of non-epithelial cells. Proc Natl Acad Sci USA 94:12425–12430
3. Yin AH, Miraglia S, Zanjani ED et al (1997) AC133, a novel marker for human hematopoietic stem and progenitor cells. Blood 90:5002–5012
4. Miraglia S, Godfrey W, Yin AH et al (1997) A novel five-transmembrane hematopoietic stem cell antigen: isolation, characterization, and molecular cloning. Blood 90:5013–5021
5. Singh SK, Hawkins C, Clarke ID et al (2004) Identification of human brain tumour initiating cells. Nature 432:396–401
6. Collins AT, Berry PA, Hyde C et al (2005) Prospective identification of tumorigenic prostate cancer stem cells. Cancer Res 65:10946–10951
7. Suetsugu A, Nagaki M, Aoki H et al (2006) Characterization of CD133+ hepatocellular carcinoma cells as cancer stem/progenitor cells. Biochem Biophys Res Commun 351:820–824
8. O'Brien CA, Pollett A, Gallinger S et al (2007) A human colon cancer cell capable of initiating tumour growth in immunodeficient mice. Nature 445:106–110
9. Eramo A, Lotti F, Sette G et al (2008) Identification and expansion of the tumorigenic lung cancer stem cell population. Cell Death Differ 15:504–514
10. Karbanová J, Missol-Kolka E, Fonseca AV et al (2008) The stem cell marker CD133 (prominin-1) is expressed in various human glandular epithelia. J Histochem Cytochem 56:977–993
11. Zhu L, Gibson P, Currle DS et al (2009) Prominin 1 marks intestinal stem cells that are susceptible to neoplastic transformation. Nature 457:603–607
12. Zacchigna S, Oh H, Wilsch-Bräuninger M et al (2009) Loss of the cholesterol-binding protein prominin-1/CD133 causes disk dysmorphogenesis and photoreceptor degeneration. J Neurosci 29:2297–2308
13. Maw MA, Corbeil D, Koch J et al (2000) A frameshift mutation in prominin (mouse)-like 1 causes human retinal degeneration. Hum Mol Genet 9:27–34
14. Fargeas CA, Florek M, Huttner WB et al (2003) Characterization of prominin-2, a new member of the prominin family of pentaspan membrane glycoproteins. J Biol Chem 278:8586–8596
15. Fargeas CA, Huttner WB, Corbeil D (2007) Nomenclature of prominin-1 (CD133) splice variants – an update. Tissue Antigens 69:602–606

16. Shmelkov SV, Jun L, St Clair R et al (2004) Alternative promoters regulate transcription of the gene that encodes stem cell surface protein AC133. Blood 103:2055–2061
17. Tabu K, Sasai K, Kimura T et al (2008) Promoter hypomethylation regulates CD133 expression in human gliomas. Cell Res 18:1037–1046
18. Huttner WB, Brand M (1997) Asymmetric division and polarity of neuroepithelial cells. Curr Opin Neurobiol 7:29–39
19. Corbeil D, Röper K, Fargeas CA et al (2001) Prominin: a story of cholesterol, plasma membrane protrusions and human pathology. Traffic 2:82–91
20. Röper K, Corbeil D, Huttner WB (2000) Retention of prominin in microvilli reveals distinct cholesterol-based lipid micro-domains in the apical plasma membrane. Nat Cell Biol 2:582–592
21. Götz M, Huttner WB (2005) The cell biology of neurogenesis. Nat Rev Mol Cell Biol 6:777–788
22. Pfenninger CV, Roschupkina T, Hertwig F et al (2007) CD133 is not present on neurogenic astrocytes in the adult subventricular zone, but on embryonic neural stem cells, ependymal cells, and glioblastoma cells. Cancer Res 67:5727–5736
23. Merkle FT, Tramontin AD, Garcia-Verdugo JM et al (2004) Radial glia give rise to adult neural stem cells in the subventricular zone. Proc Natl Acad Sci USA 101:17528–17532
24. Spassky N, Merkle FT, Flames N et al (2005) Adult ependymal cells are postmitotic and are derived from radial glial cells during embryogenesis. J Neurosci 25:10–18
25. Coskun V, Wu H, Blanchi B et al (2008) CD133+ neural stem cells in the ependyma of mammalian postnatal forebrain. Proc Natl Acad Sci USA 105:1026–1031
26. Corbeil D, Joester A, Fargeas CA et al (2009) Expression of distinct splice variants of the stem cell marker prominin-1 (CD133) in glial cells. Glia 57:860–874
27. Liu C, Sage JC, Miller MR et al (2011) Mosaic analysis with double markers reveals tumor cell of origin in glioma. Cell 146:209–221
28. Li L, Bhatia R (2011) Stem cell quiescence. Clin Cancer Res 17:4936–4941
29. Dean M, Fojo T, Bates S (2005) Tumour stem cells and drug resistance. Nat Rev Cancer 5:275–284
30. Diehn M, Cho RW, Lobo NA et al (2009) Association of reactive oxygen species levels and radioresistance in cancer stem cells. Nature 458:780–783
31. Bao S, Wu Q, McLendon RE et al (2006) Glioma stem cells promote radioresistance by preferential activation of the DNA damage response. Nature 444:756–760
32. Gaudet F, Hodgson JG, Eden A et al (2003) Induction of tumors in mice by genomic hypomethylation. Science 300:489–492
33. Yi JM, Tsai HC, Glockner SC et al (2008) Abnormal DNA methylation of CD133 in colorectal and glioblastoma tumors. Cancer Res 68:8094–8103
34. Baba T, Convery PA, Matsumura N et al (2009) Epigenetic regulation of CD133 and tumorigenicity of CD133+ ovarian cancer cells. Oncogene 28:209–218
35. You H, Ding W, Rountree CB (2010) Epigenetic regulation of cancer stem cell marker CD133 by transforming growth factor-beta. Hepatology 51:1635–1644
36. Ikushima H, Todo T, Ino Y et al (2009) Autocrine TGF-beta signaling maintains tumorigenicity of glioma-initiating cells through Sry-related HMG-box factors. Cell Stem Cell 5:504–514
37. Fukamachi H, Shimada S, Ito K et al (2011) CD133 is a marker of gland-forming cells in gastric tumors and Sox17 is involved in its regulation. Cancer Sci 102:1313–1321
38. Kim I, Saunders TL, Morrison SJ (2007) Sox17 dependence distinguishes the transcriptional regulation of fetal from adult hematopoietic stem cells. Cell 130:470–483
39. He S, Kim I, Lim MS et al (2011) Sox17 expression confers self-renewal potential and fetal stem cell characteristics upon adult hematopoietic progenitors. Genes Dev 25:1613–1627
40. Sohn J, Natale J, Chew LJ et al (2006) Identification of Sox17 as a transcription factor that regulates oligodendrocyte development. J Neurosci 26:9722–9735
41. Mak AB, Nixon AM, Moffat J (2012) The mixed lineage leukemia (MLL) fusion-associated gene AF4 promotes CD133 transcription. Cancer Res 72:1929–1934
42. Uckun FM, Herman-Hatten K, Crotty ML et al (1998) Clinical significance of MLL-AF4 fusion transcript expression in the absence of a cytogenetically detectable t(4;11)(q21;q23) chromosomal translocation. Blood 92:810–821

43. Bitoun E, Oliver PL, Davies KE (2007) The mixed-lineage leukemia fusion partner AF4 stimulates RNA polymerase II transcriptional elongation and mediates coordinated chromatin remodeling. Hum Mol Genet 16:92–106
44. Guenther MG, Lawton LN, Rozovskaia T et al (2008) Aberrant chromatin at genes encoding stem cell regulators in human mixed-lineage leukemia. Genes Dev 22:3403–3408
45. Tabu K, Kimura T, Sasai K et al (2010) Analysis of an alternative human CD133 promoter reveals the implication of Ras/ERK pathway in tumor stem-like hallmarks. Mol Cancer 9:39
46. Imai Y, Ohmori K, Yasuda S et al (2009) Breast cancer resistance protein/ABCG2 is differentially regulated downstream of extracellular signal-regulated kinase. Cancer Sci 100: 1118–1127
47. Shen H, Xu W, Luo W et al (2011) Upregulation of mdr1 gene is related to activation of the MAPK/ERK signal transduction pathway and YB-1 nuclear translocation in B-cell lymphoma. Exp Hematol 39:558–569
48. Bernhard EJ, Kao G, Cox AD et al (1996) The farnesyltransferase inhibitor FTI-277 radiosensitizes H-ras-transformed rat embryo fibroblasts. Cancer Res 56:1727–1730
49. Mohyeldin A, Garzon-Muvdi T, Quinones-Hinojosa A (2010) Oxygen in stem cell biology: a critical component of the stem cell niche. Cell Stem Cell 7:150–161
50. McCord AM, Jamal M, Williams ES et al (2009) CD133+ glioblastoma stem-like cells are radiosensitive with a defective DNA damage response compared with established cell lines. Clin Cancer Res 15:5145–5153
51. Wang FS, Wang CJ, Chen YJ et al (2004) Ras induction of superoxide activates ERK-dependent angiogenic transcription factor HIF-1alpha and VEGF-A expression in shock wave-stimulated osteoblasts. J Biol Chem 279:10331–10337
52. Matsumoto K, Arao T, Tanaka K et al (2009) mTOR signal and hypoxia-inducible factor-1 alpha regulate CD133 expression in cancer cells. Cancer Res 69:7160–7164
53. Rouault TA (2006) The role of iron regulatory proteins in mammalian iron homeostasis and disease. Nat Chem Biol 2:406–414
54. Bourseau-Guilmain E, Griveau A, Benoit JP et al (2011) The importance of the stem cell marker prominin-1/CD133 in the uptake of transferrin and in iron metabolism in human colon cancer Caco-2 cells. PLoS One 6:e25515
55. Gilbertson RJ, Rich JN (2007) Making a tumour's bed: glioblastoma stem cells and the vascular niche. Nat Rev Cancer 7:733–736
56. Jokilehto T, Jaakkola PM (2010) The role of HIF prolyl hydroxylases in tumour growth. J Cell Mol Med 14:758–770
57. Tahiliani M, Koh KP, Shen Y et al (2009) Conversion of 5-methylcytosine to 5-hydroxymethylcytosine in mammalian DNA by MLL partner TET1. Science 324:930–935
58. Klose RJ, Kallin EM, Zhang Y (2006) JmjC-domain-containing proteins and histone demethylation. Nat Rev Genet 7:715–727
59. Warburg O (1956) On the origin of cancer cells. Science 123:309–314
60. Griguer CE, Oliva CR, Gobin E et al (2008) CD133 is a marker of bioenergetic stress in human glioma. PLoS One 3:e3655
61. Tavazoie M, Van der Veken L, Silva-Vargas V et al (2008) A specialized vascular niche for adult neural stem cells. Cell Stem Cell 3:279–288

Part II
Prominin-1/CD133 and Stem Cells

Chapter 6
Prominin-1 (CD133) and the Cell Biology of Neural Progenitors and Their Progeny

Alex M. Sykes and Wieland B. Huttner

Abstract Our group discovered prominin-1 in search for markers to study the cell polarity of neural stem and progenitor cells in the developing brain. Over the past 15 years, prominin-1, also called CD133, has not only become a frequently used marker of neural stem cells and neural cancer stem cells, as is in fact the case of somatic (cancer) stem cells in general, but has also been used to understand the symmetric versus asymmetric division of the neural stem cells in the context of their apical-basal polarity. Moreover, studying prominin-1 on neural stem cells has revealed a novel fate of the midbody, that is, midbody release, and key differences in this release between normal stem cells and cancer-derived cells. Other subcellular aspects of neural stem cells, the understanding of which has been promoted by studying prominin-1, pertain to the organization of plasma membrane protrusions and the membrane microdomains they contain. Of particular relevance in this context is the primary cilium of neuroepithelial cells and its transformation into the outer segment of retinal photoreceptor cells, a process in which prominin-1 exerts a vital role.

Keywords Cell division • Cell polarity • Myelin • Neurogenesis • Plasma membrane protrusions

---

A.M. Sykes • W.B. Huttner, M.D. (✉)
Max Planck Institute of Molecular Cell Biology and Genetics,
Pfotenhauerstr. 108, 01307 Dresden, Germany
e-mail: huttner@mpi-cbg.de

D. Corbeil (ed.), *Prominin-1 (CD133): New Insights on Stem & Cancer Stem Cell Biology*, Advances in Experimental Medicine and Biology 777, DOI 10.1007/978-1-4614-5894-4_6,

6.1 Introduction

Prominin-1 was identified in the course of studies aiming to dissect the cell biology of neural stem and progenitor cells in the developing mammalian neocortex that give rise to all projection neurons. There are two principal classes of such neural stem and progenitor cells based on where mitosis occurs, (1) apical progenitors undergoing mitosis at the ventricular surface of the ventricular zone, which corresponds to the apical surface of the cortical wall, and (2) basal progenitors undergoing mitosis at an abventricular, more basal location, typically in the subventricular zone. Apical progenitors can be bipolar, exhibiting apical-basal cell polarity throughout the cell cycle including mitosis, as is characteristically the case for neuroepithelial cells and apical radial glial cells, or monopolar, exhibiting apical but not basal cell polarity, as is characteristically the case for the so-called short neural precursors. Basal progenitors can be monopolar, exhibiting basal but not apical cell polarity, as is characteristically the case for basal radial glial cells, or nonpolar, at least at mitosis, as is characteristically the case for intermediate progenitor cells and transit amplifying progenitors [1]. Over the past two decades, molecular markers of the various stem and progenitor cell types as well as their apical versus basal cell polarity have been characterized and used to dissect their cell biology [1, 2].

6.2 Discovery of Prominin-1

Prominin-1 was first identified in our group in the search for specific markers that could be used to study apical (ventricular) plasma membrane of apical progenitors. The apical membrane is a specialized compartment of the plasma membrane that is rich in microvilli and in apical progenitors also contains the primary cilium. To specifically label the apical membrane, monoclonal antibodies were raised against the mouse telencephalic neuroepithelium. Of the panel of antibodies raised, the monoclonal antibody 13A4 was ultimately selected because it specifically stained the apical, but not the basolateral, membrane of embryonic day (E) 12 apical progenitors. After screening a mouse adult kidney cDNA library, prominin-1 was subsequently identified as this apical antigen [3]. The protein was called "prominin" from the Latin word *prominere* (to stand out) to reflect its key feature to be concentrated in plasma membrane protrusions (see below). In epithelial cells exhibiting apical-basal cell polarity, prominin-1 is specifically concentrated in the apical plasma membrane [3, 4]. However, prominin-1 does not only occur in epithelial cells but also in non-epithelial cells, as was first demonstrated by David Buck and colleagues who characterized prominin-1, independently from our group, as the AC133 antigen on human hematopoietic stem cells, leading to the alternative name CD133 [3–6].

6.3 Prominin-1 and Neural Progenitors

Of the stem and progenitor cell types in the developing mammalian neocortex that give rise to cortical projection neurons (see Sect. 6.1), prominin-1 is restricted to apical progenitors [3, 7]. However, studies on regions of the developing mammalian central nervous system other than neocortex, such as postnatal cerebellum [8], on regions showing adult neurogenesis [9] (TL Walker, AM Sykes, and G Kempermann, unpublished) and on the central nervous system of various mammalian and non-mammalian vertebrates [10, 11] have shown that prominin-1, or at least *prominin-1* mRNA, is expressed in various neural progenitors. A comprehensive description of the expression of prominin-1 in the vertebrate central nervous system, however, is beyond the scope of this chapter, and the reader is referred to excellent reviews covering that topic (e.g., Ref. [12]). Below, we focus on prominin-1 expression in myelinating cells as this is of relevance for the association of prominin-1 with specific membrane microdomains, which is discussed further below.

6.4 Prominin-1 and Other Neural Cells

Prominin-1 mRNAs come as many splice variants that at the protein level result in small insertions and at least three C-terminal tail variants, which have altered PSD95/Dlg1/zo-1 (PDZ)-binding domains. It has been proposed that certain splice variants are upregulated and downregulated in tissues over development [10]. However, the actual role of the distinct splice variants is unknown. For a detailed description of prominin-1 splice variants, see Chap. 1.

A well-characterized example of a specific splice variant of prominin-1 is that found in the myelin sheath of both central and peripheral nerves of adult mice. It was observed by immunoblotting that the adult kidney and brain contain prominin-1 with different molecular weights of ~115 and ~100 kDa, respectively. In more detailed experiments over the course of brain development from E10 to adult, it was then established, using a combination of an antibody that presumably recognizes all of the known splice variants and an antibody that detects splice variant 1 (s1, 115 kDa) but not splice variant 3 (s3, 100 kDa), that s1 is initially expressed; however, over the course of brain development, s3 becomes more dominant and ultimately the only variant expressed in the adult. Upon further examination of forebrain sections, it was observed that the *prominin-1* transcript was expressed in a population of olig-2-positive oligodendrocytes. The expression of prominin-1 in oligodendrocytes was also confirmed by immunofluorescence and electron microscopy of nerve sections from the central and peripheral nervous system. In these nerves, prominin-1 was located in the myelin sheath, and it was confirmed by immunoblotting that prominin-1 was concentrated in the myelin membrane fraction. To determine that the oligodendrocytes expressed predominantly s3, cultured oligodendrocytes were analyzed, and immunoblotting confirmed that they expressed prominin-1 with the same molecular weight as the adult brain, whereas cultured astrocytes expressed only the s1 prominin-1. Lastly, to confirm these findings, in myelin-deficient mice (shiverer, rumpshaker, and jimpy), prominin-1.s3 was absent or extremely reduced in brain extracts, whereas prominin-1 in the kidney (s1) was expressed at normal levels [10].

6.5 Prominin-1 and Neural Cancer Stem Cells

Prominin-1 expression is not limited to neural stem cells as it has also been used to isolate presumptive cancer stem cells from brain tumors [13, 14]. When a prominin-1 antibody was used to sort cells from patient brain tumors, only the prominin-1-positive cells were found to be able to initiate brain tumors in a mouse model, and only 100 prominin-1-positive cells were required to generate a tumor [13]. Furthermore, prominin-1-positive glioma cells have also been found to have an increased resistance to radiation in a DNA checkpoint-dependent manner, in comparison to prominin-1-negative tumor cells [14]. Prominin-1-positive cancer stem cells have also shown to be multipotent as astrocytic, neuronal and oligodendrocytic lineages have been observed in their progeny [13, 14].

6.6 Prominin-1 and Plasma Membrane Protrusions

In terms of its subcellular localization, the hallmark of prominin-1 is its concentration in plasma membrane protrusions, irrespective of the cell type. This subcellular localization appears to reflect the preference of prominin-1 for membranes exhibiting positive curvature rather than a particular form of cytoskeleton underlying a given type of plasma membrane protrusion, as prominin-1 is found on both actin-based and microtubule-based protrusions. Thus, in neural progenitors, prominin-1 is concentrated in apical microvilli and apical primary cilia [3, 15–18]. Consistent with a preference for positive curvature membranes, prominin-1 is strikingly excluded from planar regions of the plasma membrane, even if these are adjacent to a microvillus or primary cilium [3, 15–17].

6.7 Prominin-1 and Membrane Microdomains

One explanation for the specific subcellular localization of prominin-1 in positive curvature membranes is that it is part of a novel membrane microdomain. The prominin-1 microdomain was first identified in Madin-Darby canine kidney cells transfected with mouse prominin-1 cDNA. In contrast to classical membrane microdomain markers such as caveolin and placental alkaline phosphatase (PLAP), which are insoluble in the nonionic detergent Triton X-100 [19], not only the endoplasmic reticulum (ER) form but also the plasma membrane form of prominin-1 were found to be completely soluble in Triton X-100. By contrast, when prominin-1-transfected cells were extracted with another nonionic detergent, Lubrol WX, the plasma membrane form of prominin-1 was found to be insoluble, whereas the ER form was soluble. Furthermore, ^{35}S-methionine pulse-chase experiments followed by Lubrol WX extraction revealed that the formation of Lubrol WX-insoluble, prominin-1-containing membrane microdomains occurred concomitant with exit from the trans-Golgi network. These microdomains were found to be cholesterol dependent because when cholesterol was depleted by methyl-β-cyclodextrin, the prominin-1-containing, Lubrol WX-resistant complexes became reduced in size. Interestingly, prominin-1 itself has the capacity to physically interact directly with membrane cholesterol [19]. This raises the possibility that the association of prominin-1 with Lubrol WX-resistant membrane microdomains may not simply reflect its role as cargo but, perhaps, a role as a microdomain organizer. Importantly, cholesterol depletion abolished the protrusion-specific localization of prominin-1, allowing it to diffuse into the planar regions of the plasma membrane. As membranes with high positive curvature such as microvilli and cilia are enriched in cholesterol [19], this data suggests that the association of prominin-1 with Lubrol WX-resistant, cholesterol-dependent membrane microdomains and its ability to directly interact with membrane cholesterol are key to its concentration in plasma membrane protrusions [20].

6.8 Prominin-1 and Neural Stem Cell Division

As a cell surface marker that is associated with a specific membrane microdomain, prominin-1 offers the possibility not only to isolate prominin-1-positive stem and progenitor cells and cancer stem cells but also to investigate the plasma membrane dynamics during their cell division. Thus, the specific localization of prominin-1 to apical plasma membrane protrusions in neural stem and progenitor cells has been exploited to study the symmetric versus asymmetric distribution of this membrane upon cytokinesis and to perform such analyses separately for apical progenitors undergoing symmetric proliferative divisions and apical progenitors undergoing asymmetric divisions leading to apical progenitor self-renewal and yielding basal progenitors or neurons [15]. The results of these and subsequent [21] studies are consistent with the concept that inheritance of apical plasma membrane is a prerequisite, but not necessarily sufficient, for apical progenitor divisions to be symmetric proliferative or asymmetric self-renewing. It thus appears that inheritance of certain apical plasma membrane components, though alone not sufficient, may contribute to maintaining daughter cells as apical progenitors.

With regard to neural cancer stem cells, human prominin-1 has been reported to be either symmetrically or asymmetrically distributed to the daughter cells of dividing glioma cells [22]. Interestingly, the segregation of prominin-1 correlates with Numb, when the latter showed an asymmetric distribution during cell division.

Numb is a fate-determining protein that promotes the differentiation of neural stem cells through its ability to antagonize Notch- and Hedgehog-signaling pathways [23].

Although hematopoietic stem and progenitors cells are not in the focus of this chapter, it is interesting to note that in these non-epithelial cells, which lack apical-basal cell polarity, prominin-1 can also be either symmetrically or asymmetrically distributed to the daughter cells upon cell division [24]. Placing the data on plasma membrane inheritance upon divisions of (epithelial) neural stem and progenitor cells on the one hand and (non-epithelial) hematopoietic stem and progenitor cells on the other hand into context to one another, a possible common denominator may be that daughter cell fate is linked to inheritance of membrane microdomains (e.g., prominin-1-containing membrane protrusions) rather than macrodomains (e.g., the apical membrane as a whole).

6.9 Prominin-1 and Midbody Release

Studies on prominin-1 and neural stem cell division also uncovered a novel pathway of extracellular membrane traffic – the release of the midbody from daughter cells [17, 25]. In fact, midbody release is one of several pathway by which membrane particles that also may carry prominin-1-containing membrane microdomains are released from cells into the extracellular space [16, 17, 25] (see also Chap. 3). Originating from the observation that the ventricular fluid of embryonic mouse brain contains prominin-1-positive membrane particles with a ring-like appearance and a diameter of 0.5–1 μm, referred to as P2 particles based on their sedimentation upon differential centrifugation [16], it was subsequently shown that these particles were released from the apical surface of neuroepithelial and radial glial cells and in fact constituted the central portion of the midbody, containing the remnants of the mitotic spindle and contractile ring [17]. Given that an apical, prominin-1-positive midbody will only be formed if cytokinesis bisects the apical (rather than basolateral) plasma membrane of a neuroepithelial or radial glial cell, it was assumed that the release of such midbodies into the ventricular fluid primarily reflects symmetric proliferative divisions of apical progenitors [17].

A possible link between midbody release and the mode of progenitor cell division with regard to daughter cell fate was explored by studying various stem cells and other cell lines in vitro, exploiting the fact that midbody release can occur irrespective of whether or not a cell expresses prominin-1 [25]. Interestingly, the extent of midbody release was found to be greater when stem cells were induced to undergo differentiative divisions than proliferative divisions. Even more interesting, midbody release from cancer-derived cell lines was dramatically less than from proliferating stem cells and relatively higher in cancer-derived cells that could still be stimulated to undergo differentiation, which was then accompanied by an increase in midbody release. Together, these data suggest that midbody release is somehow linked to cell differentiation. Indeed, interfering with midbody release by RNAi knockdown of components of the abscission machinery in a differentiation-responsive cell line enhanced the response to a subsequent differentiation stimulus [25].

There are at least two implications of these findings. First, inducing alterations in midbody release, specifically increasing it from cancer stem cells, may perhaps provide an approach to drive these cells toward differentiation. Second, given that prominin-1, and presumably the membrane microdomains it is associated with, clusters at the midbody, midbody release would irreversibly deplete stem cells of these microdomains. If prominin-1 and these microdomains contribute to stemness, midbody release may be part of the mechanism underlying differentiation.

6.10 Prominin-1 and the Transformation of Primary Cilia into Retinal Photoreceptor Outer Segments

As described in Chap. 3, budding of prominin-1-containing membrane from the tip of the primary cilium of neuroepithelial cells likely contributes to the generation of the 50–100-nm P4 vesicles released into the ventricular fluid of the embryonic brain [16]. This membrane budding process, and the fact that this membrane contains prominin-1, may also be of significance for the transformation of the tip of the neuroepithelial primary cilium into the outer segment of retinal photoreceptors during the development of the latter cells. Moreover, understanding the membrane dynamics that occur during this transformation, as well as during photoreceptor disc maintenance, and the role(s) of prominin-1 therein will be key to unravel the pathomechanism underlying the photoreceptor degeneration in humans and mice with mutations in the *prominin-1* gene [26–28].

While an in-depth discussion of retinal photoreceptor degeneration due to mutations in the *prominin-1* gene can be found in Chap. 4, the following considerations should be mentioned here in the present context focusing on the membrane dynamics during primary cilium to outer segment transformation and photoreceptor disc maintenance. First, retinal photoreceptor cells are the only cells known so far to exhibit a massive phenotype upon prominin-1 mutation. This may be due to the lack of redundancy of prominin proteins in these cells as the paralog of prominin-1, prominin-2, is not expressed in the retina [29]. However, in light of the findings that there are cell types that express one or the other of the two prominins, as well as tissues in which the cellular and subcellular distributions of prominin-1 and prominin-2 are distinct [11, 30, 31], an alternative, and perhaps more likely, explanation is that the membrane dynamics during primary cilium to outer segment transformation and photoreceptor disc maintenance, which are characteristic of photoreceptor cells, have a specific need for prominin-1.

Second, in prominin-1 knockout mice [28], in recessive human prominin-1 mutations that result in impaired prominin-1 trafficking to the cell surface [26], and in dominant-negative human prominin-1 mutations that cause mislocalization of endogenous unmutated prominin-1 [27], prominin-1 is lacking from (or at least reduced at) the base of the outer segment, the site of disc formation where prominin-1 is characteristically localized under normal circumstances [26]. These observations are consistent with the above notion that the photoreceptor-specific membrane dynamics during outer segment formation and maintenance are more dependent on the presence of prominin-1 than other membrane rearrangements.

6.11 Conclusions and Perspectives

Studies on prominin-1 in the nervous system, notably in neural stem and progenitor cells of the developing brain, have provided novel insights at the cellular and subcellular level. At the cellular level, these pertain to understanding neural progenitor cell polarity, their symmetric versus asymmetric divisions, and the similarities and differences between physiological stem cell and cancer stem cell divisions. At the subcellular level, studying prominin-1 has taught us new principles of plasma membrane microdomain organization, their inheritance, and their extracellular trafficking. By exploiting these insights, future studies on prominin-1 hold the promise of unraveling the mechanism underlying a major neurodegenerative disease, that is, photoreceptor degeneration, and perhaps developing novel therapeutic approaches.

Acknowledgements W.B.H. was supported by grants from the Deutsche Forschungsgemeinschaft (SFB 655, A2; TRR 83, Tp6) and the European Research Council (250197), by the Deutsche Forschungsgemeinschaft-funded Center for Regenerative Therapies Dresden and by the Fonds der Chemischen Industrie.

References

1. Fietz SA, Huttner WB (2011) Cortical progenitor expansion, self-renewal and neurogenesis – a polarized perspective. Curr Opin Neurobiol 21:23–35
2. Götz M, Huttner WB (2005) The cell biology of neurogenesis. Nat Rev Mol Cell Biol 6:777–788
3. Weigmann A, Corbeil D, Hellwig A, Huttner WB (1997) Prominin, a novel microvilli-specific polytopic membrane protein of the apical surface of epithelial cells, is targeted to plasmalemmal protrusions of non-epithelial cells. Proc Natl Acad Sci USA 94:12425–12430
4. Corbeil D, Röper K, Hannah MJ, Hellwig A, Huttner WB (1999) Selective localization of the polytopic membrane protein prominin in microvilli of epithelial cells – a combination of apical sorting and retention in plasma membrane protrusions. J Cell Sci 112:1023–1033
5. Yin AH, Miraglia S, Zanjani ED, Almeida-Porada G, Ogawa M, Leary AG, Olweus J, Kearney J, Buck DW (1997) AC133, a novel marker for human hematopoietic stem and progenitor cells. Blood 90:5002–5012
6. Miraglia S, Godfrey W, Yin AH, Atkins K, Warnke R, Holden JT, Bray RA, Waller EK, Buck DW (1997) A novel five-transmembrane hematopoietic stem cell antigen: isolation, characterization, and molecular cloning. Blood 90:5013–5021
7. Fietz SA, Kelava I, Vogt J, Wilsch-Bräuninger M, Stenzel D, Fish JL, Corbeil D, Riehn A, Distler W, Nitsch R, Huttner WB (2010) OSVZ progenitors of human and ferret neocortex are epithelial-like and expand by integrin signaling. Nat Neurosci 13:690–699
8. Lee A, Kessler JD, Read TA, Kaiser C, Corbeil D, Huttner WB, Johnson JE, Wechsler-Reya RJ (2005) Isolation of neural stem cells from the postnatal cerebellum. Nat Neurosci 8:723–729

9. Beckervordersandforth R, Tripathi P, Ninkovic J, Bayam E, Lepier A, Stempfhuber B et al (2010) In vivo fate mapping and expression analysis reveals molecular hallmarks of prospectively isolated adult neural stem cells. Cell Stem Cell 7:744–758
10. Corbeil D, Joester A, Fargeas CA, Jászai J, Garwood J, Hellwig A, Werner HB, Huttner WB (2009) Expression of distinct splice variants of the stem cell marker prominin-1 (CD133) in glial cells. Glia 57:860–874
11. Fietz SA, Lachmann R, Brandl H, Kircher M, Samusik N, Schroder R et al (2012) Transcriptomes of germinal zones of human and mouse fetal neocortex suggest a role of extracellular matrix in progenitor self-renewal. Proc Natl Acad Sci USA 109:11836–11841
12. Grosse-Gehling P, Fargeas CA, Dittfeld C, Garbe Y, Alison MR, Corbeil D, Kunz-Schughart LA (2012) CD133 as a biomarker for putative cancer stem cells in solid tumours: limitations, problems and challenges. J Pathol: 10.1002/path.4086
13. Singh SK, Clarke ID, Hide T, Dirks PB (2004) Cancer stem cells in nervous system tumors. Oncogene 23:7267–7273
14. Bao S, Wu Q, McLendon RE, Hao Y, Shi Q, Hjelmeland AB, Dewhirst MW, Bigner DD, Rich JN (2006) Glioma stem cells promote radioresistance by preferential activation of the DNA damage response. Nature 444:756–760
15. Kosodo Y, Röper K, Haubensak W, Marzesco AM, Corbeil D, Huttner WB (2004) Asymmetric distribution of the apical plasma membrane during neurogenic divisions of mammalian neuroepithelial cells. EMBO J 23:2314–2324
16. Marzesco AM, Janich P, Wilsch-Bräuninger M, Dubreuil V, Langenfeld K, Corbeil D, Huttner WB (2005) Release of extracellular membrane particles carrying the stem cell marker prominin-1 (CD133) from neural progenitors and other epithelial cells. J Cell Sci 118:2849–2858
17. Dubreuil V, Marzesco AM, Corbeil D, Huttner WB, Wilsch-Bräuninger M (2007) Midbody and primary cilium of neural progenitors release extracellular membrane particles enriched in the stem cell marker prominin-1. J Cell Biol 176:483–495
18. Wilsch-Bräuninger M, Peters J, Paridaen JT, Huttner WB (2012) Basolateral rather than apical primary cilia on neuroepithelial cells committed to delamination. Development 139:95–105
19. Röper K, Corbeil D, Huttner WB (2000) Retention of prominin in microvilli reveals distinct cholesterol-based lipid microdomains in the apical plasma membrane. Nat Cell Biol 2:582–592
20. Huttner WB, Zimmerberg J (2001) Implications of lipid microdomains for membrane curvature, budding and fission. Curr Opin Cell Biol 13:478–484
21. Konno D, Shioi G, Shitamukai A, Mori A, Kiyonari H, Miyata T, Matsuzaki F (2008) Neuroepithelial progenitors undergo LGN-dependent planar divisions to maintain self-renewability during mammalian neurogenesis. Nat Cell Biol 10:93–101
22. Lathia JD, Hitomi M, Gallagher J, Gadani SP, Adkins J, Vasanji A et al (2011) Distribution of CD133 reveals glioma stem cells self-renew through symmetric and asymmetric cell divisions. Cell Death Dis 2:e200
23. Gulino A, Di Marcotullio L, Screpanti I (2010) The multiple functions of Numb. Exp Cell Res 316:900–906
24. Bauer N, Fonseca AV, Florek M, Freund D, Jászai J, Bornhäuser M, Fargeas CA, Corbeil D (2008) New insights into the cell biology of hematopoietic progenitors by studying prominin-1 (CD133). Cells Tissues Organs 188:127–138
25. Ettinger AW, Wilsch-Bräuninger M, Marzesco AM, Bickle M, Lohmann A, Maliga Z, Karbanová J, Corbeil D, Hyman AA, Huttner WB (2011) Proliferating versus differentiating stem and cancer cells exhibit distinct midbody-release behaviour. Nat Commun 2:503
26. Maw MA, Corbeil D, Koch J, Hellwig A, Wilson-Wheeler JC, Bridges RJ et al (2000) A frameshift mutation in prominin (mouse)-like 1 causes human retinal degeneration. Hum Mol Genet 9:27–34
27. Yang Z, Chen Y, Lillo C, Chien J, Yu Z, Michaelides M et al (2008) Mutant prominin 1 found in patients with macular degeneration disrupts photoreceptor disk morphogenesis in mice. J Clin Invest 118:2908–2916

28. Zacchigna S, Oh H, Wilsch-Bräuninger M, Missol-Kolka E, Jászai J, Jansen S et al (2009) Loss of the cholesterol-binding protein prominin-1/CD133 causes disk dysmorphogenesis and photoreceptor degeneration. J Neurosci 29:2297–2308
29. Fargeas CA, Florek M, Huttner WB, Corbeil D (2003) Characterization of prominin-2, a new member of the prominin family of pentaspan membrane glycoproteins. J Biol Chem 278:8586–8596
30. Jászai J, Fargeas CA, Haase M, Farkas LM, Huttner WB, Corbeil D (2008) Robust expression of Prominin-2 all along the adult male reproductive system and urinary bladder. Histochem Cell Biol 130:749–759
31. Jászai J, Farkas LM, Fargeas CA, Janich P, Haase M, Huttner WB, Corbeil D (2010) Prominin-2 is a novel marker of distal tubules and collecting ducts of the human and murine kidney. Histochem Cell Biol 133:527–539

Chapter 7
CD133-Positive Hematopoietic Stem Cells: From Biology to Medicine

Rupert Handgretinger and Selim Kuçi

Abstract Lifelong hematopoiesis is sustained by a very small number of hematopoietic stem cells capable of self-renewal and differentiation into multiple hematopoietic lineages. The sialomucin CD34 has been, and is currently, used for the identification and purification of primitive hematopoietic progenitors. Depending on the source of stem cells, CD34 may not be expressed on all progenitor cells. An alternative stem cell marker is prominin-1 (CD133), which is expressed on a subpopulation of CD34$^+$ cells as well as on CD34$^-$ progenitor cells derived from various sources including fetal liver and bone marrow, adult bone marrow, cord blood, and mobilized peripheral blood. CD133$^+$ stem cells can reconstitute myelo- and lymphopoiesis of lethally irradiated mice, and the characterization of the CD133 expression on stem cells provides some insights into the biology of the hierarchy and functional organization of human hematopoiesis. The availability of methods for clinical large-scale isolation of CD133$^+$ cells facilitates their use in autologous and allogeneic hematopoietic stem cell transplantation and possibly in other fields of regenerative medicine.

Keywords Adult multipotent hematopoietic cells • Allogeneic stem cell transplantation • Autologous stem cell transplantation • Hematopoietic stem cells • NOD/SCID-repopulating stem cell • Stem cell expansion • Stem cell subpopulations

---

R. Handgretinger, M.D. (✉)
University Children's Hospital, Department of Hematology/Oncology,
Hoppe-Seyler-Strasse 1, 72076 Tübingen, Germany
e-mail: Rupert.Handgretinger@med.uni-tuebingen.de

S. Kuçi, M.D., Ph.D.
University Children's Hospital III, Department of Hematology/Oncology,
Theodor-Stern-Kai 7, 60590 Frankfurt am Main, Germany

D. Corbeil (ed.), *Prominin-1 (CD133): New Insights on Stem & Cancer Stem Cell Biology*,
Advances in Experimental Medicine and Biology 777, DOI 10.1007/978-1-4614-5894-4_7,

7.1 Introduction

Hematopoietic stem cells (HSCs) are defined as cells that can give rise to all blood cell lineages throughout the life span of an individual. CD34 antigen has been the most frequently used marker for identification of HSCs [1–3]. However, not all human repopulating hematopoietic cells express it [4–6]. In 1997, CD133 (or AC133 antigen) was identified and reported as a potential alternative cell surface antigen for HSCs [7, 8]. CD133, also named prominin-1, is the first member of the prominin family (see Chaps. 1 and 2). This pentaspan membrane glycoprotein is expressed on various non-hematopoietic stem cell and progenitor populations [9] (see also Chap. 15) and on a subpopulation of hematopoietic progenitor cells [10]. In this context, HSCs derived from various sources including fetal liver and bone marrow, adult bone marrow (BM), umbilical cord blood (UCB), steady-state peripheral blood (sPB), and mobilized peripheral blood (mPB) show distinct expression levels of CD133 [7, 8, 11–13]. Moreover, the proportion of CD133^{+} cells within the CD34^{+} and CD34^{-} compartments and their capacity to repopulate their hosts upon transplantation might vary among the different HSC sources. In this chapter, we will describe these phenomena in more details.

7.1.1 *CD133 Expression on Bone Marrow–Derived HSCs*

To get more insights into the identity of CD133^{+} cells, a phenotypic analysis was performed by means of flow cytometry [14]. In BM, the co-expression analysis of CD133 and other antigens on CD34^{+} cells demonstrated that the majority of CD164^{+}, CD135^{+}, CD117^{+}, CD38^{low}, CD33^{+}, and CD71^{low} cells had segregated within the CD133^{+} population [14]. In contrast, most of the CD10^{+} and CD19^{+} B cell progenitors and a fraction of CD71^{high} population were found within the CD133^{-} population. This indicated that BM-derived CD34^{+}CD133^{+} cells had been enriched in primitive and myeloid progenitor cells, while CD34^{+}CD133^{-} cells had contained B cell and late erythroid progenitors. After selection of CD133^{+} BM cells, CD10^{+} and CD71^{high} cells were almost not detectable. Remarkably, a small proportion of CD133^{+} cells (0.2–0.7%) did not co-express CD34 and harbored a CD71^{-}, CD117^{-}, CD10^{-}, CD38^{low}, CD135^{+}, HLA-DR^{high}, and CD45^{-} phenotype. Further analysis of the CD34^{+}CD38^{-} population showed that virtually all these primitive HSCs had co-expressed CD133. Among this CD34^{+}CD133^{+}CD38^{-} subset, various expressions of other markers including angiopoietin receptors TIE and TEK (CD202B) and CD135 were observed. Functionally, several lines of evidence suggest that BM-derived CD133^{+} are able to give rise to different cellular components of hematopoietic microenvironment which are critical for the development and maintenance of the hematopoietic system. Quirici and colleagues demonstrated that the culture of selected BM CD133^{+} cells on fibronectin-coated flasks with defined growth factors for three weeks had given rise to a monolayer with typical endothelial cell (EC) morphology [15]. A consequent enrichment of these ECs using *Ulex europaeus* agglutinin-1 (UEA-1) resulted in a cell fraction, which after further culture for 3 weeks, expressed several endothelial markers including typical Weibel-Palade bodies. Remarkably, these ECs were superior to fibroblasts in supporting generation of hematopoietic colonies from CD34$^+$ HSCs in vitro [15].

BM CD133$^+$ cells possess also the property to give rise to non-hematopoietic multipotent bone marrow stromal cells (BMSCs), which contribute to both the sinusoidal and endosteal niche environments as demonstrated by Bakondi and colleagues [16]. For instance, after they were ex vivo expanded, CD133$^+$ cell–derived BMSC were seeded into hydroxyapatite-tricalcium phosphate fibrin constructs that were implanted subcutaneously into mice. After 60 days, human adipocytes, osteoblasts, osteocytes, and reticular cells were observed, and remarkably they supported the murine hematopoiesis, indicating that CD133 in human BM identifies a native stem and progenitor cell population that gives rise to all cellular determinants necessary for the hematopoietic microenvironment [16]. Besides their role in hematopoiesis and vasculogenesis (see also below), Pozzobon and colleagues demonstrated that CD133$^+$ BM cells had been able to differentiate in vitro to cells of the mesodermal and ectodermal (neurogenic) lineages [17]. Under their specific and defined culture conditions, the authors observed a rapid decrease of the expression of CD133 and the stemness markers c-kit (CD117, steel factor receptor) and OCT4, while the expression of the stage-specific embryonic antigen 4 (SSEA4) remained constant [17].

In order to assess the stemness of CD133$^+$ cells, Koutna and colleagues recently compared the proliferation and differentiation potential of CD133$^+$ and CD34$^+$ derived from BM and mPB [18]. After 6 days in culture, they observed a greater number of BM CD133$^+$ cells that were nondividing and contained a higher percentage of CD38$^-$ cells than CD133$^+$ cells from mPB. Furthermore, the CD133$^+$ BM subset contained significantly more vascular endothelial growth factor receptor antigen$^+$ (VEGFR$^+$) endothelial progenitor cells which led the investigators to conclude that CD133$^+$ cell population from BM display a better proliferation capacity and a higher proportion of primitive progenitors than any other studied population, namely, BM CD34$^+$, mPB CD133$^+$, and CD34$^+$ [18].

To get more biological insights into the regulation of these cells and given that microRNAs (miRNAs) appear to play a certain role in hematopoiesis, Bissels and colleagues have elegantly investigated their expression in different subpopulations of HSCs derived from BM. miRNAs are short ribonucleic acid (average of 22 nucleotides) molecules that negatively regulate target mRNAs on a posttranscriptional level. In their experience, BM-derived donor-matched CD133$^+$ were compared with CD34$^+$CD133$^-$ and CD34$^-$CD133$^-$ at the miRNA and mRNA levels [19]. They found that between 109 (CD133$^+$) and 216 (CD34$^-$CD133$^-$), miRNAs were expressed. Interestingly, 18 miRNAs were differentially expressed between CD133$^+$ and CD34$^+$CD133$^-$ cells. Their biological role was further analyzed by co-expression of miRNAs with their bioinformatically predicted mRNA targets. In a nutshell, they found that miRNAs differentially expressed between the CD133$^+$ and CD34$^+$CD133$^-$ cells had seemed to play a role in inhibition of differentiation, prevention of apoptosis, and cytoskeletal remodeling. Interestingly, the authors demonstrated a direct regulation of CD133 by miR-142-3p and showed that this particular miRNA had had a negative influence on the overall colony-forming capacity of CD133$^+$ cells [19]. Thus, in addition to its complex regulation at the promoter level (see Chap. 5), the expression of human CD133 seems to be under the control of miRNAs.

7.1.2 CD133 Expression on Cord Blood–Derived HSCs

The biology of CD133$^+$ HSCs derived from UCB has been extensively studied because of the simple accessibility of human UCB samples by comparison to BM ones. The analysis of expression of CD133 on CD34$^+$ cells in normal UCB, BM, and mPB demonstrated no significant difference between BM and UCB in terms of CD133 expression, i.e., 51% and 36.2% of the derived CO34$^+$ cells from UCB and BM, respectively, co-expressed CD133 [11]. However, a significant difference was observed with mPB CD34$^+$ cells where 75.3% of them expressed CD133 [11]. Furthermore, most of the clonogenic cells (67%) were contained in the CD34$^+$CD133$^+$ fraction and the generation of progenitor cells in long-term culture on bone marrow stromal cells was 10–100-fold higher as compared to CD34$^+$CD133$^-$ UCB cells. In a nonobese diabetic/severe combined immunodeficiency (NOD/SCID) mouse transplant model, only the CD34$^+$CD133$^+$ cells were able to reconstitute hematopoiesis with the first human cells being detectable after 20 days with an increasing level up to 40 days posttransplantation. No engraftment was observed in mice transplanted with up to 1.2×10^5 CD34$^+$CD133$^-$ cells, whereas transplantation with solely 5,000 CD34$^+$CD133$^+$ cells engrafted in about 20% of the mice. The CD34$^+$CD133$^+$ population was additionally enriched in dendritic cell precursors, and the generated dendritic cells were functionally active. Further analysis of the CD133$^+$ G_0 cells demonstrated that a large proportion of these cells had been CD38 negative (60.4%) and possessed a high aldehyde dehydrogenase activity (ALDH^{hi}) (75.1%) when compared to their CD133$^+$ G_1 counterpart (13.5% and 4.1%, respectively) [20]. This suggests that the stem cell activity resides within the CD133$^+$ G_0 population. Transplantation of these subsets into NOD/SCID mice showed multilineage engraftment only in the mice receiving the G_0 cells, whereas no significant engraftment was observed in the mice transplanted with cells in G_1 state. Therefore, quiescent CD133$^+$ UCB cells have a phenotype consistent with HSCs highly enriched for repopulating activity. To further dissect the phenotypic trait of these cells, lineage depletion combined with ALDH^{hi} selection was used [21]. The majority of the ALDH^{hi}CD133$^+$Lin$^-$ cells co-expressed CD34 but also included CD34$^-$CD38$^-$CD133$^+$ cells. Both ALDH^{hi}CD133$^-$Lin$^-$ and ALDH^{hi}CD133$^+$Lin$^-$ cells had different clonogenic capacities in vitro. However, only the ALDH^{hi}CD133$^+$Lin$^-$ population significantly engrafted in NOD/SCID and NOD/SCIDβ2M-null mice. Limiting dilution analysis revealed a 10-fold increase in the frequency of NOD/SCID-repopulating cells compared with CD133$^+$Lin$^-$ cells, indicating that ALDH activity further enriches cells with repopulating activity. After transplantation of ALDH^{hi}CD133$^+$Lin$^-$ cells, CD34$^+$CD38$^-$ primitive hematopoietic phenotypes were maintained and facilitated repopulation in recipients of serial, secondary transplants [21].

By using CD133 and CD7 cell surface markers, UCB-derived $CD34^-CD38^-Lin^-$ and $CD34^+CD38^-Lin^-$ populations were further dissected into $CD133^-CD7^-$ and $CD133^+CD7^-$ cells [22]. The latter subpopulation comprised a very rare population at a frequency of less than 0.02% and was surprisingly highly enriched in progenitor activity. This was the only subset among the $CD34^-CD38^-Lin^-$ cells that gave rise to $CD34^+$ cells in liquid culture. After transplantation of ex vivo-cultured $CD133^+CD7^-$ cells isolated from $CD34^-CD38^-Lin^-$ cells into NOD/SCID mice, human cells were found in the bone marrow of the recipient animals, whereas transplantation of a 400-fold greater number of $CD133^-CD7^-$ did not result in detectable engraftment, suggesting the existence of a rare population of human $CD133^+CD34^-$ HSCs. In addition, UCB-derived fms-related tyrosine kinase 3 $(Flt\text{-}3)^-CD34^-$ cells were shown to have long-term repopulating potential after intra-bone marrow injection into NOD/SCID mice [23].

The proliferative capacity of UCB-derived $CD34^-$ and $CD34^+$ cells that were negative for CD33, CD38, and CD71 (Lin^-) but positive for CD133 was assessed in vitro in the presence of Flt3-ligand, stem cell factor, and thrombopoietin [24]. Interestingly, the $CD34^-Lin^-CD133^+$ cells underwent more divisions in serum-free culture compared to the $CD34^+Lin^-CD133^+$ cells. Most of the $CD34^-$ cells acquired the expression of CD34 in vitro. Hematopoietic activity of both $CD34^-$ and $CD34^+$ cells resided in the cell population expressing CD34 after 4 days in culture. Moreover, in cultures from the $CD34^-Lin^-$ cells, hematopoietic activity was seen within the dividing cells, whereas in cultures of the $CD34^+Lin^-$ cells, this activity could also be observed within the nondividing cell fraction. $CD133^+CD34^-Lin^-$ UCB HSCs acquire CD34 prior to cell division, and in vitro long-term hematopoietic activity resides within the cell fraction expressing CD34 [24].

The proliferative potential of UCB $CD133^+$ and $CD34^+$ cells was also analyzed in a serum and stromal cell-free culture system for up to several weeks [25]. $CD133^+$ cells in the G_0 phase had a high long-term culture-initiating and colony-forming cell incidence and were capable of producing a 660 million-fold expansion of nucleated cells and a 120 million-fold expansion of colony-forming unit-granulocyte-macrophage over a period of 30 weeks and were consistently superior to $CD34^+$ G_0 cells. Further analysis of freshly isolated $CD133^+$ UCB cells revealed that 98% of them co-expressed the CD34. After 2 weeks of culture using the $CD133^+CD34^-$ starting population, 7.5% of the cells co-expressed the CD34, indicating that $CD133^+$ UCB cells comprise highly enriched stem cell population.

By employing the more recently described NOD/SCID $IL2r\gamma^{-/-}$ (NSG) mouse model [26, 27], expanded $CD34^+CD133^+$ starting from UCB $CD133^+$ cells were investigated for multilineage long-term engraftment and serial reconstitution [28]. The SCID-repopulating activity resided in the $CD34^+CD133^+$ fraction of expanded cells and correlated with the cell number of $CD34^+CD133^+$ cells before and after culture. The expanded cells were able to mediate long-term hematopoiesis and serial reconstitution in NSG mice.

Determination of the gene expression pattern of $CD34^+CD133^+$ HSCs might be helpful to further characterize the "true" stem cell populations. Comparison of the genetic signature of UCB $CD34^+CD133^+$ HSCs with that of HSC-depleted UCB mononuclear cells demonstrated 139 genes differentially expressed (> threefold) in $CD34^+CD133^+$ HSCs, which fell into a variety of gene families. Solely 26 genes were expressed at over fivefold levels, including genes encoding the known phenotypic or metabolic stem cell markers, CD34, CD133, c-Kit, and ALDH [29]. Transcriptional profiling of UCB $CD34^+$ versus $CD133^+$ cells revealed many genes that were similarly expressed, but also different patterns among these two subsets were observed [30]. $CD133^+$ cells had a higher number of upregulated genes than $CD34^+$ cells, and uniquely expressed genes in $CD34^+$ or $CD133^+$ cells were associated with distinct biological functions, e.g., many transcripts overexpressed in $CD34^+$ cells were associated with the development and response to stress or external stimuli, whereas in $CD133^+$ cells, the most common transcripts were linked to the establishment and maintenance of chromatin structure, DNA metabolism, and cell cycle. Overall, the difference between the gene expression profiles of UCB $CD34^+$ and $CD133^+$ cells indicates a more primitive nature of $CD133^+$ HSCs. Such data suggest indirectly that CD133 is somehow associated with the proliferative process. In another independent study, UCB-derived $CD34^+CD38^-$ cells were further separated into slow-dividing fraction (SDF) that is associated with primitive hematopoietic function and fast-dividing fraction (FDF), which mainly proceeds to differentiation [31], and their gene expression profiles were determined. In addition, the $CD34^+CD38^-$ cell profile was compared to those of $CD34^+CD38^+$ cells. As the main result, CD133 was among the genes that were highly expressed in the SDF, and the other expressed genes suggested a primitive function with self-renewal capacity. The morphological analysis of the SDF and FDF cells revealed that the former have more plasma membrane protrusions and that CD133 is expressed on lamellipodia [31]. The majority of the SDF cells were rhodamin-123^{dull}, compatible with the notion that the SDF is associated with primitive hematopoietic function.

In order to optimize retroviral gene transfer into HSCs, UCB $CD34^-$ cells were transduced using enhanced green fluorescent protein (eGFP)–containing retrovirus gene [32]. $CD34^-CD133^+Lin^-$ could be transduced at a higher efficiency compared to the $CD34^-CD133^-Lin^-$ or the $CD34^+CD38^-Lin^-$ cells. The cells of colony-forming units (CFU) derived from $CD34^-CD133^+Lin^-$ showed expression of eGFP, and cultures seeded with $CD34^-CD133^+Lin^-$ cells had a threefold higher frequency of $eGFP^+$ cells compared to those seeded with $CD34^+CD38^-Lin^-$ cells. Therefore, $CD133^+$ cells appear to be a better target for gene transfer protocols (see below).

7.1.3 CD133 Expression on Steady-State and Mobilized Peripheral Blood–Derived HSCs

A novel subset of human adherent $CD34^-$ HSCs that gave rise to NOD/SCID-repopulating cells within highly enriched $CD133^+$ cells isolated from mPB was identified after culture of $CD133^+$ cells for 3–5 weeks with Flt3-ligand and interleukin-6 [33]. Under this culture condition, a subfraction became adherent and lacked markers of HSCs including CD133, mesenchymal stem cells, dendritic cells, and stromal fibroblasts. The adherent cells either spontaneously or after stimulation with stem cell factor gave rise to nonadherent $CD133^{+}CD34^{-}$ cells. While no in vitro hematopoietic activity was observed, a substantially higher long-term multilineage engraftment compared to that of freshly isolated $CD34^{+}$ cells could be induced after transplantation into NOD/SCID mice. In addition to the engraftment of the myeloid lineage, the nonadherent $CD133^{+}CD34^{-}$ cells reconstituted also B-, T-, and natural killer (NK) cells. Morphologically, the characterization of the adherent cell population revealed the formation of tenupodia and magnupodia. They also displayed buds that protruded up to 9 μm from the cell surface and spoonlike projections (lobopodia). Fluorescence microscopy demonstrated that the CD133 expression was exclusively confined to these buds in agreement with the preferential localization of CD133 in plasma membrane protrusions (see Chap. 1). It is noteworthy that these $CD133^{+}$-derived adherent cells can differentiate not only into cells with NOD/SCID-engrafting capacity but, under appropriate in vitro culture conditions, also in neural progenitor-like, hepatocyte-like, and skeletal muscle-like cells. Therefore, these $CD133^{+}$ cells were named adult multipotent hematopoietic cells (AMHCs) [34, 35]. It is of note that the acquisition of a polarized cell shape is not a unique feature of AMHC since cultured human mPB, UCB, and BM $CD34^{+}$ also displayed the formation of protrusions such as a leading edge at the front pole and a uropod at the rear pole during their migration and redistributed CD133 at the uropod structure [36].

The expression and proliferation potential of sPB were analyzed and compared to those of UCB and mPB [12]. The proportion of $CD133^{+}$ cells among the $CD34^{+}$ subpopulation in sPB was significantly lower than of UCB, and mPB had the highest proportion of $CD34^{+}CD133^{+}$ cells. The proliferation potential of $CD34^{+}CD133^{+}$ and $CD34^{+}CD133^{-}$ cells was examined by colony-forming analysis and by the initiation of a long-term culture-initiating cells (LTC-IC) assay. While the number of colony-forming cells was similar for the $CD34^{+}CD133^{+}$ and $CD34^{+}CD133^{-}$ subpopulation, the number of LTC-IC was much higher in the $CD34^{+}CD133^{+}$ fraction, and no LTC-IC were detected in the $CD34^{+}CD133^{-}$ fraction. Most of the $CD34^{+}CD133^{+}$-derived colonies were granulocyte-macrophage types, whereas majority of the $CD34^{+}CD133^{-}$-derived colonies were erythroid ones. Moreover, colony-forming cells were only detected in the $CD34^{+}CD133^{+}$ subpopulation after ex vivo expansion of this cell fraction. CD133 might therefore be a better marker for the identification and selection of primitive hematopoietic progenitors from sPB. A comparative analysis of the proliferative potential and clonogenicity of $CD34^{+}$ and $CD133^{+}$ cells isolated from mPB of a single donor confirmed that the $CD133^{+}$ cells contained more LTC-IC than the $CD34^{+}$ cells, while $CD34^{+}CD133^{-}$ cells harbored a significantly higher proportion of erythroid progenitor cells [37]. Such differences in lineage-committed progenitors might be relevant in certain conditions. For instance, the preferential use of $CD34^{+}$ selection might be advantageous for the patients in whom fast erythroid engraftment is necessary, whereas the use of $CD133^{+}$ selection could be advantageous when the isolated cells were further expanded ex vivo. Likewise, gene transduction protocols might be more efficient when $CD133^{+}$ cells are used to target the myeloid and macrophage lineages [37].

Finally, CD133 seems to be expressed by circulating endothelial precursors (CEPs), which are regarded as potential therapeutic agents of interest as they may participate in postnatal vasculogenesis in both physiological and pathological neovascularization [38]. The human vascular endothelial growth factor receptor-2 (VEGFR-2) is expressed on a subset of CD34$^+$ cells derived from mPB, UCB, and BM [38], and approximately 2% mPB CD34$^+$ cells were positive for VEGFR-2 and strikingly co-expressed CD133. Since CD133 is not found on mature endothelial cells, the phenotype CD34$^+$CD133$^+$VEGFR-2$^+$ identifies most likely a population of CEPs. This subset co-expresses endothelial-specific markers including VE-cadherin and E-selectin [38]. These circulating CD34$^+$CD133$^+$VEGFR-2$^+$ cells might play a role in neo-angiogenesis. The co-expression of CD133 and VEGFR-3 on CD34$^+$ cells isolated from fetal liver, UCB, and PB also identified a population of CD34$^+$ lymphatic/vascular endothelial precursor cells [39]. In order to identify circulating precursors that have the developmental potential of hemangioblasts, an ex vivo long-term culture model supporting the differentiation of both hematopoietic and endothelial cell lineages was established [40]. A cell population from sPB with a CD34$^-$Lin$^-$CD45$^-$CD133$^-$ was isolated and cultured for 6 weeks. The cultured cells were able to differentiate into both hematopoietic and endothelial lineages, and the bilineage potential of the CD34$^-$Lin$^-$CD45$^-$CD133$^-$ cells was determined at a single cell level in vitro as well as by transplantation into NOD/SCID mice. This cell population reconstituted hematopoietic tissue and generated functional endothelial cells which contributed to new vessel formation during tumor angiogenesis. Further molecular characterization of the CD34$^-$Lin$^-$CD45$^-$CD133$^-$ cells showed a stem cell profile compatible with both hematopoietic and endothelial lineage potential, and these cells might represent putative circulating hemangioblasts.

7.2 The Clinical Role of CD133$^+$ HSCs in Autologous and Allogeneic Stem Cell Transplantation

Besides the characterization of CD133$^+$ HSCs in in vitro systems and animal models, the clinical interest of these cells becomes evident by the introduction of large-scale methods for their positive immunoselection from healthy donors under Good Manufacturing Process (GMP) conditions. Thus, the CliniMACS device (Miltenyi Biotec, Bergisch-Gladbach, Germany) allows the establishment of clinical protocols for the use of CD133$^+$ cells in autologous and allogeneic transplantation [13]. With such system, the percentage of CD133$^+$ cells before and after selection was 0.75 and 94, respectively, with a median recovery of 69. Importantly, the immuno-CD133-positive selection did not influence the viability of cells, and transplantation assay into NOD/SCID mice showed a high engraftment rate depending on the number of transplanted CD133$^+$ cells. With the possibility of clinical large-scale isolation of CD133$^+$ cells from mPB as well as BM, clinical protocols or single patient treatments have been initiated in the autologous and allogeneic setting a decade ago, i.e., only 5 years after the identification and molecular cloning of CD133 [7, 8].

7.2.1 Autologous Transplantation of CD133$^+$ HSCs

An 11-year-old boy with acute lymphoblast leukemia was the first patient who received an autologous graft consisting of CD133$^+$ cells purified from mPB after a myeloablative conditioning regimen [41]. The patient's leukemic blasts were CD34$^+$ and CD133$^-$, and the positively selected CD133 cells were used to minimize the risk of reinfusing malignant blasts. The child showed a rapid recovery of neutrophils and platelets whereby he remained in remission during the follow-up period of 11.5 months [41]. A similar approach was chosen for adult patients with resistant/relapsed chronic lymphocytic leukemia. Positive selection of mPB CD133$^+$ HSCs was performed in 10 individuals and reinfused after myeloablative conditioning [42]. A median number of 2.45×10^6 CD133$^+$ cells/kg body weight with a median purity of 94.8% was transplanted. All patients showed a fast and sustained engraftment and a rapid immune reconstitution without any late infectious complications. Unfortunately, in seven out of eight evaluable patients of this study, malignant graft-contaminating cells could still be detected despite the high purity of the selected CD133$^+$ cells [42].

The possibility of immunologic purging of autologous peripheral blood stem cells based on the CD34 and CD133 expression on the malignant blasts was also evaluated in patients with acute myeloid leukemia [43]. Eradication by positive selection of CD34$^+$ and/or CD133$^+$ cells resulted in three to four log tumor cell reduction and could be achieved in about half of patients, whose blasts at diagnosis showed no, or only dim, expression of CD34 and/or CD133 antigens.

7.2.2 Allogeneic Transplantation of CD133$^+$ HSCs

The clinical large-scale isolation method of CD133$^+$ cells was first employed in 10 pediatric patients [44]. These patients received a combination of CD34$^+$ and CD133$^+$ stem cells isolated from mPB from matched unrelated donors. For safety reasons, the proportion of CD133$^+$ stem cells was gradually increased from patient to patient from 10% to 100%, and patient 10 received exclusively CD133$^+$ cells. Primary engraftment was seen in eight patients, and two patients who received CD34$^+$ and CD133$^+$ cells rejected their graft. The patient who exclusively received CD133$^+$ cells showed a complete donor engraftment [44]. A further trial was then performed in 14 patients who received purified CD133$^+$ mobilized stem cells from haploidentical donors [45]. Primary engraftment was similar to that achieved after transplantation of positively selected CD34$^+$ cells, and both methods had similar incidence of graft-versus-host disease. In another series of five patients with high-risk hematological malignancies, haploidentical peripheral CD133$^+$ cells were transplanted after intensive chemotherapy [46]. Early three-lineage engraftment was observed with a median time to recovery of neutrophils $>0.5 \times 10^9$/L and platelets $>20 \times 10^9$/L of 14 and 13.5 days, respectively. In addition, a rapid reconstitution of dendritic cells after allogeneic transplantation of peripheral mobilized CD133$^+$ cells was seen in 10 adult patients with hematological malignancies after myeloablative conditioning [47]. The dendritic cells were predominantly of donor origin, and the dynamics of dendritic cells reconstitution was similar to that seen after transplantation of T-replete grafts.

In summary, mobilized peripheral CD133$^+$ cells can be isolated for clinical studies comparable to CD34$^+$ stem cells. The limited number of patients who received allogeneic CD133$^+$ grafts so far showed a rapid and complete hematopoietic recovery and long-term engraftment of donor cells. Since CD133$^+$ cells might represent more primitive hematopoietic progenitors than CD34$^+$ cells, further clinical trials using enriched allogeneic CD133$^+$ grafts are warranted. In autologous transplantation, the differential expression of CD133 and CD34 on malignant cells can be used for reducing the number of graft-contaminating tumor cells in certain hematological diseases. In addition, the high transduction efficiency with retroviral vectors makes the CD133$^+$ cells attractive target cells for gene therapy protocols for hemoglobinopathies or immunodeficiencies.

Due to the easy access and the availability of large cell numbers, CD133$^+$ cells from mPB are currently the best choice for further clinical trials in autologous and allogeneic transplantation. CD133$^+$ cells isolated from BM are already used in clinical studies of regenerative medicine (see Chap. 15) but not in the treatment of hematological malignancies. Due to the limited cell numbers, UCB-derived CD133$^+$ stem cells might only play a role as a starting population for clinical stem cell expansion protocols in allogeneic transplantation.

7.3 Conclusion

The expression of CD133 (prominin-1) on HSCs is of considerable interest for the understanding and further exploration of the hierarchy of human hematopoiesis. The functional analysis of CD133$^+$ stem cell subpopulations will allow deeper insights into the mechanism of self-renewal of primitive stem cells and maintenance of lifelong hematopoiesis. In this context, the demonstration of its symmetric versus asymmetric distribution in dividing HSCs [10, 48, 49] and its release into the extracellular milieu in association with small membrane vesicles during the process of HSC differentiation [50] are fascinating. Further study using CD133 knockout mice, for instance, might also bring new aspects of hematopoiesis [51]. Moreover, the immunoisolation of CD133$^+$ stem cells facilitates clinical protocols in the treatment of hematological malignancies as well as in regenerative medicine.

References

1. Civin CI, Strauss LC, Brovall C, Fackler MJ, Schwartz JF, Shaper JH (1984) Antigenic analysis of hematopoiesis. III. A hematopoietic progenitor cell surface antigen defined by a monoclonal antibody raised against KG-1a cells. J Immunol 133:157–165
2. Bhatia M, Wang JC, Kapp U, Bonnet D, Dick JE (1997) Purification of primitive human hematopoietic cells capable of repopulating immune-deficient mice. Proc Natl Acad Sci U S A 94:5320–5325

3. Krause DS, Fackler MJ, Civiñ CI, May WS (1996) CD34: structure, biology, and clinical utility. Blood 87:1–13
4. Goodell MA, Rosenzweig M, Kim H, Marks DF, DeMaria M, Paradis G, Grupp SA, Sieff CA, Mulligan RC, Johnson RP (1997) Dye efflux studies suggest that hematopoietic stem cells expressing low or undetectable levels of CD34 antigen exist in multiple species. Nat Med 3:1337–1345
5. Zanjani ED, meida-Porada G, Livingston AG, Flake AW, Ogawa M (1998) Human bone marrow CD34- cells engraft in vivo and undergo multilineage expression that includes giving rise to CD34+ cells. Exp Hematol 26:353–360
6. Bhatia M, Bonnet D, Murdoch B, Gan OI, Dick JE (1998) A newly discovered class of human hematopoietic cells with SCID-repopulating activity. Nat Med 4:1038–1045
7. Yin AH, Miraglia S, Zanjani ED, meida-Porada G, Ogawa M, Leary AG, Olweus J, Kearney J, Buck DW (1997) AC133, a novel marker for human hematopoietic stem and progenitor cells. Blood 90:5002–5012
8. Miraglia S, Godfrey W, Yin AH, Atkins K, Warnke R, Holden JT, Bray RA, Waller EK, Buck DW (1997) A novel five-transmembrane hematopoietic stem cell antigen: isolation, characterization, and molecular cloning. Blood 90:5013–5021
9. Meregalli M, Farini A, Belicchi M, Torrente Y (2010) CD133(+) cells isolated from various sources and their role in future clinical perspectives. Expert Opin Biol Ther 10:1521–1528
10. Bauer N, Fonseca AV, Florek M, Freund D, Jászai J, Bornhäuser M, Fargeas CA, Corbeil D (2008) New insights into the cell biology of hematopoietic progenitors by studying prominin-1 (CD133). Cells Tissues Organs 188:127–138
11. de Wynter EA, Buck D, Hart C, Heywood R, Coutinho LH, Clayton A, Rafferty JA, Burt D, Guenechea G, Bueren JA, Gagen D, Fairbairn LJ, Lord BI, Testa NG (1998) CD34+AC133+ cells isolated from cord blood are highly enriched in long-term culture-initiating cells, NOD/SCID-repopulating cells and dendritic cell progenitors. Stem Cells 16:387–396
12. Matsumoto K, Yasui K, Yamashita N, Horie Y, Yamada T, Tani Y, Shibata H, Nakano T (2000) In vitro proliferation potential of AC133 positive cells in peripheral blood. Stem Cells 18:196–203
13. Gordon PR, Leimig T, Babarin-Dorner A, Houston J, Holladay M, Mueller I, Geiger T, Handgretinger R (2003) Large-scale isolation of CD133+ progenitor cells from G-CSF mobilized peripheral blood stem cells. Bone Marrow Transplant 31:17–22
14. Bühring HJ, Seiffert M, Bock TA, Scheding S, Thiel A, Scheffold A, Kanz L, Brugger W (1999) Expression of novel surface antigens on early hematopoietic cells. Ann N Y Acad Sci 872:25–38, Discussion
15. Quirici N, Soligo D, Caneva L, Servida F, Bossolasco P, Deliliers GL (2001) Differentiation and expansion of endothelial cells from human bone marrow CD133(+) cells. Br J Haematol 115:186–194
16. Bakondi B, Spees JL (2010) Human CD133-derived bone marrow stromal cells establish ectopic hematopoietic microenvironments in immunodeficient mice. Biochem Biophys Res Commun 400:212–218
17. Pozzobon M, Piccoli M, Ditadi A, Bollini S, Destro R, Destro R, ndre-Schmutz I, Masiero L, Lenzini E, Zanesco L, Petrelli L, Cavazzana-Calvo M, Gazzola MV, De CP (2009) Mesenchymal stromal cells can be derived from bone marrow CD133+ cells: implications for therapy. Stem Cells Dev 18:497–510
18. Koutna I, Peterkova M, Simara P, Stejskal S, Tesarova L, Kozubek M (2011) Proliferation and differentiation potential of CD133+ and CD34+ populations from the bone marrow and mobilized peripheral blood. Ann Hematol 90:127–137
19. Bissels U, Wild S, Tomiuk S, Hafner M, Scheel H, Mihailovic A, Choi YH, Tuschl T, Bosio A (2011) Combined characterization of microRNA and mRNA profiles delineates early differentiation pathways of CD133+ and CD34+ hematopoietic stem and progenitor cells. Stem Cells 29:847–857
20. Boxall SA, Cook GP, Pearce D, Bonnet D, El-Sherbiny YM, Blundell MP, Howe SJ, Leek JP, Markham AF, de Wynter EA (2009) Haematopoietic repopulating activity in human cord blood CD133+ quiescent cells. Bone Marrow Transplant 43:627–635

21. Hess DA, Wirthlin L, Craft TP, Herrbrich PE, Hohm SA, Lahey R, Eades WC, Creer MH, Nolta JA (2006) Selection based on CD133 and high aldehyde dehydrogenase activity isolates long-term reconstituting human hematopoietic stem cells. Blood 107:2162–2169
22. Gallacher L, Murdoch B, Wu DM, Karanu FN, Keeney M, Bhatia M (2000) Isolation and characterization of human CD34(−)Lin(−) and CD34(+)Lin(−) hematopoietic stem cells using cell surface markers AC133 and CD7. Blood 95:2813–2820
23. Kimura T, Asada R, Wang J, Kimura T, Morioka M, Matsui K, Kobayashi K, Henmi K, Imai S, Kita M, Tsuji T, Sasaki Y, Ikehara S, Sonoda Y (2007) Identification of long-term repopulating potential of human cord blood-derived CD34-flt3- severe combined immunodeficiency-repopulating cells by intra-bone marrow injection. Stem Cells 25:1348–1355
24. Gotze KS, Schiemann M, Marz S, Jacobs VR, Debus G, Peschel C, Oostendorp RA (2007) CD133-enriched CD34(−) (CD33/CD38/CD71)(−) cord blood cells acquire CD34 prior to cell division and hematopoietic activity is exclusively associated with CD34 expression. Exp Hematol 35:1408–1414
25. Summers YJ, Heyworth CM, de Wynter EA, Hart CA, Chang J, Testa NG (2004) AC133+ G0 cells from cord blood show a high incidence of long-term culture-initiating cells and a capacity for more than 100 million-fold amplification of colony-forming cells in vitro. Stem Cells 22:704–715
26. Shultz LD, Lyons BL, Burzenski LM, Gott B, Chen X, Chaleff S, Kotb M, Gillies SD, King M, Mangada J, Greiner DL, Handgretinger R (2005) Human lymphoid and myeloid cell development in NOD/LtSz-scid IL2R gamma null mice engrafted with mobilized human hemopoietic stem cells. J Immunol 174:6477–6489
27. Shultz LD, Ishikawa F, Greiner DL (2007) Humanized mice in translational biomedical research. Nat Rev Immunol 7:118–130
28. Drake AC, Khoury M, Leskov I, Iliopoulou BP, Fragoso M, Lodish H, Chen J (2011) Human CD34+ CD133+ hematopoietic stem cells cultured with growth factors including Angptl5 efficiently engraft adult NOD-SCID Il2rgamma−/− (NSG) mice. PLoS One 6:e18382
29. He X, Gonzalez V, Tsang A, Thompson J, Tsang TC, Harris DT (2005) Differential gene expression profiling of CD34+ CD133+ umbilical cord blood hematopoietic stem progenitor cells. Stem Cells Dev 14:188–198
30. Hemmoranta H, Hautaniemi S, Niemi J, Nicorici D, Laine J, Yli-Harja O, Partanen J, Jaatinen T (2006) Transcriptional profiling reflects shared and unique characters for CD34+ and CD133+ cells. Stem Cells Dev 15:839–851
31. Wagner W, Ansorge A, Wirkner U, Eckstein V, Schwager C, Blake J, Miesala K, Selig J, Saffrich R, Ansorge W, Ho AD (2004) Molecular evidence for stem cell function of the slow-dividing fraction among human hematopoietic progenitor cells by genome-wide analysis. Blood 104:675–686
32. Murdoch B, Gallacher L, Chadwick K, Bhatia M (2002) Characterization of retroviral gene transfer into highly purified human CD34(−) cells with primitive hematopoietic capacity. Mol Ther 5:635–643
33. Kuçi S, Wessels JT, Buhring HJ, Schilbach K, Schumm M, Seitz G, Loffler J, Bader P, Schlegel PG, Niethammer D, Handgretinger R (2003) Identification of a novel class of human adherent CD34- stem cells that give rise to SCID-repopulating cells. Blood 101:869–876
34. Kuçi S, Kuçi Z, Schmid S, Seitz G, Muller I, Dufke A, Leimig T, Murti G, Jurecic R, Schumm M, Lang P, Bruchelt G, Bader P, Klingebiel T, Niethammer D, Handgretinger R (2008) Efficient in vitro generation of adult multipotent cells from mobilized peripheral blood CD133+ cells. Cell Prolif 41:12–27
35. Handgretinger R, Gordon PR, Leimig T, Chen X, Bühring HJ, Niethammer D, Kuci S (2003) Biology and plasticity of CD133+ hematopoietic stem cells. Ann N Y Acad Sci 996:141–151
36. Giebel B, Corbeil D, Beckmann J, Hohn J, Freund D, Giesen K, Fischer J, Kogler G, Wernet P (2004) Segregation of lipid raft markers including CD133 in polarized human hematopoietic stem and progenitor cells. Blood 104:2332–2338

37. Freund D, Oswald J, Feldmann S, Ehninger G, Corbeil D, Bornhäuser M (2006) Comparative analysis of proliferative potential and clonogenicity of MACS-immunomagnetic isolated CD34+ and CD133+ blood stem cells derived from a single donor. Cell Prolif 39:325–332
38. Peichev M, Naiyer AJ, Pereira D, Zhu Z, Lane WJ, Williams M, Oz MC, Hicklin DJ, Witte L, Moore MA, Rafii S (2000) Expression of VEGFR-2 and AC133 by circulating human CD34(+) cells identifies a population of functional endothelial precursors. Blood 95:952–958
39. Salven P, Mustjoki S, Alitalo R, Alitalo K, Rafii S (2003) VEGFR-3 and CD133 identify a population of CD34+ lymphatic/vascular endothelial precursor cells. Blood 101:168–172
40. Ciraci E, Della BS, Salvucci O, Rofani C, Segarra M, Bason C, Molinari A, Maric D, Tosato G, Berardi AC (2011) Adult human circulating CD34(−)Lin(−)CD45(−)CD133(−) cells can differentiate into hematopoietic and endothelial cells. Blood 118:2105–2115
41. Koehl U, Zimmermann S, Esser R, Sorensen J, Gruttner HP, Duchscherer M, Seifried E, Klingebiel T, Schwabe D (2002) Autologous transplantation of CD133 selected hematopoietic progenitor cells in a pediatric patient with relapsed leukemia. Bone Marrow Transplant 29:927–930
42. Isidori A, Motta MR, Tani M, Terragna C, Zinzani P, Curti A, Rizzi S, Taioli S, Giudice V, D'Addio A, Gugliotta G, Conte R, Baccarani M, Lemoli RM (2007) Positive selection and transplantation of autologous highly purified CD133(+) stem cells in resistant/relapsed chronic lymphocytic leukemia patients results in rapid hematopoietic reconstitution without an adequate leukemic cell purging. Biol Blood Marrow Transplant 13:1224–1232
43. Feller N, van der Pol MA, Waaijman T, Weijers GW, Westra G, Ossenkoppele GJ, Schuurhuis GJ (2005) Immunologic purging of autologous peripheral blood stem cell products based on CD34 and CD133 expression can be effectively and safely applied in half of the acute myeloid leukemia patients. Clin Cancer Res 11:4793–4801
44. Lang P, Bader P, Schumm M, Feuchtinger T, Einsele H, Fuhrer M, Weinstock C, Handgretinger R, Kuci S, Martin D, Niethammer D, Greil J (2004) Transplantation of a combination of CD133+ and CD34+ selected progenitor cells from alternative donors. Br J Haematol 124:72–79
45. Lang P, Schumm M, Greil J, Bader P, Klingebiel T, Muller I, Feuchtinger T, Pfeiffer M, Schlegel PG, Niethammer D, Handgretinger R (2005) A comparison between three graft manipulation methods for haploidentical stem cell transplantation in pediatric patients: preliminary results of a pilot study. Klin Padiatr 217:334–338
46. Bitan M, Shapira MY, Resnick IB, Zilberman I, Miron S, Samuel S, Ackerstein A, Elad S, Israel S, Amar A, Fibach E, Or R, Slavin S (2005) Successful transplantation of haploidentically mismatched peripheral blood stem cells using CD133+−purified stem cells. Exp Hematol 33:713–718
47. Bornhäuser M, Eger L, Oelschlaegel U, Auffermann-Gretzinger S, Kiani A, Schetelig J, Illmer T, Schaich M, Corbeil D, Thiede C, Ehninger G (2005) Rapid reconstitution of dendritic cells after allogeneic transplantation of CD133+ selected hematopoietic stem cells. Leukemia 19:161–165
48. Fargeas CA, Fonseca A-V, Huttner WB, Corbeil D (2006) Prominin-1 (CD133): from progenitor cells to human diseases. Future Lipidol 1:213–225
49. Fonseca A-V, Bauer N, Corbeil D (2008) The stem cell marker CD133 meets the endosomal compartment – New insights into the cell division of hematopoietic stem cells. Blood Cells Mol Dis 41:194–195
50. Bauer N, Wilsch-Bräuninger M, Karbanová J, Fonseca AV, Strauss D et al (2011) Haematopoietic stem cell differentiation promotes the release of prominin-1/CD133-containing membrane vesicles-a role of the endocytic-exocytic pathway. EMBO Mol Med 3:398–409
51. Zacchigna S, Oh H, Wilsch-Bräuninger M, Missol-Kolka E, Jászai J et al (2009) Loss of the cholesterol-binding protein prominin-1/CD133 causes disk dysmorphogenesis and photoreceptor degeneration. J Neurosci 29:2297–2308

Chapter 8
New Insights into the Renal Progenitor Cells and Kidney Diseases by Studying CD133

Benedetta Bussolati and Giovanni Camussi

Abstract CD133$^+$ progenitor cells have been found in different segments of the human nephron. In particular, CD133-expressing cells are present in the cortex, in Bowman's capsule of the glomerulus, and in proximal convoluted tubules and in medulla, in the Henle's loop, and its thin limb segments. The collecting ducts are negative. During repair of renal injury, CD133-expressing cells are increased, suggesting a contribution in renal regeneration. An increase has also been observed in pathological conditions. CD133$^+$ cells contribute to the formation of glomerular crescents and are lining the cysts in the polycystic kidney disease. Therefore, an altered regulation of CD133$^+$ cell proliferation or differentiation could be involved in glomerular and tubular response to injury in pathological condition. In clear cell renal carcinoma, despite CD133$^+$ cells appeared to contribute to tumor vascularization, they did not display features of tumor-initiating cells.

Keywords CD133 • Crescents • Glomerulonephritides • Polycystic kidney disease • Renal progenitors • Renal repair • Renal regeneration • Renal stem cells

8.1 Introduction

CD133 belongs to a family of proteins named prominin due to their prevalent expression on plasma membrane protrusions, such as microvilli and primary cilia [1–4]. The relevant expression of prominin-1 in renal epithelial cells can be highlighted by the fact that this molecule was initially cloned from a kidney cDNA library [1]. Prominin-1 has different splice variants found to be tissue-specific and

B. Bussolati • G. Camussi, M.D. (✉)
Department of Medical Sciences, University of Torino,
Cso Dogliotti 14, 10126 Torino, Italy
e-mail: giovanni.camussi@unito.it

D. Corbeil (ed.), *Prominin-1 (CD133): New Insights on Stem & Cancer Stem Cell Biology*, Advances in Experimental Medicine and Biology 777, DOI 10.1007/978-1-4614-5894-4_8,

developmentally regulated, which might also suggest an interaction with distinct protein partners [5–7]. In particular, the mouse kidney cells express mainly the prominin-1.s1 [8]. The CD133 protein is the human ortholog of the murine prominin-1 and shares approximately 60% homology at the amino acid level [6, 9, 10]. In analogy with the mouse prominin-1, the analysis of the CD133 expression at both mRNA and protein levels showed that it was present in several terminally differentiated tissues, such as normal epithelia of the human kidney [2, 11, 12]. A great interest in CD133 expression and function within the renal tissue derives from its use as a marker of progenitor renal cells. This notion was achieved using AC133 and AC141 antibodies that bind to different glycosylation-dependent epitopes of the CD133 molecule that are selectively expressed by stem cells [3]. In accordance with stem-specific expression of the CD133 isoform detected by AC133 monoclonal antibodies (mAb), CD133 (AC133) is downregulated in the adult renal tissue in all differentiated cells [4, 13].

By studying AC133 expression, several groups identified progenitor cells within the nephron of the embryonic and adult human kidney. In addition, altered AC133 expression has been reported in glomerular and tubular diseases of the kidney. Although little is known about the role of CD133 in renal progenitors, gaining insights into the mechanisms that regulate its expression in human renal progenitors may be relevant to understand its function in physiological and pathological conditions. Unfortunately, the data obtained on human CD133$^+$ progenitors cannot be extrapolated yet to the murine system due to the lack of an AC133 antibody analogue to detect a murine stem-related isoform of prominin-1.

8.2 CD133 and Renal Progenitors of the Normal Kidney

8.2.1 CD133 Expression in the Nephron

In the human fetal kidney, at 8.5 to 9 weeks of gestational age, CD133$^+$ cells represent a conspicuous population (around 50%) and decrease thereafter, being only 0.5% to 3% of total cells in the adult human kidney [14]. In the early phases of nephrogenesis, CD133 expression was reported on renal vesicles and S-shaped bodies, structures that give rise to the Bowman's capsule and podocytes, as well as on immature tubules [14]. In human adult renal tissue, using the AC133 staining on cryostatic sections of renal biopsies, CD133 has been identified as a marker of progenitor cells within the nephron with characteristics of non-differentiated renal mesenchymal progenitors [15–19]. Cells expressing CD133 are distributed within different segments of the nephron (Fig. 8.1). In particular, in the cortex, CD133$^+$ cells were described within the Bowman's capsule of the glomerulus as well as in proximal convoluted tubules [15, 17, 18], whereas in the inner medulla, they have been observed in the Henle's loop and its thin limb segments [16, 19]. The collecting ducts were negative for CD133. These observations indicate that CD133$^+$ cells are distributed along the nephron rather than confined in a specific niche.

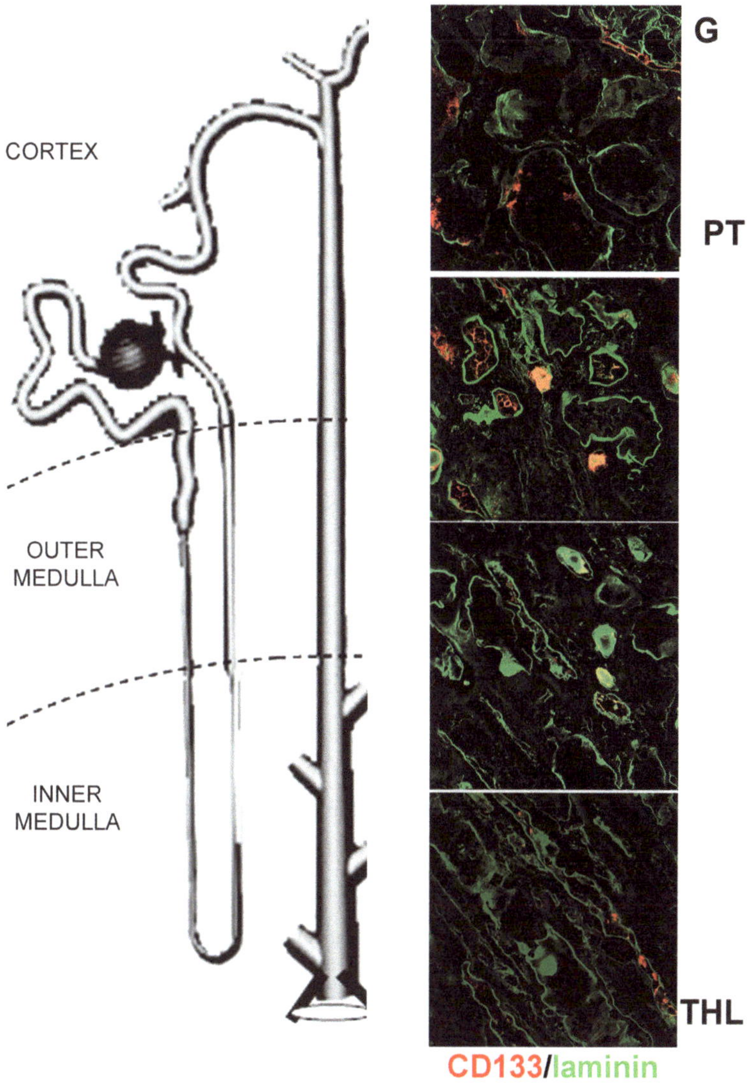

Fig. 8.1 Schematic representation of the nephron segments in the renal cortex and medulla and corresponding localization of CD133-expressing cells. The representative immunofluorescent micrographs show CD133 (AC133-1 mAb; *red*) expressed by cells lining the Bowman capsule of the glomerulus (G), by proximal tubules (PT) and by thin segments of the Henle's loop (THL). The immuno-staining for laminin (*green*) was used to evidence the renal tissue structure. Original magnification x400

Regarding CD133$^+$ cells in the Bowman's capsule, co-staining of CD133 with differentiation markers of podocytes showed a progressive acquirement of such markers and loss of CD133 starting from the urinary to the vascular pole. This may indicate that CD133$^+$ progenitors present in the Bowman's capsule differentiate into podocytes and reach the glomerular tuft via the vascular pole replacing lost podocytes [20]. When CD133⁺ cells in the tubular structures were co-labeled with classical markers of the different nephron segments, CD133⁺ cells in the cortex co-stained for the proximal tubule marker megalin [21] but not for the distal tubule marker Na/Cl co-transporter [22], suggesting their localization within the former tubular structures. In contrast, CD133⁺ cells in the medulla were negative for markers expressed by cells of the Henle's loop such as aquaporin-1 (AQ1), a marker of the thin descending segment [23]; chloride channel-1 (CLCK-1), a marker of the thin ascending segment [24]; and Tamm-Horsfall protein (THP) as a marker of the thick ascending segment [25]. The lack of co-expression of segment-specific differentiation markers in CD133⁺ cells in the inner medulla, and not in the cortex, might suggest that CD133⁺ cells within the nephron present a different degree of differentiation.

Studies performed in renal tissue undergoing ischemic injury, such as in the delayed graft function after renal transplantation, showed an increase in the number of CD133⁺ cells in respect to renal transplanted kidneys that did not undergo ischemic damage [26]. This observation strongly suggests that number of CD133⁺ cells increase during the reparative process. This might occur through the proliferation of CD133⁺ cells or alternatively through dedifferentiation of cells that survived the tissue injury. Further studies are required to show whether CD133 expression might become a marker of renal regeneration.

8.2.2 *Functional Studies on Isolated CD133⁺ Cells In Vitro*

Isolation and sorting of CD133⁺ renal progenitor cells from the different nephron segments allowed to assess their morphologic and functional properties. A genomic characterization of CD133⁺ renal progenitor cells from glomeruli and tubules of adult human kidney showed the absence of significant differences in their gene expression patterns suggesting that glomerular and tubular renal progenitors represent a genetically homogeneous population [18]. By a phenotypic point of view, CD133⁺ cells express markers typical of renal embryonic and mesenchymal stem cells, such as Six1 and Six2, Pax2 transcription factors, CD90, CD44, CD29, CD73 as surface markers, and vimentin [15]. In addition, they express stem-related transcription factors, such as Oct4A, c-Myc, and Klf-4, being the cells from the inner medulla more positive for Oct4A than those from the cortex [15, 16]. Indeed, the expression of cytokeratin by the CD133⁺ cells indicates their epithelial commitment, suggesting a role as epithelial progenitors [16, 17].

Isolated cells showed the typical properties of stem cells such as clonogenicity, ability to undergo differentiation into multiple lineages of the mesenchymal lines, such as osteogenic, adipogenic, neurogenic, and endothelial differentiation. In addition, these cells were able to differentiate into the different cell types typical of the nephron. Evidences for functional properties of epithelial cells, such as the acquirement of the transmembrane resistance, and presence of markers of podocytes, tubular cells of proximal, distal, and Henle's loop segments have been provided [15]. During differentiation, stem cell markers, including CD133, were lost.

8.2.3 Functional Studies on Isolated CD133$^+$ Cells In Vivo

Several experimental evidences indicate that CD133$^+$ progenitors can organize in renal structures in vivo. When injected subcutaneously within Matrigel in severe combined immunodeficiency (SCID) mice, undifferentiated CD133$^+$ cells showed after 10–15 days a spontaneous differentiation into tubular-like structures with morphological aspects of cell polarization and a virtual lumen containing proteinaceous materials [15]. Electron microscopy studies revealed the presence of short microvilli and tight junctions indicating the polarization of cells forming the tubular-like structures. CD133$^+$ cells derived from renal papilla, displaying higher differentiation ability, generated epithelial tubular structures with aspects of cuboidal or flat epithelium resembling the different segments of the nephron, whereas CD133$^+$ cells from renal cortex only differentiated into cuboidal structures resembling proximal tubuli [16]. In addition, these tubular structures expressed markers of the different nephron segments [16].

In parallel with these studies, murine models evaluated the integration and the functional effect of injected CD133$^+$ progenitors after renal injury. Acute renal damages were obtained in murine models of glomerular toxicity induced by Adriamycin and of myoglobin-dependent tubular necrosis induced by glycerol. In all experiments, CD133$^+$ cells migrated into the injured renal tissues, differentiated into cells of the nephron, and ameliorated renal function [15, 20]. Finally, CD133$^+$ cells from papilla integrated within developing mouse renal structures in a heterotypic metanephric organ culture [19].

Although the possible use of CD133$^+$ progenitors in human studies presents obvious limitations due to immunological barriers among individuals, the in vivo experiments reported indicate a possible physiological role of progenitors in renal regeneration after injury.

8.3 CD133 and Renal Pathology

An abnormal expression of the AC133-recognized CD133 isoform has been reported in different renal pathological settings, suggesting that CD133 might be considered as a marker of excessive proliferation of progenitors and/or of their reduced differentiation.

8.3.1 CD133 Expression in Tubular Pathology

In the tubular compartment, an increase in CD133 expression might reflect cell regeneration after injury [26]. However, an abnormal expression of CD133 as evaluated by immunofluorescence microscopy has been reported in tubular cells of patients with polycystic kidney disease [27], CD133 being strongly expressed by

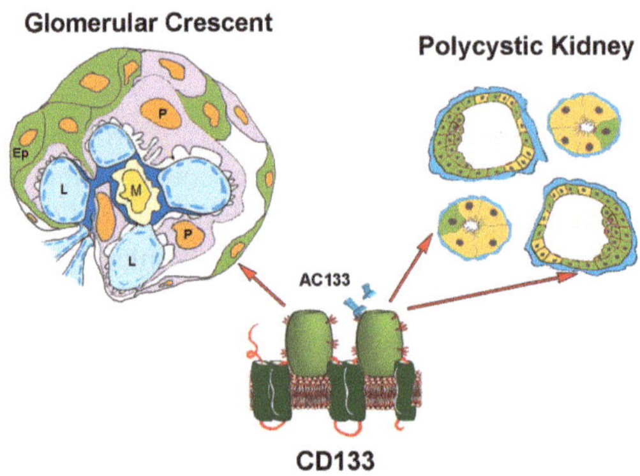

Fig. 8.2 Expression of CD133 in renal pathology. A schematic representation of the CD133 structure shows the 5-transmembrane domains and the two extracellular loops of the molecule as well as the binding of AC133 mAb on the second extracellular loop. CD133-expressing cells participate to crescent formation in glomerulonephritides and are lining the wall of cysts in polycystic kidney disease

cyst-lining cells (Fig. 8.2). This genetic disease, which is characterized by polycystin mutations, by progressive development of cysts within the renal parenchyma, and by progressive renal failure, is associated with a dedifferentiated immature phenotype of tubular cells. The expression of CD133 might be considered as a marker of immature tubular cells with an altered balance between differentiation and proliferation. Indeed, CD133$^+$ cells sorted from renal tissue of polycystic patients exhibited enhanced proliferation and were unable to differentiate in vitro. Moreover, they spontaneously formed cysts in vitro and in vivo upon injection in severe combined immunodeficiency (SCID) mice [27].

The observed alteration in the functional properties of CD133$^+$ cells might be explained by the reduction in the level of polycystins characteristic of these patients, leading to the inability to maintain planar polarity, and favoring proliferation and survival (28, 29). It can be therefore speculated that the altered differentiation capabilities of CD133$^+$ cells in polycystic kidney disease may be instrumental in generating a deregulated proliferative response after tubular injury that may result in cyst generation (27).

8.3.2 CD133 Expression in Glomerular Pathology

Under physiological condition, CD133$^+$ cells are expressed in the Bowman's capsule and could be involved in a continuous replacement of podocytes. However, the excessive proliferation of cells present in the Bowman's capsule is related to different glomerular pathologies, such as collapsing glomerulopathy and crescentic glomerulonephritis, and may lead to Bowman's space obliteration (30) (Fig. 8.2). In both pathologies, the large majority (50–75%) of cells present in the intra-glomerular lesions express CD133 (30). In analogy, CD133 is expressed by the majority of cells of the glomerular tip lesions observed at the urinary pole of the Bowman's capsule (30). Therefore, all information found in the literature indicate that $CD133^+$ cells are characterized by an increased proliferation and reduced differentiation both in polycystic kidney and glomerular crescents. Consequently, a selective pharmacological approach to modulate $CD133^+$ cell proliferation or differentiation could be of therapeutic significance.

8.3.3 CD133 Expression in Renal Carcinoma

In several tissues, CD133 is also expressed by the malignant counterpart of the resident stem cells (i.e., the cancer stem cells) (31). In renal carcinomas, a small population of $CD133^+/CD34^-$ cells (less than 1% of the total cells) is present. However, this population is not characterized by tumorigenic or tumor-initiating properties, indicating that CD133 is not a marker of renal cancer stem cells.

As $CD133^+$ cells isolated from renal carcinomas showed a comparable phenotype with that observed in the normal renal tissue, they might represent a residue of normal parenchyma or cells migrated to the tumor (32). The ability of tumor $CD133^+/CD34^-$ cells, in the presence of the tumor supernatant, to differentiate into endothelial cells could suggest that they may be involved in tumor vascularization. This was confirmed by in vivo experiments of co-transplantation of $CD133^+$ and tumor cells, at a 1:100 ratio (the same ratio present in renal carcinomas). In these co-transplants, tumor engraftment, growth, and vascularization were enhanced in comparison to tumor cells alone (32). Therefore, renal tumor-derived factors could recruit normal $CD133^+$ cells and promote their endothelial commitment, thus favoring tumor growth and dissemination. The lack of CD133 expression by renal cancer stem cells identified as a side population was recently confirmed, CD133 being solely expressed by the side population of the normal tissue (33). Moreover, the evaluation of CD133 expression in human renal clear cell carcinomas did not correlate with clinical pathological features or patient prognosis (34), implying that CD133 does not represent a useful prognostic marker for renal carcinomas. Different results were obtained by using an anti-CD133 polyclonal Ab, which did not detect the stem-related isoform recognized by the AC133 mAb. In these experiments, CD133 expression predicted progression and death in renal cell carcinoma (35).

8.4 Modulation of CD133 in Renal Progenitors

8.4.1 Modulation of $CD133^+$ Cells by Hypoxia

In neural progenitors and glioma stem cells, CD133 was recently reported to be upregulated by hypoxia (36, 37). This can be due to a transcriptional effect of hypoxia-inducible factors on CD133 synthesis as suggested by Mak and colleagues

(38). A similar effect was also reported in renal CD133⁺ cells cultured under hypoxia (1% oxygen). In this condition, the expression of CD133 is upregulated after 24-h exposure and maintained up to 72 h (16). In parallel, culture in hypoxic condition significantly modulates the progenitor/stem phenotype of $CD133^+$ cells, promoting proliferation and clonogenic potential and the expression of the stem cell transcription factor Oct4. This might be relevant in physiologic conditions, as in the papillary region the oxygen tension is very low and may be involved in maintaining the progenitor phenotype (39). In addition, hypoxia occurring after renal injury may promote CD133 functions and possibly modulate resident cells in tubuli promoting their dedifferentiation and acquirement of progenitor features (16). Hypoxia might therefore regulate stemness or differentiation of renal progenitors. The mechanism involves a balance between Oct4 and microRNA-145, a specific Oct4 microRNA (40). In this balance, hypoxia acts in the increased Oct4A expression through repression of miR145, which in turn is increased during normoxia and degrades Oct4, and hence promotes differentiation. The possibility to modulate CD133 in renal progenitors and induce regenerative programs in renal cells could be of interest for promoting renal repair after injury.

8.4.2 Involvement of CD133+ Cells in Renal Fibrosis

A recent study by Simone and colleagues showed the possible maldifferentiation of $CD133^+$ renal progenitors in myofibroblasts after delayed graft function (DGF) (41). DGF, defined as the requirement for dialysis in the first week after transplantation, is a common complication after renal transplantation; it is caused by acute kidney injury, and it is characterized by tissue inflammation and involvement of many cellular and soluble mediators. In particular, after DGF, the occurrence of oxidative stimuli such as oxygen radicals and presence of the bone morphogenetic protein-2 (BMP-2) in the tissue microenvironment promote the expression of α-smooth muscle cell actin in CD133+ cells. The acquisition of a myofibroblast marker by CD133+ cells clearly indicates their commitment toward a myofibroblastic phenotype that may be relevant in tissue fibrosis. This process of maldifferentiation could be important for progression of renal damage. An antifibrotic effect of the antioxidant N-acetyl-cysteine can be envisaged in this pathological setting.

8.4.3 Possible Role of CD133 in Renal Cells and Its Modulation

Knowledge on the function of CD133 and its stem cell-related isoform, which is recognized by AC133 mAb, is an important issue in order to understand its significance in renal pathology. However, little insight is only provided by studies on nonrenal cells. A recent study indicates that CD133 expression might be involved in cell metabolism (42). In this study, CD133 expression was shown to favor a glycolytic cell metabolism and in turn to inhibit the endocytosis of the transferrin receptor, thus resulting in the blockade of the mitochondrial activity due to iron lack. In addition, CD133 was reported to be a glucose responsive gene in myotubes (43), and its expression was related to glucose uptake in glioma cells (44). Finally, CD133 was shown to regulate the mitochondrial function implying a relationship with the anaerobic cell metabolism. In fact, inhibition of the respiratory chain complex I by rotenone caused a significant CD133 increase in glioma cells (44). Altogether, these data indicate that CD133 might be involved in the regulation of the glycolytic metabolism. The translation of this idea in renal pathophysiology is of great interest, as the acquirement of glycolytic metabolism by renal CD133+ cells after injury may promote their resistance to hypoxia and consequent survival to damage.

References

1. Weigmann A, Corbeil D, Hellwig A, Huttner WB (1997) Prominin, a novel microvilli-specific polytopic membrane protein of the apical surface of epithelial cells, is targeted to plasmalemmal protrusions of non-epithelial cells. Proc Natl Acad Sci USA 94:12425–12430
2. Fargeas CA, Florek M, Huttner WB, Corbeil D (2003) Characterization of prominin-2, a new member of the prominin family of pentaspan membrane glycoproteins. J Biol Chem 278:8586–8596
3. Yin AH, Miraglia S, Zanjani ED et al (1997) AC133, a novel marker for human hematopoietic stem and progenitor cells. Blood 90:5002–5012
4. Miraglia S, Godfrey W, Yin AH et al (1997) A novel five-transmembrane hematopoietic stem cell antigen: isolation, characterization, and molecular cloning. Blood 90:5013–5021
5. Fargeas CA, Corbeil D, Huttner WB (2003) AC133 antigen, CD133, prominin-1, prominin-2, etc.: prominin family gene products in need of a rational nomenclature. Stem Cells 21:506–508
6. Fargeas CA, Huttner WB, Corbeil D (2007) Nomenclature of prominin-1 (CD133) splice variants - an update. Tissue Antigens 69:602–606
7. Shmelkov SV, Jun L, St Clair R et al (2004) Alternative promoters regulate transcription of the gene that encodes stem cell surface protein AC133. Blood 103:2055–2061
8. Fargeas CA, Joester A, Missol-Kolka E et al (2004) Identification of novel Prominin-1/CD133 splice variants with alternative C-termini and their expression in epididymis and testis. J Cell Sci 117:4301–4311
9. Corbeil D, Röper K, Weigmann A, Huttner WB (1998) AC133 hematopoietic stem cell antigen: human homologue of mouse kidney prominin or distinct member of a novel protein family? Blood 91:2625–2626
10. Miraglia S, Godfrey W, Buck D (1998) A response to AC133 hematopoietic stem cell antigen: human homologue of mouse kidney prominin or distinct member of a novel protein family? Blood 91:4390–4391
11. Corbeil D, Röper K, Hellwig A et al (2000) The human AC133 hematopoietic stem cell antigen is also expressed in epithelial cells and targeted to plasma membrane protrusions. J Biol Chem 275:5512–5520
12. Florek M, Haase M, Marzesco AM et al (2005) Prominin-1/CD133, a neural and hematopoietic stem cell marker, is expressed in adult human differentiated cells and certain types of kidney cancer. Cell Tissue Res 319:15–26
13. Angelotti ML, Lazzeri E, Lasagni L, Romagnani P (2010) Only anti-CD133 antibodies recognizing the CD133/1 or the CD133/2 epitopes can identify human renal progenitors. Kidney Int 78:620–621

14. Lazzeri E, Crescioli C, Ronconi E et al (2007) Regenerative potential of embryonic renal multipotent progenitors in acute renal failure. J Am Soc Nephrol 18:3128–3138
15. Bussolati B, Bruno S, Grange C et al (2005) Isolation of renal progenitor cells from adult human kidney. Am J Pathol 166:545–555
16. Bussolati B, Moggio A, Collino F et al (2012) Hypoxia modulates the undifferentiated phenotype of human renal inner medullary CD133+ progenitors through Oct4/miR-145 balance. Am J Physiol Renal Physiol 302:F116–F128
17. Sagrinati C, Netti GS, Mazzinghi B et al (2006) Isolation and characterization of multipotent progenitor cells from the Bowman's capsule of adult human kidneys. J Am Soc Nephrol 17:2443–2456
18. Sallustio F, De Benedictis L, Castellano G et al (2010) TLR2 plays a role in the activation of human resident renal stem/progenitor cells. FASEB J 24:514–525
19. Ward HH, Romero E, Welford A et al (2011) Adult human CD133/1(+) kidney cells isolated from papilla integrate into developing kidney tubules. Biochim Biophys Acta 1812:1344–1357
20. Ronconi E, Sagrinati C, Angelotti ML et al (2009) Regeneration of glomerular podocytes by human renal progenitors. J Am Soc Nephrol 20:322–332
21. Cui S, Verroust PJ, Moestrup SK, Christensen EI (1996) Megalin/gp330 mediates uptake of albumin in renal proximal tubule. Am J Physiol Renal Physiol 271:F900–F907
22. Biner HL, Arpin-Bott MP, Loffing J et al (2002) Human cortical distal nephron: distribution of electrolyte and water transport pathways. J Am Soc Nephrol 13:836–847
23. Nielsen S, Frøkiaer J, Marples D et al (2002) Aquaporins in the kidney: from molecules to medicine. Physiol Rev 82:205–244
24. Uchida S (2000) In vivo role of CLC chloride channels in the kidney. Am J Physiol Renal Physiol 279:F802–F808
25. Ronco P, Brunisholz M, Geniteau-Legendre M et al (1987) Physiopathologic aspects of Tamm-Horsfall protein: a phylogenetically conserved marker of the thick ascending limb of Henle's loop. Adv Nephrol Necker Hosp 16:231–249
26. Loverre A, Capobianco C, Ditonno P et al (2008) Increase of proliferating renal progenitor cells in acute tubular necrosis underlying delayed graft function. Transplantation 85:1112–1119
27. Carvalhosa R, Deambrosis I, Carrera P et al (2011) Cystogenic potential of CD133+ progenitor cells of human polycystic kidneys. J Pathol 225:129–141
28. Calvet JP (1993) Polycystic kidney disease: primary extracellular matrix abnormality or defective cellular differentiation? Kidney Int 43:101–108
29. Harris PC (2010) What is the role of somatic mutation in autosomal dominant polycystic kidney disease? J Am Soc Nephrol 21:1073–1076
30. Smeets B, Angelotti ML, Rizzo P et al (2009) Renal progenitor cells contribute to hyperplastic lesions of podocytopathies and crescentic glomerulonephritis. J Am Soc Nephrol 20:2593–2603
31. Ferrandina G, Petrillo M, Bonanno G, Scambia G (2009) Targeting CD133 antigen in cancer. Expert Opin Ther Targets 13:823–837
32. Bruno S, Bussolati B, Grange C et al (2006) CD133+ renal progenitor cells contribute to tumor angiogenesis. Am J Pathol 169:2223–2235
33. Addla S, Brown M, Hart C et al (2008) Characterization of the Hoechst 33342 side population from normal and malignant human renal epithelial cells. Am J Physiol Renal Physiol 295:F680–F687
34. D'Alterio C, Cindolo L, Polimeno M et al (2010) Differential role of CD133 and CXCR4 in renal cell carcinoma. Cell Cycle 9:4492–4500
35. da Costa WH, Rocha RM, da Cunha IW et al (2012) CD133 immunohistochemical expression predicts progression and cancer-related death in renal cell carcinoma. World J Urol 30:553–558

36. Pistollato F, Chen HL, Schwartz PH et al (2007) Oxygen tension controls the expansion of human CNS precursors and the generation of astrocytes and oligodendrocytes. Mol Cell Neurosci 35:424–435
37. Soeda A, Park M, Lee D et al (2009) Hypoxia promotes expansion of the CD133-positive glioma stem cells through activation of HIF-1alpha. Oncogene 28:3949–3959
38. Mak AB, Blakely KM, Williams RA et al (2011) CD133 protein N-glycosylation processing contributes to cell surface recognition of the primitive cell marker AC133 epitope. J Biol Chem 286:41046–41056
39. Gunaratnam L, Bonventre JV (2009) HIF in kidney disease and development. J Am Soc Nephrol 20:1877–1887
40. Xu N, Papagiannakopoulos T, Pan G et al (2009) MicroRNA-145 regulates OCT4, SOX2, and KLF4 and represses pluripotency in human embryonic stem cells. Cell 137:647–658
41. Simone S, Cosola C, Loverre A et al (2012) BMP-2 induces a profibrotic phenotype in adult renal progenitor cells through Nox4 activation. Am J Physiol Renal Physiol 303:F23–F34
42. Bourseau-Guilmain E, Griveau A, Benoit JP, Garcion E (2011) The importance of the stem cell marker prominin-1/CD133 in the uptake of transferrin and in iron metabolism in human colon cancer Caco-2 cells. PLoS One 6:e25515
43. Yang C, Yang Y, Gupta N et al (2007) Pentaspan membrane glycoprotein, prominin-1, is involved in glucose metabolism and cytoskeleton alteration. Biochemistry 72:854–862
44. Griguer CE, Oliva CR, Gobin E et al (2008) CD133 is a marker of bioenergetic stress in human glioma. PLoS One 3:e3655

Chapter 9
CD133 Expression Strongly Correlates with the Phenotype of Very Small Embryonic-/Epiblast-Like Stem Cells

Mariusz Z. Ratajczak, Kasia Mierzejewska, Janina Ratajczak, and Magda Kucia

Abstract CD133 antigen (prominin-1) is a useful cell surface marker of very small embryonic-like stem cells (VSELs). Antibodies against it, conjugated to paramagnetic beads or fluorochromes, are thus powerful biological tools for their isolation from human umbilical cord blood, mobilized peripheral blood, and bone marrow. VSELs are described with the following characteristics: (1) are slightly smaller than red blood cells; (2) display a distinct morphology, typified by a high nuclear/cytoplasmic ratio and an unorganized euchromatin; (3) become mobilized during stress situations into peripheral blood; (4) are enriched in the $CD133^{+}Lin^{-}CD45^{-}$ cell fraction in humans; and (5) express markers of pluripotent stem cells (e.g., Oct-4, Nanog, and stage-specific embryonic antigen-4). The most recent in vivo data from our and other laboratories demonstrated that human VSELs exhibit some characteristics of long-term repopulating hematopoietic stem cells and are at the top of the hierarchy in the mesenchymal lineage. However, still more labor is needed to characterize better at a molecular level these rare cells.

Keywords CD133 • Cell isolation • Stem cells • VSELs

M.Z. Ratajczak, M.D., Ph.D. (✉) • J. Ratajczak • M. Kucia
Stem Cell Institute at James Graham Brown Cancer Center, University of Louisville, 500 S. Floyd Street, 40202 Rm. 107 Louisville, KY, USA

Department of Physiology Pomeranian Medical University, Szczecin, Poland, e-mail: mzrata01@louisville.edu

K. Mierzejewska
Department of Physiology Pomeranian Medical University, Szczecin, Poland

D. Corbeil (ed.), *Prominin-1 (CD133): New Insights on Stem & Cancer Stem Cell Biology*, Advances in Experimental Medicine and Biology 777, DOI 10.1007/978-1-4614-5894-4_9,

9.1 Introduction

One of the most challenging questions in stem cell biology is whether stem cells, which are able to differentiate into cells from multiple germ layers, can be isolated from adult tissues. From a developmental point of view, stem cells that are able to differentiate into all three germ layers (meso-, endo-, and ectoderm) are named pluripotent stem cells (PSCs), while stem cells that are already more restricted in differentiation and thus able to differentiate into cells from two or one germ layer are named multipotent stem cells. Several groups of investigators employing (1) various isolation protocols, (2) detection of surface markers, and (3) experimental in vitro and in vivo models have reported the presence of cells that possess a pluripotent or multipotent character in adult tissues, including adult bone marrow (BM) [1–9] and umbilical cord blood (UCB) [10–15].

For example, BM and UCB have each been described as a source of various stem cells that, in experimental animals, have been shown to contribute to skeletal muscle, liver, neural, and myocardial regeneration after delivery to injured tissues [9, 16–21]. Since hematopoietic stem cells (HSCs) are most abundant among the stem cells present in UCB, this unique capability of UCB-derived cells was initially (and wrongly) explained by the trans-dedifferentiation or plasticity of hematopoietic stem cells (HSCs) [22–24].

Instead, evidence has accumulated that BM and UCB contain a heterogeneous population of stem cells. For example, BM has been reported to contain multipotent adult progenitor cells (MAPCs) [25], marrow-isolated adult lineage inducible (MIAMI) cells [1], and multipotent adult stem cells (MASCs) [3], while UCB was reported to be a source of unrestricted somatic stem cells (USSCs) [15], multipotent mesenchymal cells (MSCs) [22], and omnicytes [14, 26]. Moreover, we have purified a rare population of small cells from both BM [8] and UCB [10] that exhibit embryonic stem cell characteristics, and these cells were named very small embryonic-like stem cells (VSELs). The presence of similar cells, which display PSC markers characteristic of UCB-VSELs, was recently confirmed independently by other research groups in human UCB, BM, and mobilized peripheral blood (mPB) [11–13].

VSELs isolated from UCB are phenotypically similar to VSELs initially described in adult murine BM [8]. Morphologically, they are very small in size (smaller than erythrocytes) and possess large nuclei containing an unorganized euchromatin. They express the nuclear embryonic transcription factors Oct-4 and Nanog, and the stage-specific embryonic antigen-4 (SSEA-4) is present at their surfaces. They are enriched within the $CD133^{+}$, lineage-negative (Lin^{-}), and $CD45^{-}$ cell fractions [27, 28] and co-express CD34 and CXCR4 [10, 27].

Thus, evidence has accumulated that BM, mPB, and UCB could be a source of primitive VSELs, and in this chapter, we will focus on strategies to purify such cells and their potential applications in the clinic. We envision that these cells are at the top of the hierarchy of stem cells and perhaps may even give rise to MAPCs, MIAMI cells, MASCs, MSCs, USSCs, and omnicytes. This, however, needs further experimental evidence.

9.2 Why Do PSCs Remain in Adult Tissues and Organs After Completion of Embryogenesis?

For many years, it has been accepted that adult tissues contain only tissue-committed stem cells (TCSCs), such as epidermal stem cells, HSCs, or skeletal muscle stem cells, which have a limited potential for differentiation, developing into only one type of differentiated cell [7, 20, 29, 30]. Nevertheless, we have to consider two scenarios that could occur during early embryogenesis and the development of lineage-restricted TCSCs [29–32]. In the first scenario, PSCs present in the inner cell mass of the blastocyst (and at later developmental stages in epiblasts), after giving rise to more differentiated lineage-restricted TCSCs, gradually disappear from the growing embryo and are not present in adult tissues. In the second scenario, which we believe actually takes place, some PSCs give rise to TCSCs, but some survive in adult tissues as a backup population of PSCs that renews the pool of TCSCs over time. In this second scenario, PSCs such as VSELs are precursors of TCSCs during organ and tissue rejuvenation and a source of these cells in emergency situations when organs are damaged (e.g., by heart infarct or stroke). Molecular analysis of murine BM-derived VSELs has revealed that these cells in fact express several genes that are characteristic of PSCs from the inner cell mass of the blastocyst (*Rex1*, also known as *Zfp42*) as well as from the epiblast (*Gbx2, Fgf5,* and *Nodal*) [33–35]. These cells, however, have to be protected from uncontrolled proliferation that could lead to formation of teratomas. In fact, due to epigenetic changes in genes that regulate insulin/insulin-like growth factor signaling (IIS), murine VSELs deposited in adult tissues are kept under control and in a quiescent state, which is essential for preventing their uncontrolled proliferation [33–35]. This important control mechanism affects genes that show paternal imprinting (e.g., *Igf2-H19*, *Igf2R*, and *RasGRF1*) and has been reviewed elsewhere [30]. In brief, by modifying the expression of paternally imprinted genes, VSELs downregulate pathways that are crucial in IIS and thus remain in a quiescent state [30, 33–35].

9.3 Data Supporting the Pluripotency of Murine VSELs

There are stringent criteria in vitro and in vivo for classifying a stem cell as a PSC. These criteria were established by embryologists working with embryonic stem cells (ESCs) isolated from embryos, established in vitro-immortalized ESC lines, or induced pluripotent stem cells (iPSCs). The most important in vitro criteria required to call a stem cell pluripotent include (1) undifferentiated morphology, euchromatin, and a high nuclear/cytoplasm ratio; (2) expression of PSC markers (e.g., Oct-4, Nanog, and SSEA); (3) open chromatin at the Oct-4 promoter; (4) the presence of bivalent domains; (5) female PSC reactivation of the X chromosome; and finally (6) the ability for multilineage differentiation into cells from all three germ layers. For in vivo criteria, PSCs should (1) complement blastocyst development and (2) form teratomas after inoculation in immunodeficient mice. However, it is worth mentioning that these in vivo criteria of pluripotency are not always applicable to pluripotent epiblast-derived stem cells [36, 37].

9.3.1 In Vitro Criteria That Support VSEL Pluripotency

Our morphological and molecular analysis of murine VSELs supported their pluripotent character [34, 38]. At the molecular level, these cells not only truly express acknowledged markers of pluripotency (Oct-4 and Nanog) but also possess bivalent domains, while in addition, female mouse-derived VSELs reactivate the X chromosome. Moreover, the *Oct-4* and *Nanog* promoters in VSELs, like cells isolated from ESC-derived embryoid bodies, are hypomethylated, and the chromatin immunoprecipitation (ChIP) assay revealed their association with histone H3Ac, which is a marker for open-type chromatin [33, 35, 38]. Like undifferentiated ESCs, VSELs exhibit bivalent domains in which transcriptionally active histone codes such as H3K4me3 physically coexist with repressive histone codes such as H3K27me3 at promoters for homeodomain-containing transcription factors (*Sox21, Nkx2.2, Dlx1, Lbx1h, Hlxb9, Pax5,* and *HoxA3*). Finally, we also observed that VSELs derived from female animals, like female ESCs isolated from the blastocyst ICM, reactivate one of the X chromosomes that was inactivated after fertilization, and, as a result, female PSCs display two equivalently activated X chromosomes.

9.3.2 In Vitro Data Supporting the Pluripotency of VSELs

Several coculture conditions have been successfully developed in which murine VSELs, when plated over supportive cell lines, expand/differentiate into somatic cell types [8, 39, 40]. One of these conditions is coculture over the myoblastic cell line C2C12, which allows some of the murine VSELs to differentiate and form spherical structures [41]. The cells in these spheres are immature, with large nuclei containing euchromatin and, when plated into cultures promoting tissue differentiation, expand into cells from all three germ-cell layers (e.g., neural cells, cardiomyocytes, and insulin-producing cells). Another coculture system that allows the hematopoietic specification of VSELs employs the OP9 stromal cell support [40]. Of note, this OP9-based coculture system, which we employed for VSELs, is widely employed by other investigators for hematopoietic specification of ESCs and induced pluripotent stem cells (iPSCs) [42].

9.3.3 In Vivo Multilineage Differentiation of VSELs

Our recent collaborative work indicates that VSELs may also be specified in vivo into mesenchymal stem cells (MSCs), cardiomyocytes, and HSCs [43–46].

For example, in the first study by Taichman and colleagues, VSELs isolated from (green fluorescent protein $(GFP)^+$ mice were implanted into severe combined immunodeficiency (SCID) mice, and 4 weeks later, the formation of bone-like tissues was observed [43]. Based upon the ability of uncultured VSELs to (1) differentiate in vivo into multiple mesenchymal lineages and (2) generate osseous tissues at low density, they proposed that this population of cells fulfills many of the required characteristics of MSCs. Furthermore, they have also recently obtained similar results with human VSELs (Dr. Taichman 2012).

In another collaborative study, BM-derived VSELs freshly isolated from GFP^+ transgenic mice were injected into the hearts of mice that had undergone ischemia/reperfusion injury [44]. After 35 days of follow-up, VSEL-treated mice exhibited improved global and regional left ventricular (LV) systolic function (determined by echocardiography) and attenuated myocyte hypertrophy in surviving tissue when compared with vehicle-treated controls (determined by both histology and echocardiography). Since GFP^+ VSELs were employed for transplantation, we could track the fate of injected cells in the myocardium. Although VSEL transplantation resulted in isolated new myocytes and capillaries in the infarct region, their numbers were too small to account for all of the observed benefits. Thus, it is likely that in this particular tissue injury model, some paracrine effects by transplanted VSELs played an important additional role, and we are now analyzing the profile of growth factors and cytokines secreted by these cells. In parallel, we are also optimizing our ex vivo protocol to differentiate VSELs into cardiomyocytes [45, 46].

Finally, to directly test the possibility of hematopoietic differentiation of VSELs and their relationship with HSCs, we purified these cells from murine BM, primed/cocultured them over OP9 stromal cells, and subsequently tested their hematopoietic potential by in vitro and in vivo assays [39, 40]. In parallel, cells derived from VSELs were evaluated for expression of hematopoietic genes and surface markers. While we observed that freshly isolated VSELs do not exhibit in vitro and in vivo hematopoietic potential, they may, upon coculture over OP9 stromal cells, differentiate along the hematopoietic lineage [40]. "OP9-primed" VSEL-derived cells acquired expression of several hemato/lymphopoiesis-specific genes and markers, gave rise to hematopoietic colonies in vitro, and protected lethally irradiated mice in both primary and secondary transplant models upon transplantation [40].

Recently, we also demonstrated that human UCB-derived $CD133^+$ VSELs could also be specified into the hematopoietic lineage in cocultures with OP9 stromal cells. Importantly, hematopoietic cells derived from VSELs were also able to engraft in immunodeficient mice [39]. Based on these observations, we propose that VSELs are the most primitive murine BM-residing population of stem cells that have the potential to become specified into the hematopoietic lineage, and thus, they may share some of the characteristics of long-term repopulating HSCs (LT-HSCs). Moreover, they are common precursor cells for both the hematopoietic and mesenchymal lineages in BM.

9.4 The Presence of Primitive Stem Cells in UCB-Derived VSELs as Circulating Paramedics

Someone could ask: Why are VSELs present in UCB? To address this question, we have to remember that delivery is one of the greatest stresses experienced in life, accompanied by the release of several cytokines, chemokines, and bioactive lipids, while several types of stem cells, including HSCs, VSELs, EPCs, and MSCs, are mobilized into neonatal PB and thus enriched after delivery in harvested UCB. The mobilization of these cells probably plays an important role as an endogenous mechanism to protect and repair small-scale organ and tissue damage that occurs due to hypoxia and mechanical injury during delivery. Thus, we can envision that mobilization of stem cells during delivery is a first-line stem cell therapy, which all have experienced as a neonate. It is probably an old evolutionary mechanism in mammals to increase the mobile pool of stem cells in neonatal blood to perform this important task. While HSCs are involved in immune surveillance of tissues after birth, circulating EPCs, MSCs, and in particular pluripotent VSELs play a role in regeneration of neonatal tissues. Moreover, several types of stem cells, including VSELs, are mobilized from adult tissues and organs into PB in patients after heart infarct, stroke, and severe hypoxia, and we envision that in these cases, they play the role of "circulating paramedics" that contribute to healing and regeneration of damaged tissues [47–49].

Furthermore, given the small size of $CD133^+$ VSELs, their highly migratory properties, as well as their potential to differentiate into all three germ layers, an interesting area of investigation worth pursuing is the possible role of these cells in establishing so-called fetal-maternal microchimerism. Observed following pregnancy, this phenomenon is based on bidirectional transplacental trafficking of cells present in UCB and occurs routinely during the normal course of pregnancy. While poorly understood, it is proposed to involve migration of very primitive stem cells from the fetus to the mother and from mother to fetus [26, 50]. These cells can stay in a dormant state in host tissues for many years, and it is very likely that VSELs are involved in this phenomenon. This possibility, however, requires further study.

9.5 Reports by Other Groups on VSELs and Closely Related Cells Present in Adult Tissues and Neonatal UCB

One of the most common features of VSELs is their very primitive morphology, high nuclear/cytoplasmic ratio, and relatively small size [8, 10, 27, 51]. Recently, there have been several published reports that support the existence of small, primitive VSELs or VSEL-like cells in adult tissues, and we will briefly discuss the most important of these [43, 52–57].

In one of the abovementioned reports, a population of murine VSELs was isolated from BM, and, in a set of elegant experiments, these cells were demonstrated to be able to give rise to mesenchymal and endothelial lineages [43]. The bone-forming activity of these cells, if embedded in gelatin sponges and implanted into living mice, exceeded the activity of other populations of BM-purified cells tested in the same assay. Based on this finding, it has been proposed that Sca-1$^+$Lin$^-$CD45$^-$ VSELs are at the top of the hierarchy for the mesenchymal and endothelial lineages in BM [43].

In another report, VSELs were purified from rat BM and successfully employed in regenerating damaged myocardium in an experimental rodent model of acute myocardial infarction [58]. These cells expressed SSEA-1 antigen on their surface and Oct-4 in their nuclei. Similarly, small non-hematopoietic murine BM-derived cells that correspond in size to VSELs were also shown to give rise to type 2 pneumocytes, which produce lung surfactant protein after transplantation into surfactant-deficient mice [53]. In order to determine whether surfactant-producing epithelial cells were derived from the non-hematopoietic or the hematopoietic fraction of BM cells used to treat surfactant-deficient mice, the authors employed a lineage-tracing approach in which Cre-recombinase-specific, vav-promoter-activated GFP from a ROSA-GFP construct was used as a reporter transgene [53]. Similarly, small cells with VSEL markers were also identified in murine neonatal retina [59], while in another report, it was demonstrated that small non-hematopoietic lineage-negative (lin$^-$) cells isolated from adult BM by elutriation (Fraction 25) were involved in retinal regeneration following the induction of anterior ischemic optic neuropathy and optic nerve crush in a rodent model [60]. A similar population of non-hematopoietic CD45$^-$ small stem cells harvested from BM via elutriation has recently been shown to give rise to functional insulin-producing cells in vivo in chemically induced diabetic mice [52]. Finally, several features of VSELs are displayed in so-called multilineage-differentiating, stress-enduring (Muse) cells recently isolated from murine and human BM populations [55, 61]. Of note, it has been postulated that these cells play a major role as cell populations that preferentially give rise to iPSCs when BM-derived stromal cells are induced to pluripotency by genetic manipulation [55, 61].

The presence of VSELs in human tissues has also been supported by several recent publications. For example, an interesting connection has been made between VSELs and stem cells in female [54, 56] and male gonads [57]. Cells similar to UCB-VSELs were reported by two independent groups to reside in an ovarian epithelial layer in postmenopausal ovaries [54, 56]. Moreover, it has been proposed that the ovary harbors two distinct populations of stem cells: VSELS that are pluripotent and quiescent with nuclear expression of Oct-4 and VSELs that give rise to slightly larger "progenitor" ovarian germ stem cells expressing Oct-4 in the cytoplasm [56].

Finally, VSELs or cells closely related to them have been successfully purified from human mPB [11] and UCB by other groups [12, 13, 62, 63]. A corresponding population of primitive stem cells that resembles VSELs and was named "omnicytes" was also described as circulating in UCB and is probably capable of migrating into the maternal circulatory system [14, 26].

Thus, mounting evidence supports the conclusion that small cells described as VSELs or VSEL-like cells that express developmentally primitive markers reside in adult murine and human tissues. However, to confirm that these populations of small, primitive stem cells overlap with VSELs requires further direct comparison. Most importantly, VSELs are currently purified in several laboratories worldwide, and the coming years will bring more information about their biology and in vitro and in vivo differentiation potential.

9.6 Isolation of VSELs from UCB

As mentioned before, human VSELs express CD133 as a specific cell surface antigen, and those lacking it should not consequently be referred to as VSELs. Thus, we are employing anti-CD133 monoclonal antibodies (mAbs) conjugated with fluorochromes or attached to paramagnetic immunobeads in our technical setting for the isolation of human VSELs.

9.6.1 Multiparameter Fluorescence-Activated Cell Sorting

Historically, the human UCB-derived VSELs were isolated first by our team as a minute population of Lin^-CD45^- cells that express CD133 as well as CD34 and CXCR4 [10]. Further analysis has revealed that the most primitive VSELs, which strongly express Oct-4 and SSEA-4, are enriched within the sorted $CD133^+Lin^-CD45^-$ cell fraction, and based on these observations, we concluded that CD133 is the most reliable cell surface marker for these cells (Fig. 9.1). Most recently, we identified an anti-SSEA-4 antibody that is suitable for fluorescence-activated cell sorting (FACS) and isolation of UCB-VSELs as $SSEA\text{-}4^+Lin^-CD45^-$ cells (unpublished data).

9.6.2 Enrichment for VSELs by CD133 Paramagnetic Immunobead Selection Followed by Multiparameter Sorting

As mentioned above, VSELs from human UCB were initially purified from an erythrocyte-depleted population of nucleated cells by multiparameter sorting as a population of small $CD133^+CD45^-Lin^-$ cells [10]. However, this procedure is time-consuming, and the sorting time required to process an entire UCB unit (~50 to100 ml) to isolate rare VSELs is up to 3–4 days [10]. Therefore, it has been clear from the beginning that a more rapid and less-expensive method for cell isolation has to

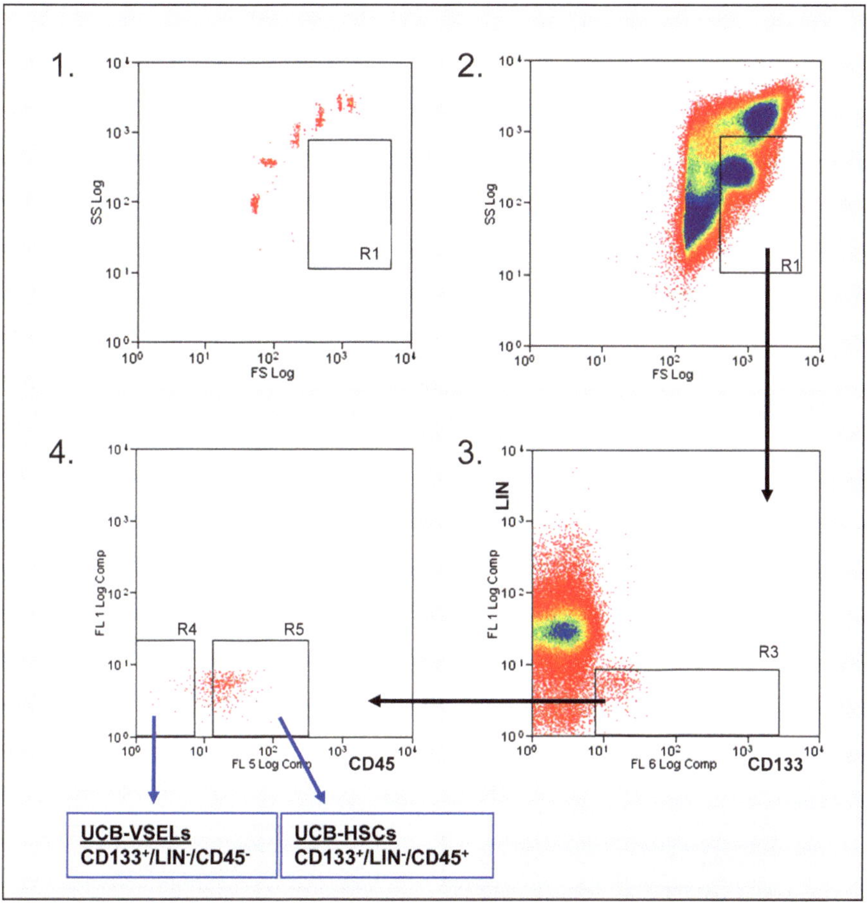

Fig. 9.1 Gating strategy for multiparameter sorting of human UCB-VSELs and UCB-HSCs by using CD133 antibody. UCB-derived cells were isolated by FACS by employing the following gating criteria. Panel **1**: All events ≥2 μm are included in gate R1 after comparison with bead particles at standard diameters of 1, 2, 4, 6, 10, or 15 μm. Panel **2**: UCB-derived total nucleated cells are visualized on a dot plot based on forward scatter (FS; x-axis) vs side scatter (SS; y-axis) signals. Panel **3**: Cells from region R1 (panel **2**) are further analyzed for Lin and CD133. Events for cells expressing Lin^- and $CD133^+$ are included in region R3. Panel **4**: The $Lin^-/CD133^+$ population from region R3 is subsequently analyzed based on CD45 antigen expression, and the $CD45^-$ and $CD45^+$ subpopulations were visualized with a dot plot showing $CD133^+Lin^-CD45^-$ (VSELs, region R4) and $CD133^+Lin^-CD45^+$ (HSCs, region R5) populations

be found. In order to establish a more efficient method for VSEL purification from UCB, we employed a three-step isolation strategy based on the removal of erythrocytes by hypotonic lysis (1st step), immunomagnetic separation of $CD133^+$ cells (2nd step), followed by FACS-based isolation of $CD133^+Lin^-CD45^-$ cells (3rd step). The entire isolation procedure takes only 2–4 h per 100 ml of UCB [27, 28].

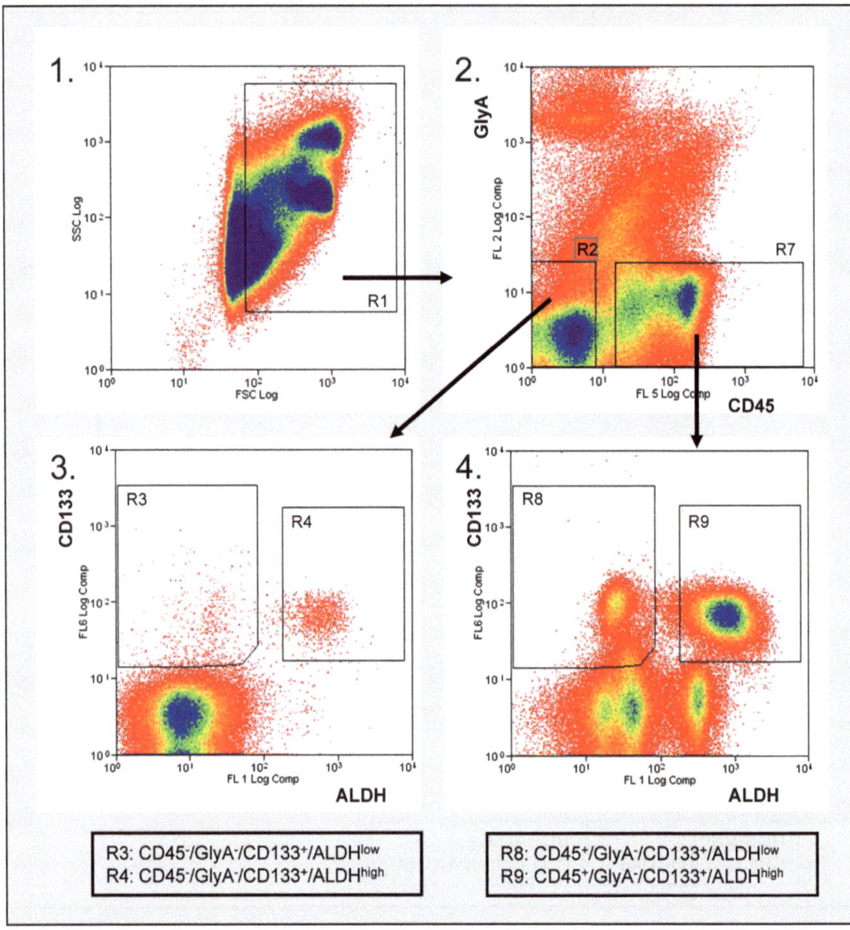

Fig. 9.2 Gating strategy for FACS analysis of UCB-VSELs and UCB-HSCs based on CD133 and CD45 expression and ALDH activity. UCB-derived $CD133^+$ VSELs were isolated by FACS by employing the following gating criteria. Panel **1**: UCB nucleated cell populations are stained using mAbs against human CD235a (GlyA), CD45, and CD133 and exposed to Aldefluor. UCB-derived total nucleated cells are visualized on a dot plot based on forward scatter (FS; x-axis) vs side scatter (SS; y-axis) signals. Panel **2**: Cells from region R1 (panel **1**) are further analyzed for $CD45^-$/$GlyA^-$ (region R2) and $CD45^+$/Gly^- (region R7) populations. Panel **3**: Cells from region R2 (panel **2**) are sorted as $CD45^-/GlyA^-/CD133^+/ALDH^{low}$ (region R3) and $CD45^-/GlyA^-/CD133^+/ALDH^{high}$ (region R4) subfractions of VSELs. Panel **4**: Cells from region R7 (panel **2**) are sorted as $CD45^+$/$GlyA^-/CD133^+/ALDH^{low}$ (region R8) and $CD45^-/GlyA^-/CD133^+/ALDH^{high}$ (region R9) HSC subpopulations

9.6.3 CD133 Cell Selection Combined with Aldefluor Staining

As an alternative strategy (Fig. 9.2), we recently proposed reducing the number of lineage factors required for purification by exposure of erythrocyte-depleted, immunomagnetic bead-selected $CD133^+$ cells to Aldefluor, followed by staining with anti-$CD133^+$ antibodies and only two lineage-specific antibodies (anti-CD45 and anti-GlyA) for the third step [39]. Aldefluor is a substrate for aldehyde dehydrogenase (ALDH), a cytosolic enzyme highly expressed in less-differentiated hematopoietic cells, and is implicated in resistance to some alkylating agents [64]. In the presence of ALDH, Aldefluor becomes modified to a fluorescent molecule that may then be used to mark ALDH-expressing cells. Inclusion of an anti-GlyA antibody is based on the well-known fact that small erythroblast $GlyA^+$ cells present in UCB do not express the CD45 antigen. Thus, selection for $CD45^-$ cells enriches for these unwanted cells. By employing this strategy, we are able to subsequently sort $CD133^+$ cells enriched for VSELs and are able to obtain ~10^3 $CD133^+/CD45^-/GlyA^-/ALDH^{low}$ and 4 × 10^3 $CD133^+/CD45^-/GlyA^-/ALDH^{high}$ UCB-VSELs from 100 ml of UCB [39]. These numbers illustrate how rare these cells are in UCB. When we compared both fractions of VSELs by real-time polymerase chain reaction (RT-PCR) gene expression analysis, we found that $CD133^+/CD45^-/GlyA^-/ALDH^{low}$ VSELs have a higher expression of *Oct-4* than the $CD133^+/CD45^-/GlyA^-/ALDH^{high}$ fraction [39].

However, we are aware that there is yet room to improve sorting and to develop alternative strategies for purifying UCB-VSELs. To accomplish this goal, one should consider the use of metabolic fluorochromes to see whether VSELs are enriched among $Rh123^{dull}$, Pyronin Y^{low}, and Hoechst 33342^{low} cells. Moreover, we expect that our current proteomics data analysis with UCB-VSELs will reveal the presence of novel positive markers that could be employed for the purification of these rare cells.

9.7 Novel Characteristics of Human VSELs

In the past two years, we investigated several experimental approaches to better characterize murine and human UCB-derived VSELs, both at the molecular and morphological levels. Our molecular gene array analysis of limited numbers of highly purified VSELs revealed that murine VSELs are somewhat heterogeneous [33–35]. In particular, while all of them express the pluripotency marker Oct-4, some of them express genes that are more closely related to genes expressed by epiblast-derived stem cells and others to those of migrating primordial germ cells. This suggests that VSELs resemble to stem cells that are expressed early during the development. Those isolated from the adult BM may differ to some extent in their differentiation potential [33–35].

Based on well-established FACS identification protocols, we also found that human $CD133^+Lin^-CD45^-$ VSELs identified in UCB, like their murine counterparts, (1) highly express telomerase (Ratajczak MZ, 2009), (2) are diploid, and (3) are viable, as shown by their ability to exclude the viability dye (7-aminoactinomycin D, 7-AAD) [30]. Importantly, we observed that the highest number of pluripotent $Oct\text{-}4^+$ VSELs reside in the population of $CD133^+Lin^-CD45^-$ UCB-derived cells [10, 27]. Moreover, some of the $CD133^+Lin^-CD45^-$ VSELs, which represent only a very small subfraction among UCB Lin^-CD45^- non-hematopoietic cells, may co-express other stem cell markers, including CD34, CXCR4, and SSEA-4 [10, 27].

Thus, VSELs represent a very rare subpopulation of stem cells among the Lin^{-} CD45^{-} non-hematopoietic UCB cell population, which also contains other stem cell types, including endothelial progenitor cells (EPCs) and MSCs. VSELs may be identified by their very small size (FSC^{low}/SSC^{low}) and co-expression of CD34, CXCR4, and notably CD133 [10, 27].

9.8 Conclusion

New data from our group and others have provided additional evidence for the existence and biological role of primitive embryonic-like stem cells in murine adult tissues and their potential role in (1) tissue and organ rejuvenation, (2) longevity, and (3) regeneration and repair of damaged tissues. These cells are kept however in a quiescent state in adult tissues by epigenetic changes in expression of some paternally imprinted genes (e.g., *Igf2, RasGrf1, Igf2R*) (Fig. 9.3), which are involved in cell proliferation in response to insulin/insulin-like growth factor signaling (IIS) [30–35]. However, in the presence of chronic elevated levels of IGF1 and insulin in blood, VSELs deposited in adult tissues may proliferate in an uncontrolled manner and become depleted much faster over time. This may contribute to the accelerated aging observed in situations with high circulating IGF-1 and insulin levels (e.g., chronic high caloric uptake or obesity). At the same time, chronic exposure to IIS may also trigger uncontrolled activation of VSELs and their malignant transformation.

The identification of UCB-derived VSELs opens up new areas of investigation and new challenges. Our data indicate that these cells can be specified along the hematopoietic lineage and, in BM and UCB, are probably equivalent to a population of LT-HSCs. The fact that many of these cells could be lost in currently employed banking procedures requires the development of better VSEL-saving banking protocols. In fact, some of the commercial UCB banks have acknowledged this fact and now employ proper volume-depletion procedures. The possibility of differentiating UCB-derived VSELs into other type of cells (e.g., neurons, MSCs, or cardiomyocytes) is currently being investigated in our laboratories. However, our data already strongly indicate that VSELs could provide a therapeutic alternative to the controversial use of human ESCs and strategies based on therapeutic cloning. Hence, while the ethical debate on the application of ESCs in therapy continues, the potential of VSELs for tissue repair and regeneration is ripe for exploration, and the expression of CD133 combined with specific antibodies offers a great advance in their prospective isolation.

We are aware that more works are needed to better characterize at a molecular level the small UCB-isolated cells. If we can confirm that a similar mechanism operates for human UCB-derived and murine BM-derived VSELs, perhaps a controlled modulation of the somatic imprint state to produce proper de novo methylation of somatic imprinted genes on the maternal and paternal chromosomes could increase the regenerative power of these cells. The coming years will determine whether these exciting possibilities are realized.

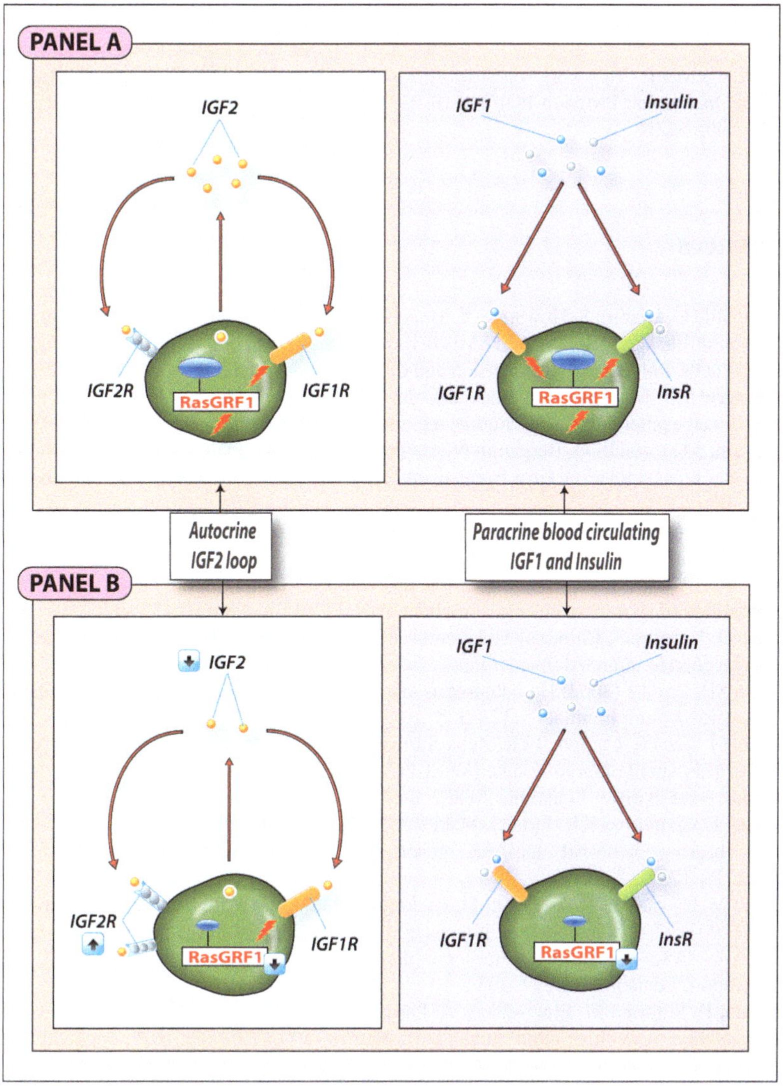

Fig. 9.3 Murine VSELs deposited in adult tissues are protected from excessive insulin/insulin-like growth factor signaling (IIS). (**a**) VSELs (*green*) are deposited in adult tissues as a backup population for tissue-committed stem cells and express non-signaling insulin-like growth factor 2 receptor (IGF2R) and two signaling receptors: insulin-like growth factor 1 receptor (IGF1R) and insulin receptor (InsR). Thus, they can be stimulated by an autocrine IGF2 loop (*left panel*) and by paracrine/endocrine IGF1 and insulin (*right panel*). (**b**) Due to changes in expression of paternally imprinted genes (e.g., *Igf2R, Igf2,* and *RasGrf1*), VSELs are protected from IIS, which would otherwise lead to their premature depletion from adult tissues as well as trigger uncontrolled proliferation and teratoma formation. *Left panel*: VSELs are protected from autocrine IIS by changes in expression of imprinted genes that are important in IIS. Downregulation of autocrine expression of IGF2, upregulation of IGF2R (a non-signaling receptor that binds IGF2 and prevents its binding to signaling IGF1R), and downregulation of RasGRF1 (which is involved in signal transduction from IGF2R) make VSELs less sensitive to autocrine/paracrine IGF2 signaling. *Right panel*: Downregulation of RasGRF1, which is involved in signaling from activated IGF1R and InsR in VSELs, also plays an important role in attenuation of IIS signaling by paracrine circulating levels of IGF1 and insulin. For the sake of simplicity, other imprinted genes (H19 and $p57^{Kip2}$) that are not involved in IIS and negatively affect VSEL proliferation are not depicted

Acknowledgements This work was supported by NIH Grant 2R01 DK074720, the EU Innovative Economy Operational Program POIG.01.01.02-00-109/09-01, and the Stella and Henry Hoenig Endowment to MZR.

References

1. D'Ippolito G, Diabira S, Howard GA, Menei P, Roos BA, Schiller PC (2004) Marrow-isolated adult multilineage inducible (MIAMI) cells, a unique population of postnatal young and old human cells with extensive expansion and differentiation potential. J Cell Sci 117:2971–2981
2. Pittenger MF, Mackay AM, Beck SC, Jaiswal RK, Douglas R, Mosca JD et al (1999) Multilineage potential of adult human mesenchymal stem cells. Science 284:143–147
3. Beltrami AP, Cesselli D, Bergamin N, Marcon P, Rigo S, Puppato E et al (2007) Multipotent cells can be generated in vitro from several adult human organs (heart, liver, and bone marrow). Blood 110:3438–3446
4. Krause DS (2008) Bone marrow-derived cells and stem cells in lung repair. Proc Am Thorac Soc 5:323–327
5. Vacanti MP, Roy A, Cortiella J, Bonassar L, Vacanti CA (2001) Identification and initial characterization of spore-like cells in adult mammals. J Cell Biochem 80:455–460
6. Orlic D, Kajstura J, Chimenti S, Jakoniuk I, Anderson SM, Li B et al (2001) Bone marrow cells regenerate infarcted myocardium. Nature 410:701–705
7. Orkin SH, Zon LI (2002) Hematopoiesis and stem cells: plasticity versus developmental heterogeneity. Nature Immunol 3:323–328
8. Kucia M, Reca R, Campbell FR, Zuba-Surma E, Majka M, Ratajczak J et al (2006) A population of very small embryonic-like (VSEL) CXCR4(+)SSEA-1(+)Oct-4+ stem cells identified in adult bone marrow. Leukemia 20:857–869
9. Mezey E, Chandross KJ, Harta G, Maki RA, McKercher SR (2000) Turning blood into brain: cells bearing neuronal antigens generated in vivo from bone marrow. Science 290:1779–1782
10. Kucia M, Halasa M, Wysoczynski M, Baskiewicz-Masiuk M, Moldenhawer S, Zuba-Surma E et al (2007) Morphological and molecular characterization of novel population of CXCR4+ SSEA-4+ Oct-4+ very small embryonic-like cells purified from human cord blood: preliminary report. Leukemia 21:297–303
11. Sovalat H, Scrofani M, Eidenschenk A, Pasquet S, Rimelen V, Henon P (2011) Identification and isolation from either adult human bone marrow or G-CSF-mobilized peripheral blood of CD34(+)/CD133(+)/CXCR4(+)/Lin(−)CD45(−) cells, featuring morphological, molecular, and phenotypic characteristics of very small embryonic-like (VSEL) stem cells. Exp Hematol 39:495–505
12. Bhartiya D, Shaikh A, Nagvenkar P, Kasiviswanathan S, Pethe P, Pawani H et al (2012) Very small embryonic-like stem cells with maximum regenerative potential get discarded during cord blood banking and bone marrow processing for autologous stem cell therapy. Stem Cells Dev 21:1–6
13. McGuckin C, Jurga M, Ali H, Strbad M, Forraz N (2008) Culture of embryonic-like stem cells from human umbilical cord blood and onward differentiation to neural cells in vitro. Nature Prot 3:1046–1055
14. Gordon MY, Levicar N, Pai M, Bachellier P, Dimarakis I, Al-Allaf F et al (2006) Characterization and clinical application of human CD34+ stem/progenitor cell populations mobilized into the blood by granulocyte colony-stimulating factor. Stem Cells 24:1822–1830
15. Kogler G, Sensken S, Airey JA, Trapp T, Muschen M, Feldhahn N et al (2004) A new human somatic stem cell from placental cord blood with intrinsic pluripotent differentiation potential. J Exp Med 200:123–135

16. Buzanska L, Machaj EK, Zablocka B, Pojda Z, Domanska-Janik K (2002) Human cord blood-derived cells attain neuronal and glial features in vitro. J Cell Sci 115:2131–2138
17. Pesce M, Orlandi A, Iachininoto MG, Straino S, Torella AR, Rizzuti V et al (2003) Myoendothelial differentiation of human umbilical cord blood-derived stem cells in ischemic limb tissues. Circ Res 93:e51–e62
18. Newsome PN, Johannessen I, Boyle S, Dalakas E, McAulay KA, Samuel K et al (2003) Human cord blood-derived cells can differentiate into hepatocytes in the mouse liver with no evidence of cellular fusion. Gastroenterology 124:1891–1900
19. Kim BO, Tian H, Prasongsukarn K, Wu J, Angoulvant D, Wnendt S et al (2005) Cell transplantation improves ventricular function after a myocardial infarction: a preclinical study of human unrestricted somatic stem cells in a porcine model. Circulation 112:I96–I104
20. Wagers AJ, Weissman IL (2004) Plasticity of adult stem cells. Cell 116:639–648
21. Di Campli C, Piscaglia AC, Pierelli L, Rutella S, Bonanno G, Alison MR et al (2004) A human umbilical cord stem cell rescue therapy in a murine model of toxic liver injury. Digestive Liver Dis 36:603–613
22. Castro RF, Jackson KA, Goodell MA, Robertson CS, Liu H, Shine HD (2002) Failure of bone marrow cells to transdifferentiate into neural cells in vivo. Science 297:1299
23. Murry CE, Soonpaa MH, Reinecke H, Nakajima H, Nakajima HO, Rubart M et al (2004) Haematopoietic stem cells do not transdifferentiate into cardiac myocytes in myocardial infarcts. Nature 428:664–668
24. Nygren JM, Jovinge S, Breitbach M, Sawen P, Roll W, Hescheler J et al (2004) Bone marrow-derived hematopoietic cells generate cardiomyocytes at a low frequency through cell fusion, but not transdifferentiation. Nature Med 10:494–501
25. Sohni A, Verfaillie CM (2011) Multipotent adult progenitor cells. Best Pract Res Clin Haematol 24:3–11
26. Mikhail MA, M'Hamdi H, Welsh J, Levicar N, Marley SB, Nicholls JP et al (2008) High frequency of fetal cells within a primitive stem cell population in maternal blood. Hum Reprod 23:928–933
27. Zuba-Surma EK, Klich I, Greco N, Laughlin MJ, Ratajczak J, Ratajczak MZ (2010) Optimization of isolation and further characterization of umbilical-cord-blood-derived very small embryonic/ epiblast-like stem cells (VSELs). Eur J Haematol 84:34–46
28. Zuba-Surma EK, Ratajczak MZ (2010) Overview of very small embryonic-like stem cells (VSELs) and methodology of their identification and isolation by flow cytometric methods. In: J Paul Robinson (managing editor) et al. (ed) Current protocols in cytometry, Chapter 9:Unit9 29
29. Ratajczak MZ, Zuba-Surma EK, Wysoczynski M, Wan W, Ratajczak J, Wojakowski W et al (2008) Hunt for pluripotent stem cell – regenerative medicine search for almighty cell. J Autoimmun 30:151–162
30. Ratajczak MZ, Shin DM, Liu R, Mierzejewska K, Ratajczak J, Kucia M et al (2012) Very small embryonic/epiblast-like stem cells (VSELs) and their potential role in aging and organ rejuvenation-an update and comparison to other primitive small stem cells isolated from adult tissues. Aging 4:235–246
31. Kucia M, Ratajczak J, Reca R, Janowska-Wieczorek A, Ratajczak MZ (2004) Tissue-specific muscle, neural and liver stem/progenitor cells reside in the bone marrow, respond to an SDF-1 gradient and are mobilized into peripheral blood during stress and tissue injury. Blood Cells Mol Dis 32:52–57
32. Ratajczak MZ, Kucia M, Reca R, Majka M, Janowska-Wieczorek A, Ratajczak J (2004) Stem cell plasticity revisited: CXCR4-positive cells expressing mRNA for early muscle, liver and neural cells 'hide out' in the bone marrow. Leukemia 18:29–40
33. Shin DM, Liu R, Klich I, Ratajczak J, Kucia M, Ratajczak MZ (2010) Molecular characterization of isolated from murine adult tissues very small embryonic/epiblast like stem cells (VSELs). Mol Cells 29:533–538
34. Shin DM, Liu R, Klich I, Wu W, Ratajczak J, Kucia M et al (2010) Molecular signature of adult bone marrow-purified very small embryonic-like stem cells supports their developmental epiblast/germ line origin. Leukemia 24:1450–1461

35. Shin DM, Liu R, Wu W, Waigel SJ, Zacharias W, Ratajczak MZ et al (2012) Global gene expression analysis of very small embryonic-like stem cells reveals that the Ezh2-dependent bivalent domain mechanism contributes to their pluripotent state. Stem Cells Dev 21:1639–1652
36. Muller FJ, Goldmann J, Loser P, Loring JF (2010) A call to standardize teratoma assays used to define human pluripotent cell lines. Cell Stem Cell 6:412–414
37. Smith KP, Luong MX, Stein GS (2009) Pluripotency: toward a gold standard for human ES and iPS cells. J Cell Physiol 220:21–29
38. Shin DM, Zuba-Surma EK, Wu W, Ratajczak J, Wysoczynski M, Ratajczak MZ et al (2009) Novel epigenetic mechanisms that control pluripotency and quiescence of adult bone marrow-derived Oct4(+) very small embryonic-like stem cells. Leukemia 23:2042–2051
39. Ratajczak J, Zuba-Surma E, Klich I, Liu R, Wysoczynski M, Greco N et al (2011) Hematopoietic differentiation of umbilical cord blood-derived very small embryonic/epiblast-like stem cells. Leukemia 25:1278–1285
40. Ratajczak J, Wysoczynski M, Zuba-Surma E, Wan W, Kucia M, Yoder MC et al (2011) Adult murine bone marrow-derived very small embryonic-like stem cells differentiate into the hematopoietic lineage after coculture over OP9 stromal cells. Exp Hematol 39:225–237
41. Kucia M, Wysoczynski M, Ratajczak J, Ratajczak MZ (2008) Identification of very small embryonic like (VSEL) stem cells in bone marrow. Cell Tissue Res 331:125–134
42. Lohle M, Hermann A, Glass H, Kempe A, Schwarz SC, Kim JB et al (2012) Differentiation efficiency of induced pluripotent stem cells depends on the number of reprogramming factors. Stem Cells 30:570–579
43. Taichman RS, Wang Z, Shiozawa Y, Jung Y, Song J, Balduino A et al (2010) Prospective identification and skeletal localization of cells capable of multilineage differentiation in vivo. Stem Cells Dev 19:1557–1570
44. Dawn B, Tiwari S, Kucia MJ, Zuba-Surma EK, Guo Y, Sanganalmath SK et al (2008) Transplantation of bone marrow-derived very small embryonic-like stem cells attenuates left ventricular dysfunction and remodeling after myocardial infarction. Stem Cells 26:1646–1655
45. Zuba-Surma EK, Guo Y, Taher H, Sanganalmath SK, Hunt G, Vincent RJ et al (2011) Transplantation of expanded bone marrow-derived very small embryonic-like stem cells (VSEL-SCs) improves left ventricular function and remodelling after myocardial infarction. J Cell Mol Med 15:1319–1328
46. Wojakowski W, Tendera M, Kucia M, Zuba-Surma E, Milewski K, Wallace-Bradley D et al (2010) Cardiomyocyte differentiation of bone marrow-derived Oct-4+CXCR4+SSEA-1+ very small embryonic-like stem cells. Int J Oncol 37:237–247
47. Wojakowski W, Tendera M, Kucia M, Zuba-Surma E, Paczkowska E, Ciosek J et al (2009) Mobilization of bone marrow-derived Oct-4+ SSEA-4+ very small embryonic-like stem cells in patients with acute myocardial infarction. J Am Coll Cardiol 53:1–9
48. Paczkowska E, Kucia M, Koziarska D, Halasa M, Safranow K, Masiuk M et al (2009) Clinical evidence that very small embryonic-like stem cells are mobilized into peripheral blood in patients after stroke. Stroke 40:1237–1244
49. Kucia M, Zhang YP, Reca R, Wysoczynski M, Machalinski B, Majka M et al (2006) Cells enriched in markers of neural tissue-committed stem cells reside in the bone marrow and are mobilized into the peripheral blood following stroke. Leukemia 20:18–28
50. Gammill HS, Nelson JL (2010) Naturally acquired microchimerism. Int J Dev Biol 54:531–543
51. Zuba-Surma EK, Kucia M, Abdel-Latif A, Dawn B, Hall B, Singh R et al (2008) Morphological characterization of very small embryonic-like stem cells (VSELs) by ImageStream system analysis. J Cell Mol Med 12:292–303
52. Iskovich S, Goldenberg-Cohen N, Stein J, Yaniv I, Fabian I, Askenasy N (2012) Elutriated stem cells derived from the adult bone marrow differentiate into insulin-producing cells in vivo and reverse chemical diabetes. Stem Cells Dev 21:86–96

53. Kassmer SH, Bruscia EM, Zhang PX, Krause DS (2012) Nonhematopoietic cells are the primary source of bone marrow-derived lung epithelial cells. Stem Cells 30:491–499
54. Virant-Klun I, Skutella T (2010) Stem cells in aged mammalian ovaries. Aging 2:3–6
55. Kuroda Y, Kitada M, Wakao S, Nishikawa K, Tanimura Y, Makinoshima H et al (2010) Unique multipotent cells in adult human mesenchymal cell populations. Proc Natl Acad Sci USA 107:8639–8643
56. Parte S, Bhartiya D, Telang J, Daithankar V, Salvi V, Zaveri K et al (2011) Detection, characterization, and spontaneous differentiation in vitro of very small embryonic-like putative stem cells in adult mammalian ovary. Stem Cells Dev 20:1451–1464
57. Bhartiya D, Kasiviswanathan S, Unni SK, Pethe P, Dhabalia JV, Patwardhan S et al (2010) Newer insights into premeiotic development of germ cells in adult human testis using Oct-4 as a stem cell marker. J Histochem Cytochem 58:1093–1106
58. Wu JH, Wang HJ, Tan YZ, Li ZH (2012) Characterization of rat very small embryonic-like stem cells and cardiac repair after cell transplantation for myocardial infarction. Stem Cells Dev 21:1367–1379
59. Liu Y, Gao L, Zuba-Surma EK, Peng X, Kucia M, Ratajczak MZ et al (2009) Identification of small Sca-1(+), Lin(−), CD45(−) multipotential cells in the neonatal murine retina. Exp Hematol 37:1096–1107
60. Goldenberg-Cohen N, Avraham-Lubin BC, Sadikov T, Goldstein RS, Askenasy N (2012) Primitive stem cells derived from bone marrow express glial and neuronal markers and support revascularization in injured retina exposed to ischemic and mechanical damage. Stem Cells Dev 21:1488–1500
61. Wakao S, Kitada M, Kuroda Y, Shigemoto T, Matsuse D, Akashi H et al (2011) Multilineage-differentiating stress-enduring (Muse) cells are a primary source of induced pluripotent stem cells in human fibroblasts. Proc Natl Acad Sci USA 108:9875–9880
62. McGuckin CP, Forraz N, Baradez MO, Navran S, Zhao J, Urban R et al (2005) Production of stem cells with embryonic characteristics from human umbilical cord blood. Cell Prolif 38:245–255
63. Seo KW, Lee SR, Bhandari DR, Roh KH, Park SB, So AY et al (2009) OCT4A contributes to the stemness and multi-potency of human umbilical cord blood-derived multipotent stem cells (hUCB-MSCs). Biochem Biophy Res Commun 384:120–125
64. Hess DA, Wirthlin L, Craft TP, Herrbrich PE, Hohm SA, Lahey R et al (2006) Selection based on CD133 and high aldehyde dehydrogenase activity isolates long-term reconstituting human hematopoietic stem cells. Blood 107:2162–2169

Part III
Prominin-1/CD133 and the Cancer Stem Cell Hypothesis

Chapter 10
New Insights into the CD133 (Prominin-1) Expression in Mouse and Human Colon Cancer Cells

Alessandro Sgambato, Maddalena Corbi, Maria Svelto, Emanuele Caredda, and Achille Cittadini

Abstract Following its discovery as a cancer stem cell marker, CD133 has been widely studied for its role in colorectal tumorigenesis. Indeed, colon cancer remains one of the major causes of cancer-related disease and death worldwide, and there is a strong need for an improvement of current diagnostic, prognostic, and therapeutic strategies. Thus, efforts have been devoted to try to understand whether CD133 might play a role in human colorectal tumorigenesis and might contribute to a better management of colon cancer patients. This chapter reviews the current knowledge on CD133 expression in normal and cancer colon tissues, both in humans and mice, discussing apparently conflicting data reported in the two species. Moreover, a great attention is devoted to the available information regarding the functional role of CD133 in colon cancer cells. Finally, the proposed clinical applications of CD133, as a prognostic and/or predictive marker as well as a target for novel antineoplastic strategies in colorectal cancer, are discussed. Overall, the available data support a potential important role of CD133 as cancer stem cell marker in colon cancer cells and warrant future studies to verify its potential use in the routine clinical management of colon cancer patients.

Keywords Cancer stem cell • CD133 • Colon cancer • Mouse model • Tumorigenesis

---

A. Sgambato, M.D., Ph.D. (✉) • M. Corbi • M. Svelto • E. Caredda • A. Cittadini
Istituto di Patologia Generale, Università Cattolica del Sacro Cuore,
Largo Francesco Vito 1, 00168 Rome, Italy
e-mail: asgambato@rm.unicatt.it

D. Corbeil (ed.), *Prominin-1 (CD133): New Insights on Stem & Cancer Stem Cell Biology*, Advances in Experimental Medicine and Biology 777, DOI 10.1007/978-1-4614-5894-4_10,

10.1 Introduction

Colorectal cancer is the third most common cancer in men and the second in women worldwide being the second leading cause of cancer-related death in Western countries and the fourth most common cause of death from cancer worldwide [1]. Several genetic and environmental factors have been implicated in its development, but the exact mechanisms responsible for the conversion of normal colon epithelium to invasive carcinoma remain still unknown. The involvement of environmental and lifestyle factors in colorectal carcinogenesis is confirmed by the increase in colorectal cancer incidence in parallel with economic development and adoption of a Western lifestyle in several countries [2]. Indeed, almost 60% of the cases occur in developed regions with the highest rates being observed in Australia, New Zealand, and Western Europe and the lowest in Africa (except Southern Africa) and south-central Asia.

Treatment of colorectal cancer includes surgery, chemotherapy, and radiation therapy. Chemotherapy, mainly used to prevent recurrences after surgery, represents the only available option in advanced and metastatic tumors. The backbone of treatment for colorectal cancer is 5-fluorouracil (5-FU) in combination with leucovorin, but several new drugs have become available in the last years including capecitabine, oxaliplatin, and irinotecan used alone or in combination. More recently, traditional chemotherapy has been supplemented with the so-called biological drugs such as monoclonal antibodies anti-VEGF and anti-EGFR [3, 4]. However, despite the emergence of these new targeted agents and the use of various therapeutic combinations, none of the treatment options available is curative especially in patients with advanced cancer, and the overall survival of patients with metastatic colorectal cancer has only slightly increased over the past decade.

On the other hand, despite the multiple therapeutic approaches available, management of early-stage colorectal cancers is also not easy and is somehow complicated by the variable behavior of the disease. Indeed, prognostic stratification is mainly based on its anatomic extent, as assessed by the TNM (for tumors/nodes/metastases) staging, which strongly affects patient's survival [5]. However, TNM is not able to accurately predict clinical outcome of patients especially with early-stage lesions which can display an inconsistent progression, as it is in the case of stage I colorectal cancers which display an overall 5-year survival around 80–90% and are considered to have very good prognosis. However, a small subset of stage I colorectal cancer patients display relapses and disease progression, and available prognostic markers are not able to identify this subgroup of patients, which might benefit from a more aggressive (i.e., adjuvant therapies) treatment [5]. Thus, the development of new prognostic markers able to identify early-stage colorectal cancer patients at high risk of progression is a major topic in colorectal cancer research especially as new treatment options become available. On this basis, a better understanding of the molecular mechanisms that underlie the development and progression of colon cancer appears critical to improve management of colorectal cancer patients.

Colorectal cancer originates from epithelial cells lining the gastrointestinal tract, which undergo sequential mutations in specific DNA sequences that disrupt normal mechanisms of proliferation and self-renewal.

The intestinal tract consists of the small intestine (duodenum, jejunum, and ileum) and the large intestine or colon. The absorptive epithelium of the small intestine is ordered into villi and crypts of Lieberkühn, finger-like invaginations supported by the lamina propria mucosae and representing the functional unit of the intestine. Terminally differentiated cells (enterocytes, enteroendocrine cells, and goblet cells) located in the top third of the crypt are continually extruded into the lumen and are replaced by new cells derived from stem cells located at the bottom of the crypt. A fourth differentiated type, the Paneth cell, resides at the bottom of crypts and was proposed to perform a specialized function other than absorption of digested nutrients, the predominant task of the small intestinal epithelium [6]. Paneth cells and villi are absent in the colon that displays a flat surface epithelium. Differentiated colon epithelial cells are replaced approximately every 5 days and are maintained by stem cells [7]. In both the mouse small intestine and colon, the stem cells lie toward the bottom of the crypt in the proliferative zone and are responsible for generating all epithelial cell types along the crypt–villus axis. It is thought that four to six stem cells are present in each crypt even if the precise number of stem cells per crypt and what controls their number remain uncertain [8].

Stem cells display the unique ability to generate a large spectrum of different cell types despite retaining an undifferentiated phenotype themselves. This property is mainly due to their ability to divide asymmetrically. An asymmetric cell division generates two different daughter cells: one that retains stem cell features, including the ability to divide asymmetrically, and remains in the stem cell compartment (or "niche") and another cell with a restricted developmental potential which replicates relatively rapidly (transit-amplifying cells, TA) and gives rise to terminally differentiated cell types after one or more additional cell divisions. TA cells spend approximately 2 days in the crypt, in which they divide 4–5 times before they terminally differentiate into one of the specialized intestinal epithelial cell types: the absorptive enterocytes or colonocytes, the mucous-secreting goblet cells, and hormone-secreting enteroendocrine cells. While differentiating, cells migrate upward the crypt, and few days after terminal differentiation, they reach the tip of the crypt, undergo spontaneous apoptosis, and are shed into the lumen (Fig. 10.1). The Wnt pathway has been shown to play a pivotal role in regulating proliferation and differentiation of epithelial cells in the intestinal crypts [9, 10].

Studies on colon cancer development have provided a detailed knowledge of the sequence of genetic mutations occurring during the process with the formulation of a model of multistep carcinogenesis, which has been subsequently extended, to other epithelial tumors [11]. Almost all cancers of the colon are derived from a dysfunctional regulation of the Wnt/beta-catenin pathway, essential for the development of the normal colonic mucosa [12]. Studies on hereditary colon cancer syndromes have further contributed to our understanding of human colorectal tumorigenesis. About 80% of patients with FAP (familial adenomatous polyposis), a hereditary colon cancer syndrome, display loss or mutation in the *APC* (adenomatous polyposis coli) gene, which encodes a protein that normally controls β-catenin levels by binding to the cytoplasmic β-catenin protein and, in the absence of Wnt ligand, promoting its phosphorylation and degradation by the proteasome. When *APC* is mutated or

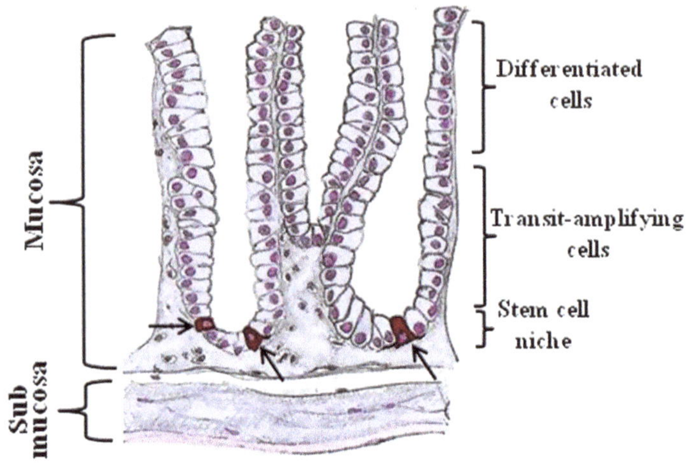

Fig. 10.1 Schematic organization of normal colon crypt. Stem cells (in *red*, *arrows*) are localized at the crypt base in the so-called stem cell niche. By asymmetric division, stem cells generate daughter cells with a limited development potential, which reproduce relatively quickly (transit-amplifying cells) and give rise to terminally differentiated functional cells

absent, the cytosolic β-catenin is stabilized, and the protein can then migrate into the nucleus where it serves as a coactivator for the TCF family of transcription factors which activate the transcription of oncogenes such as c-myc and cyclin D1 [13]. The study of juvenile polyposis syndrome, a condition that predisposes to hamartomatous gastrointestinal polyp formation, has revealed the important role of SMAD/BMP (Sma- and Mad-related protein/bone morphogenetic protein) in intestinal architecture [14]. The observation that during adenoma development, in patients with FAP, APC mutations cause expansion of the crypt base cell population [15] together with the well-known observation of multiple intestinal cell types with different degrees of differentiation within a colon cancer [16] has supported for colon cancer a hierarchical model of development as opposed to the stochastic model.

According to the hierarchical model, also known as the cancer stem cell (CSC) model of tumorigenesis, the tumor would be organized as a normal tissue with only a rare subpopulation of undifferentiated cells displaying the unique biological properties necessary for tumor initiation, maintenance, and spreading [17]. Like normal stem cells, CSC is able to both self-renew and give rise to a differentiated progeny and would be located in a niche with mesenchymal cells that would ensure their survival in a secure environment, regulating their activation or quiescence through secretion of soluble factors [18]. The discovery of stem cells in the majority of normal tissues including colon crypts supports the hypothesis that normal stem cells might represent a possible target for tumorigenic mutations due to both their long life and their ability to self-renew and therefore might be the origin of CSC. However, the origin of CSC remains so far unclear, and it cannot be excluded that they might derive from more mature cells that reacquire stem cell properties during tumor formation rather than being the direct progeny of mutated stem cells [19].

A precise phenotypic, functional, and molecular characterization of colon CSC will be essential to shed lights on their origin. To this aim, it is important to be able to unequivocally identify CSC, and this explains the great effort devoted worldwide to the identification of specific CSC markers.

Indeed, phenotypic characterization of colon CSCs is still controversial [20–22]. Nevertheless, CD133 is presently considered a useful marker to identify CSC in colorectal cancers, and its detection has been used to identify CSC in colorectal cancer samples [23–26] although the co-expression of other molecules, such as CD44, CD166, ALDH1, Lgr5, and EpCAM, has been suggested to identify the CSC pool more precisely than CD133 expression alone [27].

10.2 CD133 in Human Colorectal Normal and Cancer Tissues

10.2.1 The Identification of CD133 as a CSC Marker in Colon Cancer

CD133 (also known as prominin-1) is a member of the prominin family of proteins and is a surface protein with five transmembrane domains initially discovered as a marker of hematopoietic stem cells [28] and of bone marrow-derived circulating endothelial progenitors involved in postnatal angiogenesis, inflammation, and tissue regeneration [29, 30]. More recently, CD133 has been also used to mark and to selectively enrich both normal and cancer stem cells from a variety of tissues [31].

Two studies originally identified CD133 as a marker of CSC in colorectal tissues [20, 21]. Following the previous identification of CD133 as a marker of cancer neural stem cells, Ricci-Vitiani and colleagues investigated the existence of a CD133 expressing (CD133+) cell population in colon cancer and identified the presence of rare cells expressing CD133, but not cytokeratin 20 (CK20), an intermediate filament protein whose presence is essentially restricted to differentiated intestinal cells, within the bulk of colorectal cancer tissues following tissue dissociation and immunophenotype analysis by flow cytometry using an antihuman CD133 antibody targeting the AC133 epitope (Miltenyi Biotech) [21, 28]. They also confirmed these findings by immunostaining colon cancer samples with the same antibody and demonstrating the presence of rare CD133+ cells in tumor as well as in normal colon tissues [21]. They then compared the ability of tumor-derived CD133+ and CD133– cells to engraft and give rise to subcutaneous tumors in severe combined immunodeficient (SCID) mice reporting that colon cancer CD133+ cells readily reproduced the original tumor at histological level, whereas CD133– cells displayed a significantly reduced ability to form tumors [21]. Moreover, staining and microscopic analysis indicated that CD133+-derived tumor xenografts consistently reproduced the primary tumor at histological level, including specific features infrequently observed in colorectal cancer. When CD133+ were isolated from primary xenografts and subsequently serially transplanted for several generations into secondary mice, CD133+ cells did not lose their tumorigenic potential but rather increased their aggressiveness, as indicated by the acquisition of a progressively faster tumor growth without significant phenotypic alterations. This observation suggested that CD133+ cells resident in the colon tumor mass displayed a virtually unlimited growth potential as also confirmed but their ability to grow for more than a year in vitro as undifferentiated tumor spheres in serum-free medium, maintaining the ability to reproduce the same morphological and antigenic pattern of the original tumor when injected in SCID mice [21].

O'Brien and colleagues also isolated, based on CD133 expression, tumorigenic and non-tumorigenic fractions of cells within human colon cancer samples [20]. They used subrenal capsule transplantation into pre-irradiated nonobese diabetic/severe combined (NOD/SCID) mice to identify human colon cancer-initiating cells (CC-IC) and, in agreement with Ricci-Vitiani, demonstrated an enrichment of more than 200-fold of CC-IC in the suspension of CD133+ cells isolated from colon cancer samples with the same antibody used by Ricci-Vitiani, compared to unfractionated tumor cells. These authors also observed a higher percentage of CD133+ cells in liver metastases compared to primary tumors, as subsequently confirmed by our group [32], and demonstrated that tumors generated from CD133+ cells were phenotypically similar to the original tumors [20]. It is noteworthy, however, that O'Brien and colleagues also calculated that not all CD133+ cells are CC-ICs and suggested that the use of additional markers, in combination with CD133, might be necessary to more exactly identify the fraction of tumor-initiating cells in colon cancer [20].

10.2.2 Conflicting Results on the Role of CD133 as a CSC Marker in Colon Cancer

These initial observations fostered a multitude of studies on the role of CD133 in colon cancer, and the majority of them confirmed the ability of CD133 to identify a subpopulation of CSC within colon cancer, but discordant observations have been also reported. Thus, Shmelkov and colleagues analyzed by immunohistochemistry the expression of CD133 on fresh-frozen tissues postfixed in paraformaldehyde using the antihuman CD133/clone AC133 (from Miltenyi) and reported a diffuse expression of CD133 on the luminal surface of the majority of differentiated epithelial cells, a pattern overlapping with the expression of epithelial cell markers and suggesting that CD133 is a marker of epithelial mature rather than stem cells in normal human colon [22]. They also reported a wide expression of CD133 on epithelial cancer cells in all primary colon tumors examined, while this was not true for liver metastatic colon lesions since most of them did not express detectable CD133+ cells. In addition, they demonstrated that both CD133+ and CD133– cells from colon metastatic cancers were able to form colon-spheres and to reproduce the tumor in NOD/SCID mice. Based on these observations, they concluded that CD133+ cells are not linked to CC-ICs and that the inability of the CD133– cells to generate tumors in immunodeficient mice reported in previous works [20, 21] may be due to the possibility of a predominant presence among these cells of non-tumorigenic stromal and inflammatory cells devoid of tumor initiation ability [22].

It is very difficult to conciliate these contrasting findings although some explanations might be suggested [33]. One possibility proposed by Liao and colleagues is that the reported discrepancies about the different tumor-initiating ability of CD133 positive and negative cells can be explained by the different abundance of CD133 expression on tumor cells [34]. Thus, in an attempt to reconcile Shmelkov findings [22] with the proposed role of CD133 as CSC marker in colon cancer, they proposed that all colon cancer cells might indeed express CD133 but at different levels suggesting that quantitative rather than qualitative analysis of CD133 abundance on tumor cells might provide more accurate information about their tumor initiation ability [34]. To support this hypothesis, they demonstrated that it is possible to identify populations of colon cancer cells expressing different levels of CD133 (high, medium, low), which relate to self-renewal, proliferation, and tumorigenicity capacities [34]. In their study, the CD133 low subgroup isolated from the SW620 human colon cancer cell line still retains the ability to initiate tumor growth, although with a lower efficiency than cells expressing CD133 at higher level, and it will be important to verify whether the same results can be obtained with cells isolated from human primary colorectal tumors.

Other studies also reported that CD133 is not expressed only on undifferentiated stem cells questioning its ability to identify CSC in colon cancer. Thus, Dalerba and colleagues demonstrated that CD44 was able to identify CSC in primary colon cancer tissues collected from surgical specimens or xenografts established in NOD/SCID mice and that other markers, such as EpCAM and CD166, but not CD133, could be used for further enrichment of the CSC fraction [27]. Similar results were also reported by Chu and colleagues who were not able to demonstrate an increased tumorigenic ability of CD133+ cells isolated from primary tumor xenografts obtained directly by injection of human primary colon cancer cells into immunodeficient mice [35]. They identified CD44 as a reliable marker of colon CSC and demonstrated that this cell population can be enriched by other markers, such as ALDH, but not CD133 [35]. Finally, Haraguchi and colleagues demonstrated, through studies on both colon cancer cell lines and primary tumors, that the simultaneous presence of CD133 and CD44 was able to identify colon CSC better than each of the two markers alone [36]. Overall, these data cast doubts on the proposed role of CD133 as a CSC marker in colon cancer but might well be conciliated hypothesizing that multiple types of colon CSC may exist or, more likely, that they can express multiple markers whose surface expression might not be stable but change due to growth conditions [37] and the interaction with a changeable surrounding microenvironment [38, 39].

10.2.3 CD133 Functions in Colon Cells

Despite the enormous interest for this molecule, the function and ligand of CD133 remain unknown, and its involvement in essential properties of stem cells such as self-renewal and differentiation remains to be defined. We previously reported that membrane expression of the CD133 molecule, as assessed by flow cytometry and Western blot analysis using the antihuman CD133/clone AC133 monoclonal antibody (mAb), decreased significantly during colon differentiation, using as a model the sodium butyrate-induced differentiation of HT29 human colon cancer cells. This decrease was likely due to posttranslational mechanisms since no changes were detected in the methylation status of the *CD133* gene promoter as well as in its mRNA expression level. However, it is of note that we could not detect any significant change in the expression of the CD133 protein during HT29 differentiation by using the antibody directed against the C-terminal region of the molecule (from Abgent) [40]. Similar results were reported by Feng and colleagues who also demonstrated that while downregulation of CD133 is associated with differentiation, CD133 depletion using specific siRNA is not able per se to induce cellular differentiation of colon cancer cells [41]. Our results were in agreement with previous studies reporting a downregulation of CD133 expression, evaluated using the AC133 antibody, upon differentiation of the Caco-2 colon carcinoma cell line [42]. It was also demonstrated that the AC133 epitope can be downregulated independently from the presence of corresponding mRNA in the Caco-2 cells [42, 43], and its general tissue distribution was much less widespread than the corresponding CD133 mRNA [43, 44]. The finding that methylation of *CD133* gene promoter did not change during differentiation was also in agreement with the observation that downregulation of the CD133 protein does not always involve changes in the DNA methylation pattern in colorectal cancer cells [45]. A surprising finding of our study was that the AC133 epitope might be downregulated independently from the CD133 protein or mRNA, an observation subsequently confirmed by Kemper who reported that detection of the AC133 epitope on the cell surface of colon CSC decreased upon differentiation in the absence of corresponding changes in CD133 promoter activity, mRNA, splice variant, protein expression, or even cell surface expression of CD133 [46].

No other indications are available regarding CD133 functions in colon normal and cancer cells. Indeed, Feng and colleagues were not able to detect any appreciable effect on the proliferation, cell cycle distribution, colony formation, and differentiation of HCT116 and HT-29 colon cancer cells following CD133 downregulation [41], a finding confirmed by Horst who also did not detect any significant effect on the proliferation, migration, invasion, and colony formation of Caco-2 and LoVo colon cancer cells by depletion of CD133 using specific siRNA [24]. Others and we demonstrated, however, that cell growth conditions could affect CD133 expression [37, 47], which might also be affected by the interaction with the surrounding microenvironment [38, 39], but the definitive functional significance of these observations remains to be elucidated.

10.2.4 Prognostic Significance of CD133 Expression in Human Colon Cancers

Despite the debate on the uncertain role of CD133 as marker of CSC in colon cancer, a great attention has been devoted to evaluate its prognostic role in colon cancer being the CSCs responsible for metastasis, treatment resistance, and recurrence [48]. To identify a potential relationship between CD133 and prognosis in colorectal cancer patients, many studies have evaluated CD133 by immunohistochemistry in colon cancer samples and correlated the results obtained with clinicopathological characteristics of patients, as well as with the clinical evolution of the disease, reporting somehow conflicting results.

Kojima and colleagues detected CD133 expression in 29 of 189 colorectal cancer patients (15.3%) using the anti-CD133/clone AC133 mAb antibody (Miltenyi) [25]. They reported expression of the protein exclusively on the cell membrane at the luminal surface of cancer glands with a weak and equivocal cytoplasmic expression in cancer cells. Of the 29 tumors with CD133 expression, 21 cases (11.1%) showed CD133 overexpression (luminal expression in more than 10% of cancer area), and there was no difference between invasive and surface tumoral areas. Moreover, all CD133-positive cases were well- or moderately differentiated adenocarcinoma, and only tumor areas exhibiting ductal structures displayed a positive expression, while poorly differentiated adenocarcinomas were never positive for CD133. Analyzing the relationship between CD133 expression and the clinical outcome of patients among well- and moderately differentiated adenocarcinomas, CD133 overexpression was associated with a worse outcome and a higher incidence of distant metastases but could not be considered an independent risk factor in terms of patient survival and risk of recurrence [25].

Horst and colleagues analyzed the expression of CD133 by immunohistochemistry in colorectal cancer samples collected from patients who underwent intentionally curative surgical resection [23]. They only included moderately differentiated (G2) colorectal adenocarcinomas with T-categories T2 or T3 and no nodal or distant metastases at the time of diagnosis. To verify the specificity of the detection method, they initially compared tumor sections stained with three different anti-CD133 primary antibodies: the AC133 clone from Miltenyi Biotech, a rabbit mAb (clone C24B9) from Cell Signaling Technology, and an anti-CD133 goat polyclonal antibody from Santa Cruz Biotechnology. They found that the three antibodies gave comparable staining patterns with CD133 being detected at the luminal surface of colorectal cancer glands with shedding into the lumina. They also showed that CD133+ cells were CK20- and EpCAM+, which is in agreement with findings obtained using tumor cell suspensions [21, 27]. CD133 expression was scored on the basis of CD133+ tumor glands: tumors showing more than 50% positive glands (either apical staining or staining of intraglandular cellular debris) were considered as $CD133^{High}$, whereas those with less than 50% stained glands were considered $CD133^{Low}$. When comparing the CD133 status with clinicopathological variables, no correlation was observed with age, gender, or T-category of the tumor. On the other hand, a significant correlation was observed between CD133 expression and clinical outcome since patients with CD133^{High} tumors displayed a significantly shorter 5- and 10-year survival than those with CD133^{Low} tumors and CD133^{High} expression confirmed to be an independent prognostic marker strongly associated with patient survival in a multivariate Cox regression analysis. The authors noted that CD133 expression was restricted only to tumor glands and was never detected on tumor buds, small clusters of undifferentiated cells ahead of the invasive front of the lesion showing no lumen formation, which are recognized as an independent prognostic factor significantly related to both lymph node and distant metastases in colorectal cancer [49]. Thus, they suggested that budding tumor cells might contain a population of CSCs that are not characterized by CD133 expression, as subsequently demonstrated by Hostettler and colleagues [50]. Afterward, Horst and colleagues evaluated CD133 expression, using the rabbit mAb (clone C24B9; Cell Signaling), in a matched case–control collection of 54 pairs of colon cancers with and without synchronous liver metastases [24]. They found that in normal mucosa adjacent to tumors staining for CD133 was always negative and that high CD133 expression (≥50% of CD133 positive glands) was significantly associated with synchronous liver metastases. They also demonstrated, comparing tissue samples of both tumor sites, that CD133 expression in primary colon cancers and corresponding liver metastases were similar. On this basis, they concluded that CD133 is a valuable prognostic marker for assessing the risk of disease progression, metastasis, and death of colon cancer patients.

CD133 was confirmed to be an independent prognostic indicator for colorectal cancer patients in a study by Ong and colleagues who analyzed by immunostaining with the mAb AC133 clone (Miltenyi) a tissue microarray containing 501 primary colorectal cancer cases [51]. Scoring was performed using an automated imaging and scoring platform, and receiver-operating characteristic (ROC) analysis was used to determine cutoff scores for positive expression. They found that CD133 expression was associated with significantly worse survival of colorectal cancer patients. In agreement with previous studies, these authors found a strong inverse correlation between the expression of CD133 and CK20 with CD133 staining being mainly localized at the glandular-luminal surface of tumor epithelial cells and in the intraglandular debris shed by tumor cells and absent or very mild in normal colonic mucosa [21, 23]. This study for the first time reported that CD133 expression is also associated with tumor response to 5-fluorouracil-based chemotherapy. In fact, the authors compared the survival rates of stage III and stage IV patients treated with or without chemotherapy noting that CD133– tumors gained significant benefit from chemotherapy, while no survival benefit was observed for patients with CD133+ tumors. In the same study, other candidate CSC markers in colorectal cancer, such as Oct-4 and Sox-2, were also analyzed and, although Oct-4 was associated with worse survival of patients, only CD133 expression was a predictive marker for survival benefit from 5-fluorouracil-based adjuvant chemotherapy [51]. Our group also analyzed CD133 expression by immunostaining in colorectal cancer patients and obtained both membrane and cytoplasmic staining using different anti-CD133 antibodies. In agreement with previous studies, scattered positive cells were detected at

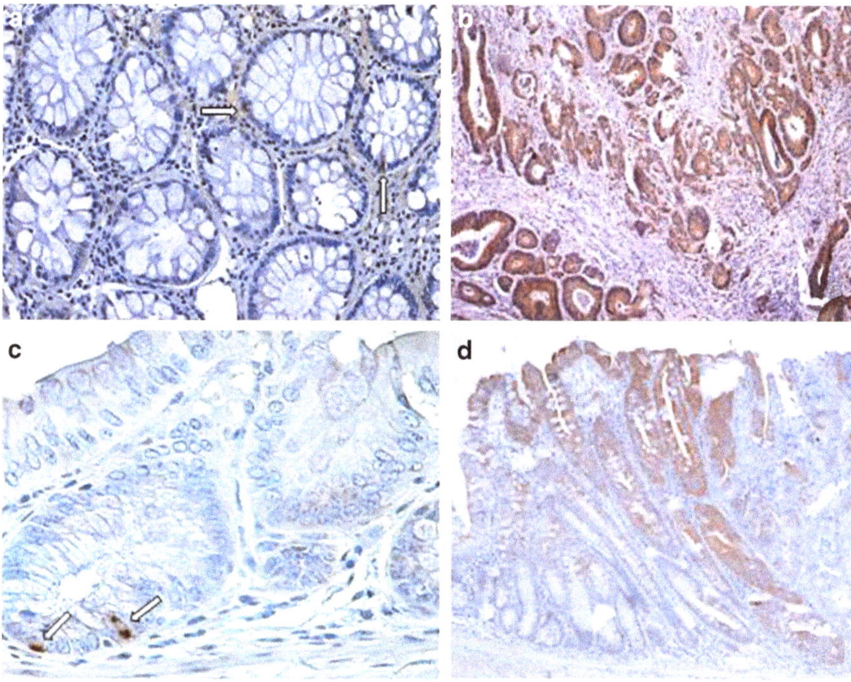

Fig. 10.2 Examples of CD133 immunohistochemical staining in human normal colonic mucosa (**a** and **c**) and colorectal samples (**b** and **d**). Note the rare (*arrowed*) positive signal for CD133 in normal mucosa (**a**, ×200 and **c**, ×400), while a widely positive immunostaining is detected in cancer samples (×200). A rabbit mAb (clone C24B9; Cell Signaling Technology; 1:150) was used for the staining

the bases of the crypts in normal colonic mucosa, while a diffuse and heterogeneous staining was frequently evident in tumor cells (Fig. 10.2). Increased expression of CD133 confirmed to be an independent prognostic parameter at a multivariate analysis associated with an increased risk of recurrence and death ([52] and unpublished data).

A significant association between high CD133 expression and worse prognosis was also reported by Li and colleagues who used a different mouse mAb from Abcam and were able to detect CD133 expression not only on the apical membrane but also on the basal surface of tumor cells [26]. Moreover, intense staining was observed within the tumoral area, and the authors suggested that $CD133^+$ cells could likely proliferate and invade the surrounding tissue, giving rise to cells with weak or negative expression of CD133. Positive staining was observed both on well-differentiated and poorly differentiated tumors with ductal structures rather than those without ductal structures [26]. The same mAb from Abcam was also used in a recent work reporting a significant association between CD133 expression and histological grade, depth of invasion, metastasis, and overall 5-year survival. Membrane and cytoplasmic staining of cancer cells was reported in this study with positivity detected predominantly around the lumens of cancerous glands in well-differentiated carcinomas [53]. Saigusa and colleagues used the same mAb to investigate the relationship between CD133 expression and chemoradiotherapy [54]. To this aim, they analyzed by immunohistochemistry the expression of CD133 in patients with colorectal cancer who had received preoperative chemoradiotherapy. They reported distinct patterns of CD133 expression in colon cancer specimens in glandular-luminal surface, in intraluminal cells, and in cytoplasm [54]. Overall, positivity for CD133 was higher in colorectal cancer samples of patients who had undergone preoperative chemoradiotherapy compared to untreated patients. Moreover, histopathological responders were significantly more frequent in cases negative for CD133 expression compared to cases with CD133 expression in both luminal surface and cytoplasm. They also demonstrated that irradiation of HT29 human colon cancer cells induced a dose-dependent increase of CD133 expression further confirming a relationship between CD133 and resistance to chemoradiation and warranting further studies to verify whether CD133 might serve as a predictive marker for the response to chemoradiotherapy in colorectal cancer patients. Comparable results were reported using the same mAb by Jao and colleagues who found that cytoplasmic CD133 was related to local recurrence and survival in patients with rectal cancer after neoadjuvant chemoradiotherapy and suggested that cytoplasmic CD133 expression might help to identify surviving cancer cells in areas with nearly total regression after chemoradiotherapy [55].

Besides the in situ localization of CD133 detected by immunohistochemistry, several studies have also evaluated the expression level of CD133 mRNA in patients with colorectal cancer. Artells and colleagues analyzed the relationship between CD133 mRNA levels and clinicopathological characteristics, clinical outcome, and response to adjuvant chemotherapy in colorectal cancer patients [56]. CD133 mRNA levels were estimated in matched tumor and normal colon tissues and were found overall higher in tumor than in normal tissue samples. Higher levels of CD133 mRNA correlated with a shorter disease-free and overall survival both in all patients and in the subgroup receiving adjuvant chemotherapy [56]. The association between high CD133 expression and resistance to chemotherapeutic drugs, such as 5-FU and oxaliplatin, has been previously reported in colon cancer cell lines in vitro [57] although it remains to elucidate whether CD133 is simply a marker of resistance or itself is involved in the process of drug resistance [58].

Assessment of CD133 mRNA, together with carcinoembryonic antigen (CEA) and CK20, has been also suggested as a sensitive method that could replace the conventional cytological examination, for the detection of free cancerous cells in the peritoneum of colorectal cancer patients undergoing surgical resection. Detection of CD133, as well as CEA and/or CK20 mRNA, in peritoneal washings showed a significant correlation with lymph node metastases and tumor stage and confirmed to be an independent prognostic factor of poor prognosis both in terms of overall and peritoneal recurrence-free survival. The sensitivity and specificity of CD133 for peritoneal recurrence was higher than those of CEA and CK20 suggesting the metastatic ability of CD133+ free cancer cells [59]. Similar results were obtained by Iinuma and colleagues who looked for circulating tumor cells using a real-time

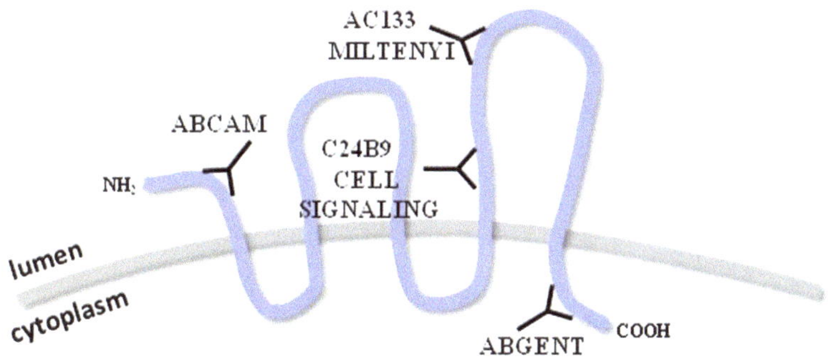

Fig. 10.3 Schematic representation of CD133 molecule and predicted epitopes recognized by various anti-CD133 antibodies

reverse transcription polymerase chain reaction assay to detect the levels of CD133 mRNA, in the peripheral blood of patients who had undergone curative surgery for colorectal cancer [60]. They found that simultaneous detection of peripheral blood CD133, CK, and CEA mRNA had a strong prognostic significance in stage II and III colorectal cancer patients both in terms of overall and disease-free survival [60].

Several studies failed to detect any relationship between CD133 expression levels and disease progression or survival in colorectal cancer patients. Choi and colleagues reported a significant association of CD133 expression levels with tumor stage, but not with patients' survival [61], a finding confirmed by Lugli and colleagues [62].

In conclusion, contrasting results have been reported in the literature regarding CD133 staining in colorectal cancer cells (in term of distribution and percentage of positive cells) as well as its prognostic significance. Several factors might explain such discrepancies: (1) non-homogeneity of patient cohorts, (2) mixed tumor stages, (3) different criteria used to classify positive staining, (4) different cutoff used to discriminate positive and negative tumors, and (5) different antibodies used for the analysis. Indeed, the antibodies used in various studies have a different specificity and recognize different epitopes on the CD133 molecule (Fig. 10.3). We believe that this can be a highly confusing factor since, as we and others have previously demonstrated, different results can be obtained by using different antibodies [40, 63, 64] (Table 10.1). The observation that high CD133 expression has been reported to be a negative prognostic factor for colorectal cancer in several studies using different antibodies, however, strongly suggests an important prognostic significance of its detection. It remains to be defined the functional significance of CD133 in colon cancer cells and whether CD133 expression levels (detected at the protein and/or mRNA level) strictly relate to the presence of CSC in the tumor since, as reported, multiple types of colon CSC may exist which might express other markers besides CD133 (Table 10.2). Moreover, it has been reported that CD133 expression is upregulated in colorectal cancer cells that have a hyperactivated Ras–Raf–MEK–ERK pathway, and since mutations of genes involved in this pathway have been related to poor prognosis, it has been suggested that the negative prognostic

Table 10.1 Different types of antibodies used for the immunohistochemical detection of CD133 molecule in colon cancer samples

Reference	Antibody	Epitope	Type of staining
Kojima et al. [25]	mAb (clone AC133, Miltenyi Biotech)	Extracellular	Luminal cell surface of cancer glands with weak and equivocal cytoplasmic staining
Horst et al. [23]	mAb (clone AC133, Miltenyi Biotech)	Extracellular	Luminal cell surface of cancer glands and intraglandular debris
Shmelkov et al. [22]	mAb (clone AC133, Miltenyi Biotech)	Extracellular	Luminal surface of differentiated epithelial normal and cancer cells
Ong et al. [51]	mAb (clone AC133, Miltenyi Biotech)	Extracellular	Glandular-luminal surface of cancer epithelial cells and intraglandular debris
Horst et al. [23, 24]	mAb (clone C24B9, Cell Signaling)	Extracellular	Luminal cell surface of cancer glands and intraglandular debris
Lugli et al. [62]	mAb (clone C24B9, Cell Signaling)	Extracellular	Membranous staining
Saigusa et al. [54]	mAb (clone C24B9, Cell Signaling)	Extracellular	Luminal and intraluminal cytoplasmic staining
Jao et al. [55]	mAb (clone C24B9, Cell Signaling)	Extracellular	Apical and/or endoluminal surfaces, in the cytoplasm and in the lumen
Li et al. [26]	mAb (Abcam)	Extracellular	Luminal and basal surface of cancer cells
Xi and Zhao [53]	Polyclonal (Abcam)	–	Membrane and cytoplasm of cancer cells, in lumen of cancerous glands only in well-differentiated areas
Horst et al. [23]	Polyclonal (Santa Cruz)	–	Luminal cell surface of cancer glands and intraglandular debris
Choi et al. [61]	Polyclonal (Santa Cruz)	–	Cytoplasmic staining

Table 10.2 Cell surface molecules suggested as CSC marker in colon cancer

Reference	Markers identified	Method
Ricci-Vitiani et al. [21]	CD133	Flow cytometry, xenografts
O'Brien et al. [20]	CD133	Xenografts
Dalerba et al. [27]	EpCAM^{high} CD44+ CD166	Flow cytometry, xenografts
Haraguchi et al. [36]	CD133 CD44	Flow cytometry, xenografts
Chu et al. [35]	CD44 ALDH	Xenografts
Liao et al. [34]	CD133^{high}	Flow cytometry, xenografts

significance of CD133 expression might be related to associated mutations or activity status of the Ras–Raf pathway rather to a direct influence of CD133 on tumor evolution and progression [65]. Regardless of the functional significance of CD133 expression in cancer cells, however, we believe that efforts are warranted to reach standardization for CD133 detection in clinical samples and for staining classification since CD133 detection might become a valuable support in the routine clinical management of colorectal cancer patients both as a prognostic and a predictive marker.

10.3 CD133 in Mouse Colorectal Normal and Cancer Tissues

The use of mouse models is very important in research. They are invaluable tools for understanding pathological mechanisms and for the assessment of diagnostic, prognostic, and therapeutic aspects related to human diseases. In particular, through the development of genetically modified models, it is possible to investigate specific fields of interest. Thus, mice have been also used to investigate the role of CD133 in normal colorectal tissues and to analyze its involvement in the process of tumor development. However, they have been mostly used as recipients for tumorigenicity assays and to answer to the question of whether CD133 has to be considered a marker of colon CSC or not, and few studies have directly analyzed the role of CD133 expressing cells in murine colorectal tumorigenesis.

One of the first papers analyzing the expression of CD133 in mouse normal and cancer colon tissues unexpectedly reported that CD133 is present in the majority of the differentiated and progenitor epithelial cells of the normal colonic mucosa along the entire crypt, including its transient amplifying and basal cell layers, as well as in spontaneous primary colon tumors in mice [22]. These findings were strikingly in contrast with the studies in human reporting that CD133 is only expressed on a little subset of cells both in normal and cancer colorectal tissues [20, 21] (see previous paragraph). However, while in the majority of studies on the role of CD133 as CSC marker it has been mainly analyzed (as mentioned) at the protein level using specific anti-CD133 antibodies, in this study the distribution of CD133 was primarily analyzed using an in vivo model based on a knock-in mouse in which the expression of lacZ reporter gene was driven by the endogenous CD133 promoter [22]. It is of note that the lacZ reporter gives indications about the transcriptional activity of the gene under study which does not necessarily reflect the expression and/or function of the corresponding protein which might well be regulated at a posttranscriptional level. Thus, the results obtained that relate to CD133 mRNA expression might not fully correspond to the expression of the corresponding protein. It is noteworthy that in the same study, the authors did confirm their genetic findings using an anti-mouse prominin-1 mAb (clone 13A4 from eBioscience) to analyze by immunohistochemistry the expression of CD133 on fresh-frozen colon tissues postfixed in paraformaldehyde and reported a ubiquitous expression of the CD133 protein in mouse colon normal and tumor tissues [22]. They found a broad expression of CD133 on the luminal surface and in the center of the crypts. Thus, CD133 expression was higher toward the upper portion of the intestinal gland, where, as previously described, differentiated cells are mainly represented (see Fig. 10.1). To analyze CD133 expression in murine colon cancers, these authors took advantage of an $IL10^{-/-}$/CD133 lacZ mice in which chronic inflammation results in spontaneous tumor formation in the colon. Using this model, they reported that CD133 is widely expressed by all cancer epithelial cells being not restricted to a rare subset of tumor cells. Overall, these data demonstrate that both CD133 mRNA and protein are present in the majority of differentiated cells and stem/progenitors of the colonic mucosa as well as in the majority of colon cancer cells, and the discrepancies with studies in humans remain unclear and difficult to be conciliated although some explanations might be suggested [33].

Similar results were reported by Zhu and colleagues using an inducible-Cre-nuclear LacZ reporter allele knocked into the CD133 locus [19]. Using this model, they were also able to demonstrate expression of CD133 in various developing and differentiated mouse tissues including goblet and columnar epithelial cells lining the colon, while it was more restricted in the small intestine being mainly expressed at the base of crypts tightly overlapping with Lgr5, a known marker of small intestinal stem cells. In the same murine model, the tumorigenesis of small intestine, which is induced by the endogenous activation of the Wnt signaling (a phenomenon that mimics human colonic tumorigenesis) in CD133+ cells, was associated with an initial marked expansion of CD133+ cells which replaced normal mucosa architecture across the entire small intestine giving rise to neoplastic lesions whose cells mostly lost expression of CD133. On the other hand, endogenous activation of the Wnt signaling in colon CD133+ cells did not induce appearance of any neoplastic lesion, as expected considering that CD133 marks differentiated cells in colon [19].

To get further insights into the role(s) played by CD133 in colon tumorigenesis, our group analyzed the expression of CD133 during the entire process of colon tumorigenesis in mouse. We took advantage of a mouse model of colitis-related cancer in which colorectal cancer is induced by the combined administration of azoxymethane, an initiating agent, and dextran sodium sulfate, a chemical irritant which causes chronic inflammation and subsequent proliferation of adenomatous polyps [66] giving rise to cancers in about 90–100% of cases. Mice were sacrificed at various intervals, and CD133 expression was evaluated by immunohistochemistry on routinely fixed and paraffin-embedded colon samples using a rabbit mAb (clone C24B9; Cell Signaling Technology). Comparable results were obtained using a polyclonal antibody (Santa Cruz) as well as the monoclonal AC133 antibody (Miltenyi) [67].

In normal tissue, rare scattered positive cells were detectable at the bottom of crypts, a pattern comparable to what previously reported for human colon [20, 21] in contrast with the findings reported by Shmelkov's group [22]. The percentage of positive cells significantly increased in dysplastic lesions compared to normal tissue and appeared then to progressively decrease in the passage from dysplasia to adenoma to cancer although always remaining higher than in normal tissue. In tumor tissues, CD133 expressing cells were also more represented at the periphery of tumors and less evident at the tumoral center [67]. The distribution of CD133 expressing cells observed in normal and diseased colon tissues is consistent with the hypothesis that normal stem cells reside at the basis of the crypts and that CD133 expression is lost in more differentiated cells. Moreover, the observation that CD133 expressing cells represent the majority of cells in the earliest (i.e., dysplastic) phases of the process and then progressively decrease at later stages (adenoma and cancer) is in agreement with data from Zhu's group which reported a similar behavior for CD133 expressing cells in small intestine tumorigenesis [19]. The open question is whether CD133 is indeed a stem cell marker in colon tissue and how to explain the discrepant data reported in the literature.

Our findings support an important involvement of the CD133 molecule in colon tumorigenesis and especially in the early phases of the process, and we hypothesize that an aberrant increase of the CD133 expressing cells normally resident within adult colon tissue might initiate and sustain the process. It will be of interest to determine the factor(s) responsible for this phenomenon, which might be consequent to an expansion of the pool of CD133 expressing stem cells, following tumor-initiating mutations, or to the expression of the CD133 molecule on cells normally not expressing it, which might eventually acquire stemness properties including the ability to initiate tumors. We also believe, as suggested by recent evidence, that interaction of cells with the surrounding microenvironment likely plays an important role in the process [38, 39].

10.4 Final Remarks

Overall, the data available in the literature support a potential role of CD133 as a marker of CSC in colon cancer, although the significance of its relative amount as well as of other markers remains to be defined. This is not a futile argument considering that CSC is expected to represent only a small number of cells within a tumor having the ability to initiate both the primary disease and its recurrences and metastases. In this view, the availability of reliable markers is essential for the isolation of CSC and a better understanding of their role in tumorigenesis. Moreover, targeting these cells is essential to eradicate cancer and prevent disease recurrences and metastasis. Therefore, it is of paramount importance to be able to identify CSC within tumors and, thus, to definitively assess whether CD133 can mark CSC in colon cancer especially now that therapies specifically targeting CD133 are becoming available [68, 69].

High CD133 expression, as assessed by immunohistochemistry, has been associated with disease aggressiveness and evolution [23, 24, 26], treatment resistance [51, 54, 56], and survival [23] in colorectal cancer, and more recently, the prognostic value of circulating CD133 mRNA levels in colorectal cancer has been also reported [56, 59, 60].

All these findings suggest a potentially important role of CD133 expressing cells in the initiation and evolution of human colorectal tumorigenesis and support its role as CSC marker in human colon cancer. The conflicting results reported in the literature, however, cannot be overlooked and deserve further investigation. It is likely that identification and isolation of CSC from colon cancer can be more accurately accomplished by using multiple cell surface markers associated with cancer stemness such as CD44, CD166, and Lgr5 besides CD133 (Table 10.2). Regarding CD133, as previously reported by us and others [33, 47], we believe that discrepancies can be, at least in part, explained by the different methods (i.e., protein vs mRNA analysis) and/or different antibodies used to evaluate CD133 (Table 10.1). Moreover, cell growth conditions [37, 47], hypoxia [70], and microenvironment [38, 39] have been also shown to affect CD133 expression, and further studies are warranted to identify the underlying molecular mechanisms and how much they might affect the results reported both in vitro and in vivo. On this basis, we previously suggested the need for a consensus panel aimed to define standardized procedures and reagents to evaluate expression of this molecule especially in view of its potential clinical applications, and we believe this need is still valid [47].

In conclusion, giving its presence on CSC in colon as well as many other human cancers, CD133 has all the features to become a valuable target for tumor-targeted therapy and contribute to the development of more successful anticancer therapies. Therefore, further studies on the function of CD133 in normal colon epithelial cells and its role in human colorectal tumorigenesis are warranted and strongly recommended without any doubts.

References

1. Ferlay J, Shin HR, Bray F, Forman D, Mathers C, Parkin DM (2010) Estimates of worldwide burden of cancer in 2008: GLOBOCAN 2008. Int J Cancer 127:2893–2917
2. Center MM, Jemal A, Smith RA, Ward E (2009) Worldwide variations in colorectal cancer. CA Cancer J Clin 59:366–378
3. Meyerhardt JA, Mayer RJ (2005) Systemic therapy for colorectal cancer. N Engl J Med 352:476–487
4. Omura K (2008) Advances in chemotherapy against advanced or metastatic colorectal cancer. Digestion 77(Suppl 1):13–22
5. O'Connell MJ (2004) Current status of adjuvant therapy for colorectal cancer. Oncology (Williston Park) 18:751–755
6. Porter EM, Bevins CL, Ghosh D, Ganz T (2002) The multifaceted Paneth cell. Cell Mol Life Sci 59:156–170

7. Reya T, Clevers H (2005) Wnt signalling in stem cells and cancer. Nature 434:843–850
8. Bjerknes M, Cheng H (1999) Clonal analysis of mouse intestinal epithelial progenitors. Gastroenterology 116:7–14
9. Van de Wetering M, Sancho E, Verweij C, de Lau W, Oving I, Hurlstone A et al (2002) The β-catenin/TCF-4 complex imposes a crypt progenitor phenotype on colorectal cancer cells. Cell 111:241–250
10. Van der Flier LG, Sabates-Bellver J, Oving I, Haegebarth A, De Palo M, Anti M, Van Gijn ME, Suijkerbuijk S, Van de Wetering M, Marra G, Clevers H (2007) The intestinal Wnt/TCF signature. Gastroenterology 132:628–632
11. Fearon ER, Vogelstein B (1990) A genetic model for colorectal cancer tumorigenesis. Cell 61:759–767
12. Pinto D, Gregorieff A, Begthel H, Clevers H (2003) Canonical Wnt signals are essential for homeostasis of the intestinal epithelium. Genes Dev 17:1709–1713
13. Korinek V, Barker N, Morin PJ, van Wichen D, de Weger R, Kinzler KW, Vogelstein B, Clevers H (1997) Constitutive transcriptional activation by a beta-catenin-Tcf complex in APC−/− colon carcinoma. Science 275:1784–1787
14. Haramis AP, Begthel H, van den Born M, van Es J, Jonkheer S, Offerhaus GJ, Clevers H (2004) De novo crypt formation and juvenile polyposis on BMP inhibition in mouse intestine. Science 303:1684–1686
15. Boman BM, Walters R, Fields JZ, Kovatich AJ, Zhang T, Isenberg GA, Goldstein SD, Palazzo JP (2004) Colonic crypt changes during adenoma development in familial adenomatous polyposis: immunohistochemical evidence for expansion of the crypt base cell population. Am J Pathol 165:1489–1498
16. Dexter DL, Spremulli EN, Fligiel Z, Barbosa JA, Vogel R, VanVoorhees A, Calabresi P (1981) Heterogeneity of cancer cells from a single human colon carcinoma. Am J Med 71:949–956
17. Vermeulen L, Todaro M, de Sousa MF, Sprick MR, Kemper K, Perez Alea M, Richel DJ, Stassi G, Medema JP (2008) Single-cell cloning of colon cancer stem cells reveals a multilineage differentiation capacity. Proc Natl Acad Sci USA 105:13427–13432
18. Ailles LE, Weissman IL (2007) Cancer stem cells in solid tumors. Curr Opin Biotechnol 18:460–466
19. Zhu L, Gibson P, Currle DS, Tong Y, Richardson RJ, Bayazitov IT, Poppleton H, Zakharenko S, Ellison DW, Gilbertson RJ (2009) Prominin 1 marks intestinal stem cells that are susceptible to neoplastic transformation. Nature 457:603–607
20. O'Brien CA, Pollett A, Gallinger S, Dick JE (2007) A human colon cancer cell capable of initiating tumour growth in immunodeficient mice. Nature 445:106–110
21. Ricci-Vitiani L, Lombardi DG, Pilozzi E, Biffoni M, Todaro M, Peschle C, De Maria R (2007) Identification and expansion of human colon-cancer-initiating cells. Nature 445:111–115
22. Shmelkov SV, Butler JM, Hooper AT, Hormigo A, Kushner J, Milde T et al (2008) CD133 expression is not restricted to stem cells, and both CD133+ and CD133- metastatic colon cancer cells initiate tumors. J Clin Invest 118:2111–2120
23. Horst D, Kriegl L, Engel J, Kirchner T, Jung A (2008) CD133 expression is an independent prognostic marker for low survival in colorectal cancer. Br J Cancer 99:1285–1289
24. Horst D, Scheel SK, Liebmann S, Neumann J, Maatz S, Kirchner T, Jung A (2009) The cancer stem cell marker CD133 has high prognostic impact but unknown functional relevance for the metastasis of human colon cancer. J Pathol 219:427–434
25. Kojima M, Ishii G, Atsumi N, Fujii S, Saito N, Ochiai A (2008) Immunohistochemical detection of CD133 expression in colorectal cancer: a clinicopathological study. Cancer Sci 99:1578–1583
26. Li CY, Li BX, Liang Y, Peng RQ, Ding Y, Xu DZ, Zhang X, Pan ZZ, Wan DS, Zeng YX, Zhu XF, Zhang XS (2009) Higher percentage of CD133+ cells is associated with poor prognosis in colon carcinoma patients with stage IIIB. J Transl Med 7:56
27. Dalerba P, Dylla SJ, Park IK, Liu R, Wang X, Cho RW, Hoey T, Gurney A, Huang EH, Simeone DM, Shelton AA, Parmiani G, Castelli C, Clarke MF (2007) Phenotypic characterization of human colorectal cancer stem cells. Proc Natl Acad Sci U S A 104:10158–10163

28. Yin AH, Miraglia S, Zanjani ED, Almeida-Porada G, Ogawa M, Leary AG, Olweus J, Kearney J, Buck DW (1997) AC133, a novel marker for human hematopoietic stem and progenitor cells. Blood 90:5002–5012
29. Asahara T, Murohara T, Sullivan A, Silver M, van der Zee R, Li T, Witzenbichler B, Schatteman G, Isner JM (1997) Isolation of putative progenitor endothelial cells for angiogenesis. Science 275:964–967
30. Shaked Y, Ciarrocchi A, Franco M, Lee CR, Man S, Cheung AM, Hicklin DJ, Chaplin D, Foster FS, Benezra R, Kerbel RS (2006) Therapy-induced acute recruitment of circulating endothelial progenitor cells to tumors. Science 313:1785–1787
31. Singh SK, Hawkins C, Clarke ID, Squire JA, Bayani J, Hide T, Henkelman RM, Cusimano MD, Dirks PB (2004) Identification of human brain tumour initiating cells. Nature 432:396–401
32. Puglisi MA, Sgambato A, Saulnier N, Rafanelli F, Barba M, Boninsegna A, Piscaglia AC, Lauritano C, Novi ML, Barbaro F, Rinninella E, Campanale C, Giuliante F, Nuzzo G, Alfieri S, Doglietto GB, Cittadini A, Gasbarrini A (2009) Isolation and characterization of CD133+ cell population within human primary and metastatic colon cancer. Eur Rev Med Pharmacol Sci 13(Suppl 1):55–62
33. LaBarge MA, Bissell MJ (2008) Is CD133 a marker of metastatic colon cancer stem cells? J Clin Invest 118:2021–2024
34. Liao Y, Hu X, Huang X, He C (2010) Quantitative analyses of CD133 expression facilitate researches on tumor stem cells. Biol Pharm Bull 33:738–742
35. Chu P, Clanton DJ, Snipas TS, Lee J, Mitchell E, Nguyen ML, Hare E, Peach RJ (2009) Characterization of a subpopulation of colon cancer cells with stem cell-like properties. Int J Cancer 24:1312–1321
36. Haraguchi N, Ohkuma M, Sakashita H, Matsuzaki S, Tanaka F, Mimori K, Kamohara Y, Inoue H, Mori M (2008) CD133+CD44+ Population Efficiently Enriches Colon Cancer Initiating Cells. Annals Surg Oncol 15:2927–2933
37. Jaksch M, Múnera J, Bajpai R, Terskikh A, Oshima RG (2008) Cell cycle-dependent variation of a CD133 epitope in human embryonic stem cell, colon cancer, and melanoma cell lines. Cancer Res 68:7882–7886
38. Chao C, Carmical JR, Ives KL, Wood TG, Aronson JF, Gomez GA, Djukom CD, Hellmich MR (2012) CD133+ colon cancer cells are more interactive with the tumor microenvironment than CD133– cells. Lab Invest 92:420–436
39. Yang Z, Wang Z, Fan Y, Zheng Q (2012) Expression of CD133 in SW620 colorectal cancer cells is modulated by the microenvironment. Oncol Lett 4:75–79
40. Sgambato A, Puglisi MA, Errico F, Rafanelli F, Boninsegna A, Rettino A, Genovese G, Coco C, Gasbarrini A, Cittadini A (2010) Post-translational modulation of CD133 expression during sodium butyrate-induced differentiation of HT29 human colon cancer cells: implications for its detection. J Cell Physiol 224:234–241
41. Feng HL, Liu YQ, Yang LJ, Bian XC, Yang ZL, Gu B, Zhang H, Wang CJ, Su XL, Zhao XM (2010) Expression of CD133 correlates with differentiation of human colon cancer cells. Cancer Biol Ther 9:216–223
42. Corbeil D, Röper K, Hellwig A, Tavian M, Miraglia S, Watt SM, Simmons PJ, Peault B, Buck DW, Huttner WB (2000) The human AC133 hematopoietic stem cell antigen is also expressed in epithelial cells and targeted to plasma membrane protrusions. J Biol Chem 275:5512–5520
43. Florek M, Haase M, Marzesco AM, Freund D, Ehninger G, Huttner WB, Corbeil D (2005) Prominin-1/CD133, a neural and hematopoietic stem cell marker, is expressed in adult human differentiated cells and certain types of kidney cancer. Cell Tissue Res 319:15–26
44. Miraglia S, Godfrey W, Yin AH, Atkins K, Warnke R, Holden JT, Bray RA, Waller EK, Buck DW (1997) A novel five-transmembrane hematopoietic stem cell antigen: isolation, characterization, and molecular cloning. Blood 90:5013–5021
45. Yi JM, Tsai HC, Glöckner SC, Lin S, Ohm JE, Easwaran H et al (2008) Abnormal DNA methylation of CD133 in colorectal and glioblastoma tumors. Cancer Res 68:8094–8103

46. Kemper K, Sprick MR, de Bree M, Scopelliti A, Vermeulen L, Hoek M, Zeilstra J, Pals ST, Mehmet H, Stassi G, Medema JP (2010) The AC133 epitope, but not the CD133 protein, is lost upon cancer stem cell differentiation. Cancer Res 70:719–729
47. Sgambato A, Errico F, Caredda E, Puglisi MA, Cittadini A (2011) Divergent expression of CD133 in different studies: the need for a consensus panel? Int J Cancer 128:2240–2250
48. Pardal R, Clarke MF, Morrison SJ (2003) Applying the principles of stem-cell biology to cancer. Nat Rev Cancer 3:895–902
49. Hase K, Shatney C, Johnson D, Trollope M, Vierra M (1993) Prognostic value of tumor "budding" in patients with colorectal cancer. Dis Colon Rectum 36:627–635
50. Hostettler I, Zlobec I, Terracciano L, Lugli A (2010) ABCG5-positivity in tumor buds is an indicator of poor prognosis in node-negative colorectal cancer patients. World J Gastroenterol 16:732–739
51. Ong CW, Kim LG, Kong HH, Low LY, Iacopetta B, Soong R, Salto-Tellez M (2010) CD133 expression predicts for non-response to chemotherapy in colorectal cancer. Mod Pathol 23:450–457
52. Reggiani Bonetti L, Migaldi M, Caredda E, Ponz de Leon M, Di Gregorio C, Barresi V, Scannone D, Danese S, Cittadini A and Sgambato A (2012) Increased expression of CD133 is a strong predictor of poor outcome in stage I colorectal cancer patients. Scand J Gastroenterol 47: 1211–1217
53. Xi HQ, Zhao P (2011) Clinicopathological significance and prognostic value of EphA3 and CD133 expression in colorectal carcinoma. J Clin Pathol 64:498–503
54. Saigusa S, Tanaka K, Toiyama Y, Yokoe T, Okugawa Y, Kawamoto A, Yasuda H, Morimoto Y, Fujikawa H, Inoue Y, Miki C, Kusunoki M (2010) Immunohistochemical features of CD133 expression: association with resistance to chemoradiotherapy in rectal cancer. Oncol Rep 24:345–350
55. Jao SW, Chen SF, Lin YS, Chang YC, Lee TY, Wu CC, Jin JS, Nieh S (2012) Cytoplasmic CD133 Expression is a Reliable Prognostic Indicator of Tumor Regression After Neoadjuvant Concurrent Chemoradiotherapy in Patients with Rectal Cancer. Ann Surg Oncol. [Epub ahead of print]
56. Artells R, Moreno I, Díaz T, Martínez F, Gel B, Navarro A, Ibeas R, Moreno J, Monzó M (2010) Tumour CD133 mRNA expression and clinical outcome in surgically resected colorectal cancer patients. Eur J Cancer 46:642–649
57. Dallas NA, Xia L, Fan F, Gray MJ, Gaur P, van Buren G 2nd, Samuel S, Kim MP, Lim SJ, Ellis LM (2009) Chemoresistant colorectal cancer cells, the cancer stem cell phenotype, and increased sensitivity to insulin-like growth factor-I receptor inhibition. Cancer Res 69:1951–1957
58. Neuzil J, Stantic M, Zobalova R, Chladova J, Wang X, Prochazka L, Dong L, Andera L, Ralph SJ (2007) Tumour-initiating cells vs. cancer 'stem' cells and CD133: what's in the name? Biochem Biophys Res Commun 355:855–859
59. Nakamura K, Iinuma H, Aoyagi Y, Shibuya H, Watanabe T (2010) Predictive value of cancer stem-like cells and cancer-associated genetic markers for peritoneal recurrence of colorectal cancer in patients after curative surgery. Oncology 78:309–315
60. Iinuma H, Watanabe T, Mimori K, Adachi M, Hayashi N, Tamura J, Matsuda K, Fukushima R, Okinaga K, Sasako M, Mori M (2011) Clinical significance of circulating tumor cells, including cancer stem-like cells, in peripheral blood for recurrence and prognosis in patients with Dukes' stage B and C colorectal cancer. J Clin Oncol 29:1547–1555
61. Choi D, Lee HW, Hur KY, Kim JJ, Park GS, Jang SH, Song YS, Jang KS, Paik SS (2009) Cancer stem cell markers CD133 and CD24 correlate with invasiveness and differentiation in colorectal adenocarcinoma. World J Gastroenterol 15:2258–2264
62. Lugli A, Iezzi G, Hostettler I, Muraro MG, Mele V, Tornillo L, Carafa V, Spagnoli G, Terracciano L, Zlobec I (2010) Prognostic impact of the expression of putative cancer stem cell markers CD133, CD166, CD44s, EpCAM, and ALDH1 in colorectal cancer. Br J Cancer 103:382–390

63. Hermansen S, Christensen K, Jensen S, Kristensen B (2011) Inconsistent immunohistochemical expression patterns of four different CD133 antibody clones in glioblastoma. J Histochem Cytochem 59:391–407
64. Mak AB, Blakely KM, Williams RA, Penttilä PA, Shukalyuk AI, Osman KT, Kasimer D, Ketela T, Moffat J (2011) CD133 N-glycosylation processing contributes to cell-surface recognition of the primitive cell marker AC133. J Biol Chem 286:41046–41056
65. Kemper K, Versloot M, Cameron K, Colak S, de Sousa E, Melo F, de Jong JH, Bleackley J, Vermeulen L, Versteeg R, Koster J, Medema JP (2012) Mutations in the Ras-Raf Axis underlie the prognostic value of CD133 in colorectal cancer. Clin Cancer Res 18:3132–3141
66. Tanaka T, Kohno H, Suzuki R, Yamada Y, Sugie S, Mori H (2003) A novel inflammation-related mouse colon carcinogenesis model induced by azoxymethane and dextran sodium sulfate. Cancer Sci 94:965–973
67. Arena V, Caredda E, Cufino V, Stigliano E, Scaldaferri F, Gasbarrini A, Cittadini A, Sgambato A (2011) Differential CD133 expression pattern during mouse colon tumorigenesis. Anticancer Res 31:4273–4275
68. Waldron NN, Kaufman DS, Oh S, Inde Z, Hexum MK, Ohlfest JR, Vallera DA (2011) Targeting tumor-initiating cancer cells with dCD133KDEL shows impressive tumor reductions in a xenotransplant model of human head and neck cancer. Mol Cancer Ther 10:1829–1838
69. Sun J, Zhang C, Liu G, Liu H, Zhou C, Lu Y, Zhou C, Yuan L, Li X (2012) A novel mouse CD133 binding-peptide screened by phage display inhibits cancer cell motility in vitro. Clin Exp Metastasis 29:185–196
70. Matsumoto K, Arao T, Tanaka K, Kaneda H, Kudo K, Fujita Y, Tamura D, Aomatsu K, Tamura T, Yamada Y, Saijo N, Nishio K (2009) mTOR signal and hypoxia-inducible factor-1 alpha regulate CD133 expression in cancer cells. Cancer Res 69:7160–7164

Chapter 11
Prominin-1 (CD133) Expression in the Prostate and Prostate Cancer: A Marker for Quiescent Stem Cells

Davide Pellacani, Emma E. Oldridge, Anne T. Collins, and Norman J. Maitland

Abstract The origin and phenotype of stem cells in human prostate cancer remains a subject of much conjecture. In this scenario, CD133 has been successfully used as a stem cell marker in both normal prostate and prostate cancer. However, cancer stem cells have been identified without the use of this marker, opening up the possibility of a CD133 negative cancer stem cell. In this chapter, we review the current literature regarding prostate cancer stem cells, with specific reference to the expression of CD133 as a stem cell marker to identify and purify stem cells in normal prostate epithelium and prostate cancer.

Keywords AC133 • Cancer stem cells • CD133 • Prominin-1 • Prostate • Prostate cancer • Stem cells

11.1 Introduction

The prostate is a small extra peritoneal gland, which sits under the bladder and in front of the rectum. The major function of the prostate is to produce a slightly alkaline fluid, which constitutes 20% of the ejaculate and contains polyamines and proteins, such as prostatic acid phosphatase (PAP) and prostate-specific antigen (PSA). It also functions as an endocrine gland, rapidly metabolizing testosterone to the more effective dihydrotestosterone [1].

D. Pellacani • E.E. Oldridge • A.T. Collins • N.J. Maitland, Ph.D. (✉)
YCR Cancer Research Unit, Department of Biology, University of York, Wentworth Way, YO10 5DD York, UK
e-mail: n.j.maitland@york.ac.uk

D. Corbeil (ed.), *Prominin-1 (CD133): New Insights on Stem & Cancer Stem Cell Biology*, Advances in Experimental Medicine and Biology 777, DOI 10.1007/978-1-4614-5894-4_11,

11.1.1 Cellular Architecture of the Prostate Epithelium

Prostate tissue is organized as tubular-alveolar glands comprised of an epithelial parenchyma surrounded by fibromuscular stroma. The mature human prostate epithelium is bilayered and is mainly composed of three kinds of cells: luminal cells, basal cells, and neuroendocrine cells. Terminally differentiated luminal cells, that is, the exocrine compartment of the prostate, are the most abundant cell type in normal and hyperplastic epithelium and secrete both PSA and PAP into the glandular lumen. These cells are dependent on androgens for their survival [2] and consequently express high levels of androgen receptor (AR) [3]. The relatively undifferentiated basal cells are in direct contact with the basement membrane, which separates the epithelial and the stromal compartments. They lack secretory activity, express low or undetectable levels of AR, and are not dependent on androgens for their survival [4]. Rare neuroendocrine cells (NE) are scattered throughout the basal layer. NE cells are terminally differentiated but androgen-insensitive [5], and release neuroendocrine peptides, such as bombesin, calcitonin, and parathyroid hormone-related peptide [6], which support epithelial growth and viability.

11.1.2 Prostate Cancer

Prostate cancer (PCa) is the most commonly diagnosed cancer in men in the western world. It is generally regarded to be slow growing and originates from a pre-neoplastic lesion termed prostate intraepithelial neoplasia (PIN) [7]. At the cellular level, PCa is characterized by a drastic reduction of the basal cell content (<1% of the cells) and a concomitant expansion of the AR^+ luminal cell compartment [8], which becomes highly proliferative [9]. This change in cellular composition is accompanied by a progressive degradation of the prostatic architecture, resulting in loss of the glandular structure and destruction of the basement membrane [10, 11].

In the case of localized PCa, surgical removal of the prostate (radical prostatectomy) is still the elected therapy and is curative in the majority of cases [12], but is not without side effects. Patients with more advanced and metastatic PCa are usually treated with androgen ablation therapy, targeted toward the AR^+ secretory luminal cancer cells that constitute the bulk of the tumor [13]. Although initially PCa responds to antiandrogens, around 25% of the cases relapse and develop castration-resistant prostate cancer (CRPC) [14] with a median life expectancy of 2 years [15].

11.2 Prostate Stem Cells and the Epithelial Hierarchy

Similar to many other epithelia in the human body, there is strong evidence that homeostasis of the prostate epithelium is governed by a hierarchy of cells with different proliferative potentials [16]. In the model originally proposed by Isaacs and

Coffey [17], undifferentiated stem cells reside in the basal compartment of the prostate epithelium and give rise to terminally differentiated luminal cells through an amplifying progeny. Observations supporting this hypothesis go back to the mid-1980s, where experiments of prostate regression/regeneration in castrated rats showed the existence of long-lived, androgen-independent cells which had the ability to reconstitute a fully differentiated epithelium upon androgen reintroduction [18, 19]. Evidence linking basal to luminal cells in the same hierarchy comes also from histological studies showing the presence of intermediate cells which express basal and luminal markers [20–22], suggesting that prostate epithelial cells are in a continuum of differentiation stages within a hierarchical system. In situ lineage-tracing studies of human prostate tissues show that all the prostate epithelial cell types have a common clonal origin [23] and confirm a basal stem cell phenotype [24]. In the last decade, a solid body of evidence, using both in vitro and in vivo models, confirmed that prostate stem cells have a basal phenotype [25, 26], possess a high proliferative potential [27], can differentiate to luminal cells [28, 29], and can reconstitute prostatic-like acinar structure in vivo [27, 30]. Notably, Leong and colleagues were able to reconstitute functional prostatic structures from a single Lin^{-}Sca-1^{+}/CD133^{+}/CD44^{+}/CD117^{+} cell [31].

Two independent reports however seem to contradict this hypothesis, showing the presence of a luminal progenitor in the mouse prostate. Castration-resistant Nkx3.1-expressing cells (CARNs) are rare luminal cells which (after castration) can self-renew in vivo and reconstitute prostatic ducts in a single cell transplantation experiment [32]. Moreover, lineage-tracing experiments recently showed that the basal and luminal compartments of the mouse prostate have independent stem cells, which regenerate only the compartment of origin during two consecutive rounds of castration/regeneration [33]. Both studies show evidence for an androgen-independent luminal progenitor cell that is responsible for regenerating androgen-dependent luminal cells, in contrast to the basal cell phenotype seen in humans (and in the original mouse experiment from Leong and colleagues [31]). Although of great importance, these results could reflect a difference in the prostate biology between mice and humans, as all the evidence points to a single lineage for basal, luminal, and neuroendocrine cells in the human prostate. In fact, mouse and human prostate show differences at the anatomical and cellular levels: the mouse prostate is composed of four distinct lobes, not recognizable in the human prostate, which is subdivided into zones [34]. Moreover, the mouse prostate epithelium lacks a complete basal layer, with scattered basal cells intercalated by luminal cells in direct contact with the basement membrane [35].

11.2.1 Epithelial Hierarchy in Prostate Cancer and Cancer Stem Cell Hypothesis

As discussed previously, prostate cancer is characterized by an expansion of the luminal cell compartment which has acquired the ability to proliferate [9]. However, several independent reports support the idea that a small subpopulation of aberrant basal cells persist within PCa [36–38]. As normal prostate stem cells reside within the basal layer, the idea emerged that rare undifferentiated basal cells within PCa could contain stem-like cancer cells which are able to self-renew and generate aberrantly differentiated cancer cells. There is now a wealth of evidence that tumors are hierarchically arranged, a pattern shared between leukemias [39, 40], breast [41], brain [42], colon [43], pancreas [44], liver [45], lung [46], and endometrium [47], where cancer stem cells (CSCs) have been successfully selected using the same markers that identify the correspondent normal stem cells. In the last few years, the identification of CSCs has been one of the key research topics in prostate cancer; however, the field is still debating the precise phenotype and origin of prostate CSCs. In fact, depending on the markers, the models and the kind of assay used to test for "stemness," several groups have identified stem-like cells with both a basal and luminal origin.

The evidence that cancer stem cells reside in the luminal compartment comes from studies in transgenic mice, where luminal cells targeted with oncogenic transgenes were able efficiently to generate tumors in mice. For example, the selective deletion of PTEN (frequently mutated in human cancers) and in cells expressing PSA [48], cytokeratin 8 [33], or in CARNs [32], resulted in tumor formation. Moreover, Germann and colleagues identified, in an androgen-dependent human xenograft, a castration-resistant, partially differentiated, luminal cell type, expressing ALDH1A1 and NANOG and the luminal markers NKX3-1 and CK18, and low levels of AR [49]. These cells survived castration in a quiescent state and started proliferating and differentiating after androgen replacement, regenerating the tumor mass [49].

In contrast, several independent studies using various model systems support a basal/undifferentiated cell origin for prostate CSCs. Normal prostate stem cells are long-lived, facilitating the accumulation of genetic and epigenetic mutations over the lifetime of an individual, while there is less opportunity for mutations to accumulate in post mitotic luminal cells. In this scenario, mutated stem cells give rise to aberrantly differentiated luminal cells with the ability to proliferate. It is indeed possible that a continuous activation of the stem cells due to chronic inflammation signals [50] could favor epigenetic and genetic instability, first resulting in proliferative inflammatory atrophy (PIA), followed by PIN, and consecutively PCa.

From a histological point of view, metastases and high grade cancers often include rare cells expressing basal cell markers, such as high-molecular-weight cytokeratins [37]. Moreover, it has been proposed that the castrate-resistant state results from clonal expansion of androgen-independent cells that are present at a frequency of 1 per 10^5–10^6 androgen-responsive cells [51].

Confirmation of a putative basal cancer stem cell came from our laboratory: where cells selected from human PCa biopsies with a $CD44^+/\alpha_2\beta_1integrin^{hi}/CD133^+$ phenotype were able to self-renew in vitro and differentiate to an $AR^+/PAP^+/CK18^+$ luminal phenotype [52]. Findings from other laboratories support this idea: the $CD44^+$ population from tumor xenografts and cell lines (AR^-, $OCT\text{-}4^+$, and $BMI\text{-}1^+$) has enhanced proliferative potential and tumor-initiating ability in vivo compared to $CD44^-$ cells [53]. The side population, determined using Hoechst 33342 dye efflux and selected from primary PCa tissues, exhibits a basal phenotype and has sphere-forming features [54]. Goldstein et al. (2010) reported that basal epithelial cells (but not luminal) from both mouse and human prostate were able to initiate tumors in immunodeficient mice when infected with AKT, ERG, and AR overexpressing vector and recombined with fetal urogenital sinus mesenchyme [55, 56]. In a conditional PTEN knockout mouse model for prostate cancer, an expansion of a basal stem/progenitor cell phenotype was observed after induction of PTEN deletion, with consequent tumor initiation [57]. More recently, it was shown that basal Lin^{-}/$Sca\text{-}1^{high}$/$CD49f^{high}$ cells have the capacity to form tumor-like spheroids in vitro and are tumorigenic in vivo [58]. In this model, the stromal component (cancer-associated fibroblasts) also played a crucial role in modulating the CSCs and stimulating tumor formation. Moreover, $TRA\text{-}1\text{-}60^{+}$/$CD151^{+}$/$CD166^{+}$ cells, isolated from human prostate xenografts, expressed basal cell markers and exhibited stem-like cell characteristics, recapitulating the cellular hierarchy of the original tumor in serial transplantation experiments [59]. The latter phenotype is shared with stem cells selected on the basis of CD133 expression from primary human tissues [52, 60].

An easy argument to resolve this discrepancy between basal and luminal stem cell phenotype could be that different cell types can exhibit CSC properties (e.g. regenerative potential and fate) depending on the environment (stem cell niche), the model used, and the type of experiment conducted. In agreement with this hypothesis is a recent paper from the Blanpain group on mouse breast stem cells, showing two separate lineages for myoepithelial and luminal cells in intact mammary tissue [61]. However, basal/myoepithelial stem cells were able to regenerate both basal and luminal cells, showing a greater potency and plasticity than luminal cells.

These observations are in accordance with a hierarchical model where a proportion of the regeneration potential is maintained in partially differentiated progenitor cells. Under physiological conditions, these progenitor cells (transit amplifying) are able to proliferate and differentiate to generate luminal cells, maintaining the epithelial turnover, while the stem cells remain in a quiescent state and do not need to be activated. The undifferentiated basal stem cells however seem to have more potential and are able to regenerate complete prostatic structures from even a single cell [31]. This hypothesis could explain also the current discrepancy observed in the prostate CSC field. It is possible that the genomic and phenotypic aberrancies accumulate in cancer progenitor cells conferring them with partial stem cell properties, such as self-renewal and regeneration, while the stem cells remain dormant and potentially reactivate only in response to treatment [62].

11.3 Prominin-1 Expression in the Prostate

The pentaspan membrane glycoprotein prominin-1 (CD133) is encoded (in humans) by the gene *PROM1* located on chromosome 4 [63]. Antibodies directed against the glycosylated form of this protein (AC133) have been used to select cells with stem cell proprieties from numerous tissues and tumors (reviewed in [64]). Using this marker, Richardson and colleagues were able to enrich for cells with stem cell characteristics from the $\alpha_2\beta_1$integrin^{hi} fraction of basal prostate epithelial cells derived from primary tissues [27]. These cells possessed a high in vitro proliferative potential and were able to reconstitute prostatic acini in immunocompromised mice. Since this key publication, numerous other groups have used CD133 to select cells with stem cell features from many prostate model systems.

11.3.1 CD133 Expression in Human and Mouse Prostate Tissues

In 2004, Richardson and colleagues showed for the first time that a rare subpopulation of CD133$^+$ cells was present in the basal layer of the human prostate [27]. These cells are randomly scattered throughout the acinus and are either found alone or clustered at budding regions or branching points. More recently, Missol-Kolka and colleagues analyzed CD133 expression in the mouse prostate, finding it widely expressed on the luminal side of the epithelium [65]. This apparent inconsistency poses critical questions on the biology of CD133 in mouse and human prostate and on the use of antibodies specific for different protein epitopes. It is important to remember that the most used antibodies for marking and selecting stem cells from many different human adult tissues are directed against the AC133 or the 293C3/AC141 epitopes of the CD133 molecule [64]. These epitopes are located in a glycosylated portion of CD133 in the second extracellular loop [66] and seem to be particularly dependent on protein folding and glycosylation [67–69]. Other human-specific antibodies against the CD133 polypeptide have been developed, but these show a less restricted expression of CD133 [70, 71], suggesting that the corrected glycosylation and protein folding of CD133 is necessary for a precise marking of the stem cells, while throughout the body, CD133 seems to be expressed in many cell types in addition to the adult stem cells. To clarify this discrepancy, the Corbeil group compared the expression of CD133 in the human prostate using two different antibodies: AC133 and 80B258 (against the polypeptide chain) [65]. Interestingly, they revealed that in the basal layer of the prostate epithelium, only a small subpopulation of cells was marked by the 80B258 antibody and that this population seemed to coincide with the stem cell population stained by the AC133 antibody. However, the 80B258 antibody also showed positivity in a proportion of the luminal cells, typically with an apical membrane staining. These results mirrored the expression in the mouse prostate, clearly stating that CD133 expression in the prostate is indeed not restricted to the rare basal epithelial cells with stem cell features but that its expression is reacquired by terminally differentiated luminal cells but with a different conformation/glycosylation pattern. Understanding precisely how CD133 expression and posttranslational modifications are regulated throughout the prostate epithelial hierarchy (discussed in Sect. 11.4) and, more generally, how the AC133 epitope is regulated/masked are still unanswered questions in this field.

11.3.2 Histological Expression of CD133 in Prostate Cancer Tissues

In 2005, our laboratory published the first identification of stem-like cells in prostate cancer [52]. We reported that $CD44^{+}/\alpha_2\beta_1integrin^{hi}/CD133^{+}$ cells from primary prostate cancers had the ability to self-renew and differentiate in vitro. Although the final confirmation that these cells have tumor-initiating capacity in vivo was not presented at the time, this publication indeed generated a wide interest in the role of CD133 expression in prostate cancer. Several publications showed that CD133 (AC133) positive cells exist within prostate cancer [54, 72, 73]; however, the percentage of positive cells varied considerably between publications. Eaton and colleagues showed that CD133 was expressed (at low frequency, <1%) in half of the primary cancers tested, and its expression was increased in matched metastasis [73]. Miki and colleagues reported that $CD133^{+}$ cells within prostate cancer tissues lacked nuclear AR expression, suggesting an undifferentiated phenotype [72]. In contrast with these reports, Missol-Kolka and colleagues reported no CD133 expression in 18 prostate cancer samples. This was a surprising result as, in this study, they used antibodies against the CD133 polypeptide chain, which has already been shown to be widely expressed in normal tissues [65].

11.4 Regulation of CD133 Expression in the Prostate

The histological expression pattern of CD133 in the normal prostate suggests that the gene is expressed specifically in basal stem cells (AC133 epitope) and, in a different isoform, in terminally differentiated luminal cells, while the vast majority of the basal and intermediate cells remain CD133 negative. This implies a very dynamic and tight control for CD133 expression throughout the prostate epithelial hierarchy (Fig. 11.1).

CD133 expression can be controlled at multiple steps including transcriptional regulation, alternative transcription initiation sites, alternative splicing, and post-translational modifications [74, 75].

In the prostate, transcriptional regulation, together with posttranslational and conformational changes, seems to be essential for the correct expression pattern of CD133. Within basal prostate cells, CD133 mRNA is expressed at high levels only in rare AC133-positive cells [76, 77], while the majority of the cells do not express this gene. CD133 is then reexpressed in luminal cells but in a non-AC133 reactive form. This dynamic transcriptional regulation fits with a model of changes in the activation of transcription factors and chromatin remodeling around the five independent promoters which regulate CD133 expression in a tissue-specific manner, producing transcripts containing an alternative first exon [78].

In prostate, expression is initiated by promoter P1, generating a transcript that includes exon 1A [78]. We have confirmed this result in prostate epithelial cell lines

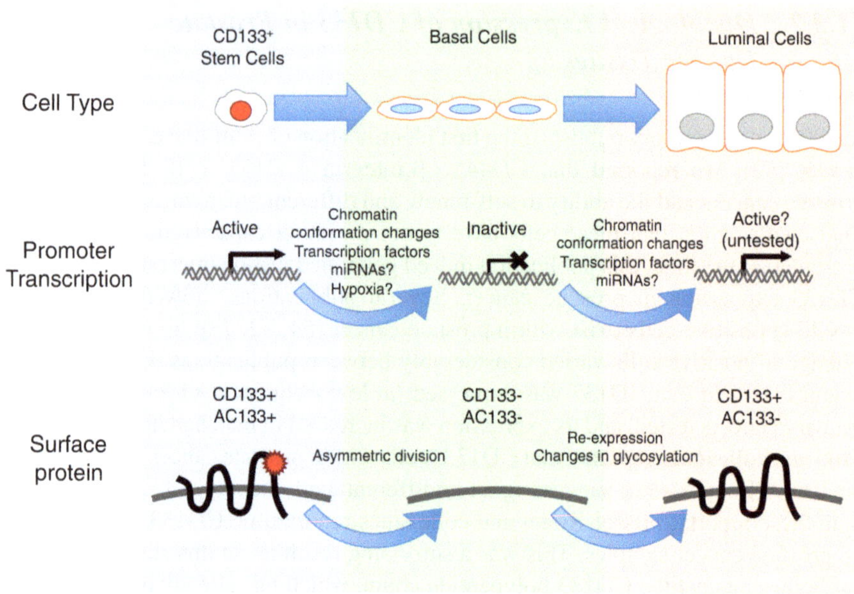

Fig. 11.1 Schematic representation of the proposed tight regulation of CD133 throughout the normal prostate epithelial hierarchy

(Fig. 11.2) by specific amplification of each alternative first exon by RT-PCR. The transcription factors which specifically regulate this promoter in prostate are still uncertain; however, insights can be drawn from studies conducted in other tissues. A strict relationship has been shown between hypoxia, hypoxia-inducible factor (HIF)1α and HIF2α transcription factors, and CD133 expression, although sometimes with opposite effects depending on the tissues studied [66, 79–81]. The overall consensus is however that hypoxic conditions stimulate CD133 expression and promote the expansion of CD133-expressing cells. This is accompanied by the upregulation of many other stem cell features in both prostate [82] and other tissues [80, 83–86]. Iida and colleagues showed that hypoxia induces CD133 expression in lung cancer cells. This induction is mediated by OCT4 and SOX2, both of which are induced by HIF1α and HIF2α, via their direct interaction with the P1 promoter [80]. Other studies reported the involvement of several other pathways and transcription factors in regulating CD133, such as the Ras/ERK pathway [87], the mTOR pathway [81], the TGF-β pathway (through DNA methylation) [88], and AF4 transcription factor [89]. However, CD133 transcriptional activation seems to be extremely tissue-specific, and more detailed studies on prostate and prostate cancer are still required.

Another widely reported mechanism, which regulates CD133 transcription, is the hypermethylation of its promoter. The CpG island present around promoters P1, P2, and P3 of CD133 is frequently hypermethylated in a number of normal tissues and tumors, inhibiting CD133 transcription [90–93]. Although the same mechanism is present in some prostate cell lines [76], this seems to be a result of culture adaptation

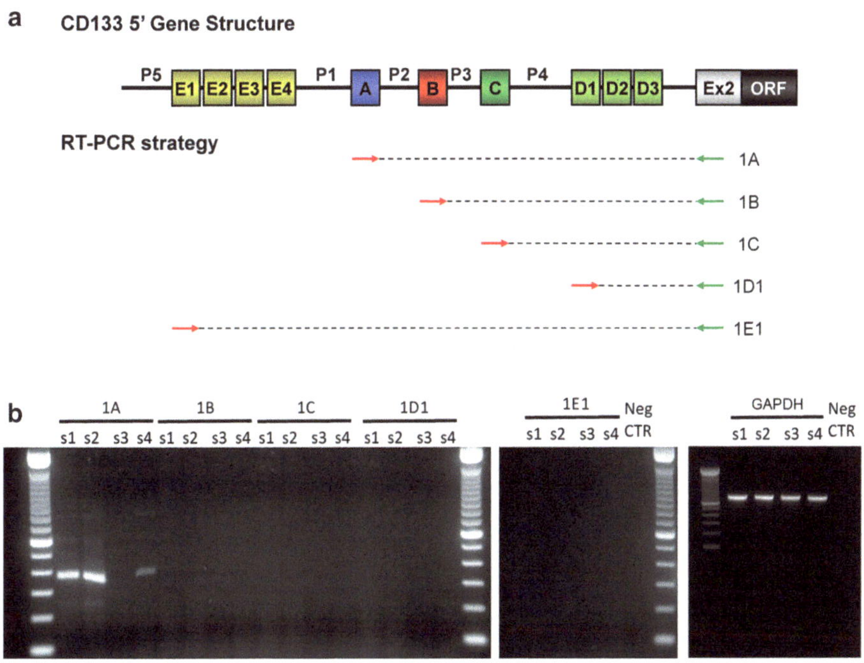

Fig. 11.2 (**a**) CD133 5' gene structure (*top panel*) (A,B,C,D1-3, E1-4 = alternative first exons; P1-P5 = promoters; Ex2 = Exon 2; ORF = open reading frame) and RT-PCR strategy for specific amplification of the alternative first exons (*bottom panel*). (**b**) Amplification of CD133 alternative first exons by RT-PCR in RC-165 N/hTERT and DU145 cell lines treated with 1 µM 5-Aza-2'-deoxycytidine for 96 h. s1 = RC-165 N/hTERT DMSO; s2 = RC-165 N/hTERT 5-Aza-2'-deoxycytidine; s3 = DU145 DMSO; s4 = DU145 5-Aza-2'-deoxycytidine

of these cells. In both prostate primary cultures and prostate tissues, CD133 regulation is independent of DNA methylation, and CD133 is almost always hypomethylated. The results in Fig. 11.3 clearly show that promoter-specific hypermethylation of CD133 to be the result of cellular adaptation to serial passaging (in this case, of a xenografted primary tumor) (Fig. 11.3) and that this process is not restricted to CD133. We have recently demonstrated that a more dynamic regulation involving chromatin condensation and a switch in histone marks seems to play the major role in regulating CD133 transcription in the prostate. The presence of active or inactive chromatin marks correlated perfectly with CD133 expression in prostate cell lines, while treatment with histone deacetylase inhibitors induced CD133 expression both in prostate cell lines and primary epithelial cultures with no involvement of DNA methylation [76]. Interestingly, our results also indicate that the tissue-specific choice of CD133 first exon is not under the control of DNA methylation in cell lines. Treatment of prostate cell lines with the demethylating agent 5-Aza-2'-deoxycytidine induced a marked reexpression of CD133 mRNA and AC133 protein [76]. Figure 11.2 shows that DU145 cells treated with 1 µM 5-Aza-2'-deoxycytidine for 96 h specifically reexpressed a CD133 isoform-containing exon 1A (Fig. 11.2),

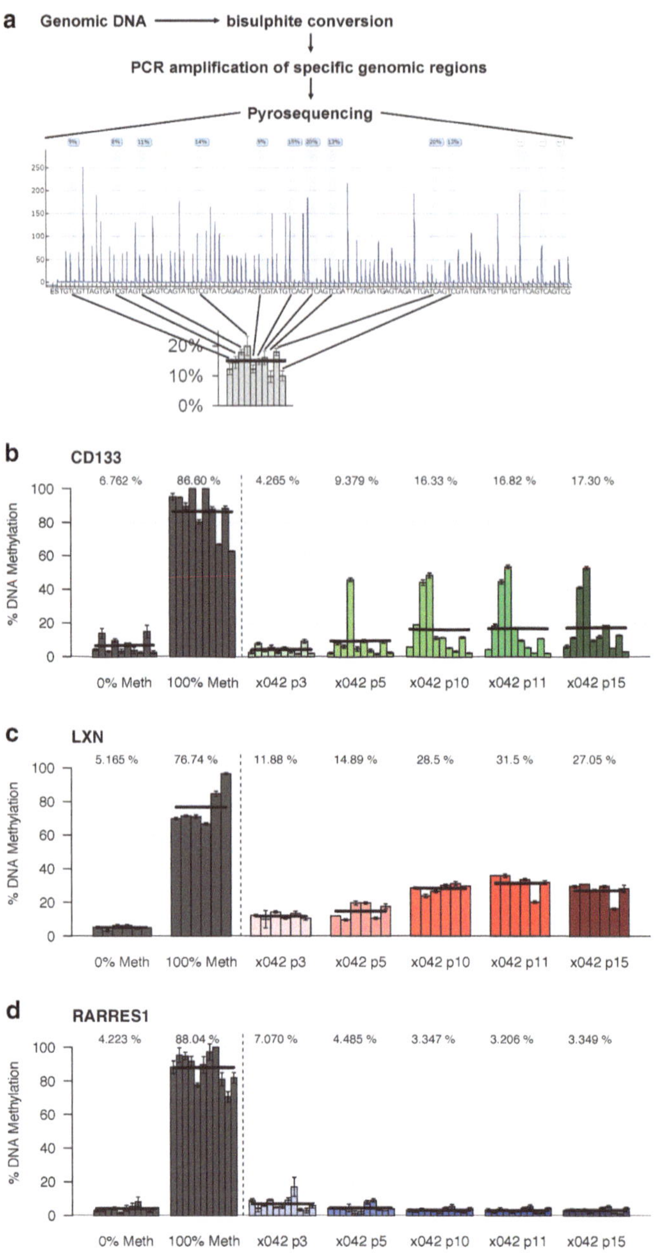

Fig. 11.3 Schematic representation of the pyrosequencing methylation analysis workflow (**a**). In brief: genomic DNA is bisulphite converted and the genomic region of interest is amplified by PCR; the PCR product is then sequenced by pyrosequencing [108] and a pyrogram is generated. DNA methylation percentages from each CpG site in the region are summarized in a bar plot with a horizontal line representing the average. Pyrosequencing methylation analysis of CD133 (**b**), LXN (**c**), and RARRES1 (**d**) performed in prostate primary xenografts throughout several passages spanning 1.5 years of culture (from passage 3 to passage 15) (bars = single CpG sites; $n=3$ technical replicas; mean ± SD; line = average of all the CpG sites, 0% Meth = 0% methylation control, 100% Meth = 100% methylation control)

while the cell line RC-165 N/hTERT (which lack a hypermethylated *CD133* promoter) was used as a control.

The switch between an AC133-positive CD133 isoform in prostate basal stem cells and an AC133-negative isoform in luminal cells is indeed another riddle in prominin-1 biology. What is the difference between these two isoforms and how are they regulated? Yu and colleagues proposed that alternative splicing could regulate this. They showed that CD133-2 isoform, which lacks a small exon of 27 nucleotides, was expressed in hematopoietic stem cells and recognized by the anti-AC133 monoclonal antibodies [94]. On the other hand, Mak and colleagues showed that N-glycosylation processing was necessary for the correct recognition of the AC133 epitope, indicating that changes in glycosylation patterns or conformation alone could lead to a AC133-negative form of CD133 [69].

11.5 CD133 as a Stem Cell Marker in Prostate and Prostate Cancer

As discussed above, our laboratory was able to select cells with stem cell characteristics using the AC133 antibody from both benign and malignant prostate tissues [27, 52]. $\alpha_2\beta_1$integrin^{hi}/CD133$^+$ cells from benign prostatic tissue display a basal phenotype and are quiescent (Ki67$^-$), but have a much higher ability to form colonies and higher proliferative potential than $\alpha_2\beta_1$integrin^{hi}/CD133$^-$ cells. When implanted subcutaneously in immunocompromised mice, CD133$^+$ cells were able to form acini-like structure that recapitulated the entire spectrum of prostate differentiation.

In cancer tissues, CD44$^+$/$\alpha_2\beta_1$integrin^{hi}/CD133$^+$ cells had a very similar phenotype to their benign counterpart, displaying basal cell markers such as cytokeratin 5 and 14 [52]. These cells also showed a high colony forming ability and a proliferative potential much higher than their benign counterpart and were able to differentiate in vitro to luminal-like cancer cells. Indeed, cultures generated from these cells were more invasive in vitro compared to BPH cultures and displayed the frequent prostatic gene fusion TMPRSS2-ERG [36, 60]. Furthermore, $\alpha_2\beta_1$integrin^{hi}/CD133$^+$ prostate CSCs have a distinct gene expression profile relative to both their normal and differentiated counterpart [60], showing differential expression of genes associated with inflammation, cellular adhesion, and metastasis.

Since 2005, CD133 has been used by many other groups as a prostate (and PCa) stem cell marker in various models, sometimes however with inconsistent results. In hTERT immortalized prostate cell lines, AC133-positive cells displayed stem cell characteristics in vitro that mirrored cells from patient tissues [72]. However, another group showed that side population was a much better CSC marker in this model compared to CD133 [95].

Wei and colleagues also reported that CD44$^+$/$\alpha_2\beta_1$integrin^{hi}/CD133$^+$ cells are present in the DU145 cell line and possess cancer stem cell features in vitro and in vivo [96]. This result was then confirmed by Dubrovska, showing also that prostate cancer cell lines grown in sphere-forming conditions increased the number of $CD44^+$/$CD133^+$ cells, in vitro and in vivo tumorigenic potential [97]. Moreover, the Shay group showed that $CD133^+$ cells from DU145 have higher telomerase activity [98]. However, others failed to see the same stem cell characteristic in $CD133^+$ DU145 cells [99]. This inconsistency in studies with the same, long established cell line may reflect heterogeneity and selection of dominant clones as seen in several other cellular models of cancer, rather than genuine changes in $CD133^+$ cell content [100].

Interestingly, Vander Griend and colleagues [101] demonstrated that a small subpopulation of $CD133^+$ cells was present in several cancer cell lines, which possessed a higher clonogenic potential. However, this subpopulation expressed AR, which is in stark contrast to all other reports showing that $CD133^+$ cells from human normal and cancerous tissues and cell lines have a basal phenotype [27, 52, 72, 77, 97, 102]. Moreover, many other groups failed to find a $CD133^+$ subpopulation in these cell lines [76, 99], confirming that CD133 can be highly dysregulated after long-term culture in vitro and the marker should be used with caution in cell lines.

The ultimate evidence for CSCs is reconstitution of a tumor in a recipient animal, which is identical to the parental tumor and that can be serially xenotransplanted indefinitely. Preliminary data from our laboratory showed that CD133+ cells selected form primary PCa xenografts were indeed able to form tumors from as few as 10 cells [36]. $\alpha_2\beta_1$integrin^{hi}/$CD133^+$ cells selected from the BPH-1 cell line were also able to form tumors in mice when recombined with cancer-associated fibroblasts; however, they had a much lower tumor-initiating potential compared to $\alpha_2\beta_1$integrin^{hi}/$CD133^-$ cells [103]. Serial transplantation (not performed in the latter study) is important to distinguish between bona fide stem cells and highly proliferating progenitor cells. Indeed, more work needs to be carried out in order to better assess the potential of CD133 as a bona fide stem cell marker in human prostate cancer; that is, identify whether it is a permanent marker or (more likely) one whose expression is condition and context dependent.

11.6 Conclusions

As discussed above, human prostate epithelial stem cells are quiescent and reside in the basal layer. Many groups were able to separate cells with prostate-regenerating capabilities from this compartment using several different markers. Indeed, basal cells expressing CD133 are quiescent stem cells able to regenerate differentiated luminal cells in vitro and in vivo [27, 28], confirming the validity of CD133 as a prostate stem cell marker.

However, the prostate cancer field is still debating on the nature of CSCs. In human prostate cancer, a solid body of evidence indicates that cells with a basal or partially differentiated (intermediate) phenotype have cancer stem cell characteristics and that their normal counterpart can function as a cell-of-origin for prostate cancer. In this scenario, the use of CD133 as a cancer stem cell marker is still somewhat uncertain: although several reports show that $CD133^+$ cells from various models have cancer stem cell features, other groups were able to isolate prostate CSCs without the use of CD133 but with basal markers, such as CD44, ALDH, and $\alpha_2\beta_1$integrin. This could be partially explained by the tight regulation needed for the correct expression of CD133, which is indeed influenced by niche and environmental conditions (especially after long-term culture in vitro and serial passaging in vivo), resulting in unstable expression of CD133 which is expressed only in the appropriate microenvironment. It is important to remember though that CD133 expression remains at the moment only a stem cell marker and not a stem cell feature, as sometimes reported in the literature. CD133 function is largely unknown, especially in the prostate, where it is expressed as two distinct isoforms in basal and luminal cells. CD133 is localized to plasma membrane protrusions where it interacts with membrane cholesterol [104]. These membrane microdomains could be enriched in components involved in maintaining stem cell properties, and their loss, perhaps through asymmetric cell division, might promote cell differentiation [105]. Interestingly, CD133 has been shown to segregate with the template DNA in asymmetric cell division in lung cancer cells, while differentiation markers were expressed only in the daughter cell [106]. Interestingly, the frequency of asymmetric cell division was enhanced by environmental factors such as cell-cell contact, serum, and hypoxia, all of which seem to play an important role in CD133 regulation. Moreover, a recent report showed that CD133 is a suppressor of differentiation in neuroblastoma [107], giving a first insight into CD133 function and linking this marker functionally to the maintenance of an undifferentiated (stem) phenotype.

It is likely that the definition of CD133 function and regulation in relation to prostate stem cells and CSCs will be of primary importance and should provide novel insights into the nature of the tumor initiation process.

Acknowledgements This work was supported by Yorkshire Cancer Research program grant (DP, EEO, ATC, and NJM), project grant Y256 (DP), and The Freemasons' Grand Charity (DP).

References

1. Kumar VL, Majumder PK (1995) Prostate gland: structure, functions and regulation. Int Urol Nephrol 27:231–243
2. Kyprianou N, Isaacs JT (1988) Activation of programmed cell death in the rat ventral prostate after castration. Endocrinology 122:552–562
3. Sar M, Lubahn DB, French FS, Wilson EM (1990) Immunohistochemical localization of the androgen receptor in rat and human tissues. Endocrinology 127:3180–3186
4. Bonkhoff H, Remberger K (1993) Widespread distribution of nuclear androgen receptors in the basal cell layer of the normal and hyperplastic human prostate. Virchows Archiv 422:35–38
5. Bonkhoff H, Stein U, Remberger K (1995) Endocrine-paracrine cell types in the prostate and prostatic adenocarcinoma are postmitotic cells. Hum Pathol 26:167–170
6. Rumpold H, Heinrich E, Untergasser G, Hermann M, Pfister G, Plas E et al (2002) Neuroendocrine differentiation of human prostatic primary epithelial cells in vitro. Prostate 53:101–108

7. Bostwick DG (1996) Progression of prostatic intraepithelial neoplasia to early invasive adenocarcinoma. Eur Urol 30:145–152
8. Grisanzio C, Signoretti S (2008) p63 in prostate biology and pathology. J Cell Biochem 103:1354–1368
9. De Marzo AM, Meeker AK, Epstein JI, Coffey DS (1998) Prostate stem cell compartments: expression of the cell cycle inhibitor p27Kip1 in normal, hyperplastic, and neoplastic cells. Am J Pathol 153:911–919
10. Gleason DF (1966) Classification of prostatic carcinomas. Cancer Chemother Rep 50:125–128
11. Bostwick DG (1994) Grading prostate cancer. Am J Clin Pathol 102:S38–S56
12. Hammad FT (2008) Radical prostatectomy. Ann N Y Acad Sci 1138:267–277
13. Nagle RB, Ahmann FR, McDaniel KM, Paquin ML, Clark VA, Celniker A (1987) Cytokeratin characterization of human prostatic carcinoma and its derived cell lines. Cancer Res 47:281–286
14. Feldman BJ, Feldman D (2001) The development of androgen-independent prostate cancer. Nat Rev Cancer 1:34–45
15. Yap TA, Zivi A, Omlin A, De Bono JS (2011) The changing therapeutic landscape of castration-resistant prostate cancer. Nat Rev Clin Oncol 8:597–610
16. Oldridge EE, Pellacani D, Collins AT, Maitland NJ (2012) Prostate cancer stem cells: Are they androgen-responsive? Mol Cell Endocrinol 360:14–24
17. Isaacs JT, Coffey DS (1989) Etiology and disease process of benign prostatic hyperplasia. Prostate 2:33–50
18. English HF, Santen RJ, Isaacs JT (1987) Response of glandular versus basal rat ventral prostatic epithelial cells to androgen withdrawal and replacement. Prostate 11:229–242
19. Evans GS, Chandler JA (1987) Cell proliferation studies in the rat prostate: II. The effects of castration and androgen-induced regeneration upon basal and secretory cell proliferation. Prostate 11:339–351
20. Hudson DL, Guy AT, Fry P, O'Hare MJ, Watt FM, Masters JR (2001) Epithelial cell differentiation pathways in the human prostate: identification of intermediate phenotypes by keratin expression. J Histochem Cytochem 49:271–278
21. Verhagen AP, Ramaekers FC, Aalders TW, Schaafsma HE, Debruyne FM, Schalken JA (1992) Colocalization of basal and luminal cell-type cytokeratins in human prostate cancer. Cancer Res 52:6182–6187
22. Verhagen AP, Aalders TW, Ramaekers FC, Debruyne FM, Schalken JA (1988) Differential expression of keratins in the basal and luminal compartments of rat prostatic epithelium during degeneration and regeneration. Prostate 13:25–38
23. Blackwood JK, Williamson SC, Greaves LC, Wilson L, Rigas AC, Sandher R et al (2011) In situ lineage tracking of human prostatic epithelial stem cell fate reveals a common clonal origin for basal and luminal cells. J Pathol 225:181–188
24. Gaisa NT, Graham TA, McDonald SA, Poulsom R, Heidenreich A, Jakse G et al (2011) Clonal architecture of human prostatic epithelium in benign and malignant conditions. J Pathol 225:172–180
25. Collins AT, Habib FK, Maitland NJ, Neal DE (2001) Identification and isolation of human prostate epithelial stem cells based on alpha(2)beta(1)-integrin expression. J Cell Sci 114:3865–3872
26. Burger PE, Gupta R, Xiong X, Ontiveros CS, Salm SN, Moscatelli D et al (2009) High aldehyde dehydrogenase activity: a novel functional marker of murine prostate stem/progenitor cells. Stem Cells 27:2220–2228
27. Richardson GD, Robson CN, Lang SH, Neal DE, Maitland NJ, Collins AT (2004) CD133, a novel marker for human prostatic epithelial stem cells. J Cell Sci 117:3539–3545
28. Frame FM, Hager S, Pellacani D, Stower MJ, Walker HF, Burns JE et al (2010) Development and limitations of lentivirus vectors as tools for tracking differentiation in prostate epithelial cells. Exp Cell Res 316:3161–3171

29. Schmelz M, Moll R, Hesse U, Prasad AR, Gandolfi JA, Hasan SR et al (2005) Identification of a stem cell candidate in the normal human prostate gland. Eur J Cell Biol 84:341–354
30. Burger PE, Xiong X, Coetzee S, Salm SN, Moscatelli D, Goto K et al (2005) Sca-1 expression identifies stem cells in the proximal region of prostatic ducts with high capacity to reconstitute prostatic tissue. Proc Natl Acad Sci USA 102:7180–7185
31. Leong KG, Wang B-E, Johnson L, Gao W-Q (2008) Generation of a prostate from a single adult stem cell. Nature 456:804–808
32. Wang X, Kruithof-de Julio M, Economides KD, Walker D, Yu H, Halili MV et al (2009) A luminal epithelial stem cell that is a cell of origin for prostate cancer. Nature 461:495–500
33. Choi N, Zhang B, Zhang L, Ittmann M, Xin L (2012) Adult murine prostate basal and luminal cells are self-sustained lineages that can both serve as targets for prostate cancer initiation. Cancer Cell 21:253–265
34. Shen MM, Wang X, Economides KD, Walker D, Abate-Shen C (2008) Progenitor cells for the prostate epithelium: roles in development, regeneration, and cancer. Cold Spring Harb Symp Quant Biol 73:529–538
35. Taylor RA, Toivanen R, Risbridger GP (2010) Stem cells in prostate cancer: treating the root of the problem. Endocr Relat Cancer 17:R273–R285
36. Maitland NJ, Frame FM, Polson ES, Lewis JL, Collins AT (2011) Prostate cancer stem cells: do they have a basal or luminal phenotype? Horm Cancer 2:47–61
37. Googe PB, McGinley KM, Fitzgibbon JF (1997) Anticytokeratin antibody 34 beta E12 staining in prostate carcinoma. Am J Clin Pathol 107:219–223
38. van Leenders GJ, Aalders TW, Hulsbergen-van de Kaa CA, Ruiter DJ, Schalken JA (2001) Expression of basal cell keratins in human prostate cancer metastases and cell lines. J Pathol 195:563–570
39. Bonnet D, Dick JE (1997) Human acute myeloid leukemia is organized as a hierarchy that originates from a primitive hematopoietic cell. Nat Med 3:730–737
40. Lapidot T, Sirard C, Vormoor J, Murdoch B, Hoang T, Caceres-Cortes J et al (1994) A cell initiating human acute myeloid leukaemia after transplantation into SCID mice. Nature 367:645–648
41. Al-Hajj M, Wicha MS, Benito-Hernandez A, Morrison SJ, Clarke MF (2003) Prospective identification of tumorigenic breast cancer cells. Proc Natl Acad Sci USA 100:3983–3988
42. Singh SK, Hawkins C, Clarke ID, Squire JA, Bayani J, Hide T et al (2004) Identification of human brain tumour initiating cells. Nature 432:396–401
43. Ricci-Vitiani L, Lombardi DG, Pilozzi E, Biffoni M, Todaro M, Peschle C et al (2007) Identification and expansion of human colon-cancer-initiating cells. Nature 445:111–115
44. Li C, Heidt DG, Dalerba P, Burant CF, Zhang L, Adsay V et al (2007) Identification of pancreatic cancer stem cells. Cancer Res 67:1030–1037
45. Ma S, Chan K-W, Hu L, Lee TK-W, Wo JY-H, Ng IO-L et al (2007) Identification and characterization of tumorigenic liver cancer stem/progenitor cells. Gastroenterology 132:2542–2556
46. Eramo A, Lotti F, Sette G, Pilozzi E, Biffoni M, Di Virgilio A et al (2008) Identification and expansion of the tumorigenic lung cancer stem cell population. Cell Death Differ 15:504–514
47. Rutella S, Bonanno G, Procoli A, Mariotti A, Corallo M, Prisco MG et al (2009) Cells with characteristics of cancer stem/progenitor cells express the CD133 antigen in human endometrial tumors. Clin Cancer Res 15:4299–4311
48. Ma X, Ziel-van der Made AC, Autar B, van der Korput HA, Vermeij M, van Duijn P et al (2005) Targeted biallelic inactivation of Pten in the mouse prostate leads to prostate cancer accompanied by increased epithelial cell proliferation but not by reduced apoptosis. Cancer Res 65:5730–5739
49. Germann M, Wetterwald A, Guzmán-Ramírez N, van der Pluijm G, Culig Z, Cecchini MG et al (2012) Stem-like cells with luminal progenitor phenotype survive castration in human prostate cancer. Stem Cells 30:1076–1086

50. Maitland NJ, Collins AT (2008) Inflammation as the primary aetiological agent of human prostate cancer: a stem cell connection? J Cell Biochem 105:931–939
51. Craft N, Chhor C, Tran C, Belldegrun A, DeKernion J, Witte ON et al (1999) Evidence for clonal outgrowth of androgen-independent prostate cancer cells from androgen-dependent tumors through a two-step process. Cancer Res 59:5030–5036
52. Collins AT, Berry PA, Hyde C, Stower MJ, Maitland NJ (2005) Prospective identification of tumorigenic prostate cancer stem cells. Cancer Res 65:10946–10951
53. Patrawala L, Calhoun T, Schneider-Broussard R, Li H, Bhatia B, Tang S et al (2006) Highly purified CD44+ prostate cancer cells from xenograft human tumors are enriched in tumorigenic and metastatic progenitor cells. Oncogene 25:1696–1708
54. Brown MD, Gilmore PE, Hart CA, Samuel JD, Ramani VAC, George NJ et al (2007) Characterization of benign and malignant prostate epithelial Hoechst 33342 side populations. Prostate 67:1384–1396
55. Goldstein AS, Huang J, Guo C, Garraway IP, Witte ON (2010) Identification of a cell of origin for human prostate cancer. Science 329:568–571
56. Lawson DA, Zong Y, Memarzadeh S, Xin L, Huang J, Witte ON (2010) Basal epithelial stem cells are efficient targets for prostate cancer initiation. Proc Natl Acad Sci USA 107:2610–2615
57. Wang S, Garcia AJ, Wu M, Lawson DA, Witte ON, Wu H (2006) Pten deletion leads to the expansion of a prostatic stem/progenitor cell subpopulation and tumor initiation. Proc Natl Acad Sci USA 103:1480–1485
58. Liao C-P, Adisetiyo H, Liang M, Roy-Burman P (2010) Cancer-associated fibroblasts enhance the gland-forming capability of prostate cancer stem cells. Cancer Res 70:7294–7303
59. Rajasekhar VK, Studer L, Gerald W, Socci ND, Scher HI (2011) Tumour-initiating stem-like cells in human prostate cancer exhibit increased NF-κB signalling. Nat Commun 2:162
60. Birnie R, Bryce SD, Roome C, Dussupt V, Droop A, Lang SH et al (2008) Gene expression profiling of human prostate cancer stem cells reveals a pro-inflammatory phenotype and the importance of extracellular matrix interactions. Genome Biol 9:R83
61. Van Keymeulen A, Rocha AS, Ousset M, Beck B, Bouvencourt G, Rock J et al (2011) Distinct stem cells contribute to mammary gland development and maintenance. Nature 479:189–193
62. Lang SH, Frame FM, Collins AT (2009) Prostate cancer stem cells. J Pathol 217:299–306
63. Jászai J, Fargeas CA, Florek M, Huttner WB, Corbeil D (2007) Focus on molecules: prominin-1 (CD133). Exp Eye Res 85:585–586
64. Mizrak D, Brittan M, Alison MR (2008) CD133: molecule of the moment. J Pathol 214:3–9
65. Missol-Kolka E, Karbanová J, Janich P, Haase M, Fargeas CA, Huttner WB et al (2011) Prominin-1 (CD133) is not restricted to stem cells located in the basal compartment of murine and human prostate. Prostate 71:254–267
66. Campos B, Herold-Mende CC (2011) Insight into the complex regulation of CD133 in glioma. Int J Cancer 128:501–510
67. Kemper K, Sprick MR, de Bree M, Scopelliti A, Vermeulen L, Hoek M et al (2010) The AC133 epitope, but not the CD133 protein, is lost upon cancer stem cell differentiation. Cancer Res 70:719–729
68. Miraglia S, Godfrey W, Yin AH, Atkins K, Warnke R, Holden JT et al (1997) A novel five-transmembrane hematopoietic stem cell antigen: isolation, characterization, and molecular cloning. Blood 90:5013–5021
69. Mak AB, Blakely KM, Williams RA, Penttila P-A, Shukalyuk AI, Osman KT et al (2011) CD133 protein N-glycosylation processing contributes to cell surface recognition of the primitive cell marker AC133 epitope. J Biol Chem 286:41046–41056
70. Karbanová J, Missol-Kolka E, Fonseca A-V, Lorra C, Janich P, Hollerová H et al (2008) The stem cell marker CD133 (Prominin-1) is expressed in various human glandular epithelia. J Histochem Cytochem 56:977–993
71. Bidlingmaier S, Zhu X, Liu B (2008) The utility and limitations of glycosylated human CD133 epitopes in defining cancer stem cells. J Mol Med 86:1025–1032

72. Miki J, Furusato B, Li H, Gu Y, Takahashi H, Egawa S et al (2007) Identification of putative stem cell markers, CD133 and CXCR4, in hTERT-immortalized primary nonmalignant and malignant tumor-derived human prostate epithelial cell lines and in prostate cancer specimens. Cancer Res 67:3153–3161
73. Eaton CL, Colombel M, van der Pluijm G, Cecchini MG, Wetterwald A, Lippitt J et al (2010) Evaluation of the frequency of putative prostate cancer stem cells in primary and metastatic prostate cancer. Prostate 70:875–882
74. Shmelkov SV, St Clair R, Lyden D, Rafii S (2005) AC133/CD133/Prominin-1. Int J Biochem Cell Biol 37:715–719
75. Fargeas CA, Huttner WB, Corbeil D (2007) Nomenclature of prominin-1 (CD133) splice variants – an update. Tissue Antig 69:602–606
76. Pellacani D, Packer RJ, Frame FM, Oldridge EE, Berry PA, Labarthe M-C et al (2011) Regulation of the stem cell marker CD133 is independent of promoter hypermethylation in human epithelial differentiation and cancer. Mol Cancer 10:94
77. Shepherd CJ, Rizzo S, Ledaki I, Davies M, Brewer D, Attard G et al (2008) Expression profiling of CD133+ and CD133− epithelial cells from human prostate. Prostate 68:1007–1024
78. Shmelkov SV, Jun L, St Clair R, McGarrigle D, Derderian CA, Usenko JK et al (2004) Alternative promoters regulate transcription of the gene that encodes stem cell surface protein AC133. Blood 103:2055–2061
79. Soeda A, Park M, Lee D, Mintz A, Androutsellis-Theotokis A, McKay RD et al (2009) Hypoxia promotes expansion of the CD133-positive glioma stem cells through activation of HIF-1alpha. Oncogene 28:3949–3959
80. Iida H, Suzuki M, Goitsuka R, Ueno H (2012) Hypoxia induces CD133 expression in human lung cancer cells by up-regulation of OCT3/4 and SOX2. Int J Oncol 40:71–79
81. Matsumoto K, Arao T, Tanaka K, Kaneda H, Kudo K, Fujita Y et al (2009) mTOR signal and hypoxia-inducible factor-1 alpha regulate CD133 expression in cancer cells. Cancer Res 69:7160–7164
82. Ma Y, Liang D, Liu J, Axcrona K, Kvalheim G, Stokke T et al (2011) Prostate cancer cell lines under hypoxia exhibit greater stem-like properties. PLoS One 6:e29170
83. Yeung TM, Gandhi SC, Bodmer WF (2011) Hypoxia and lineage specification of cell line-derived colorectal cancer stem cells. Proc Natl Acad Sci USA 108:4382–4387
84. Mathieu J, Zhang Z, Zhou W, Wang AJ, Heddleston JM, Pinna CMA et al (2011) HIF Induces Human Embryonic Stem Cell Markers in Cancer Cells. Cancer Res 71:4640–4652
85. Bar EE (2011) Glioblastoma, cancer stem cells and hypoxia. Brain Pathol 21:119–129
86. Bar EE, Lin A, Mahairaki V, Matsui W, Eberhart CG (2010) Hypoxia increases the expression of stem-cell markers and promotes clonogenicity in glioblastoma neurospheres. Am J Pathol 177:1491–1502
87. Tabu K, Kimura T, Sasai K, Wang L, Bizen N, Nishihara H et al (2010) Analysis of an alternative human CD133 promoter reveals the implication of Ras/ERK pathway in tumor stem-like hallmarks. Mol Cancer 9:39
88. You H, Ding W, Rountree CB (2010) Epigenetic regulation of cancer stem cell marker CD133 by transforming growth factor-beta. Hepatology 51:1635–1644
89. Mak AB, Nixon AML, Moffat J (2012) The mixed lineage leukemia (MLL) fusion-associated gene AF4 promotes CD133 transcription. Cancer Res 72:1929–1934
90. Baba T, Convery PA, Matsumura N, Whitaker RS, Kondoh E, Perry T et al (2009) Epigenetic regulation of CD133 and tumorigenicity of CD133+ ovarian cancer cells. Oncogene 28:209–218
91. Pleshkan VV, Vinogradova TV, Sverdlov ED (2008) Methylation of the prominin 1 TATA-less main promoters and tissue specificity of their transcript content. Biochim Biophys Acta 1779:599–605
92. Tabu K, Sasai K, Kimura T, Wang L, Aoyanagi E, Kohsaka S et al (2008) Promoter hypomethylation regulates CD133 expression in human gliomas. Cell Res 18:1037–1046

93. Yi JM, Tsai H-C, Glöckner SC, Lin S, Ohm JE, Easwaran H et al (2008) Abnormal DNA methylation of CD133 in colorectal and glioblastoma tumors. Cancer Res 68:8094–8103
94. Yu Y, Flint A, Dvorin EL, Bischoff J (2002) AC133-2, a novel isoform of human AC133 stem cell antigen. J Biol Chem 277:20711–20716
95. Zhou J, Wang H, Cannon V, Wolcott KM, Song H, Yates C (2011) Side population rather than CD133(+) cells distinguishes enriched tumorigenicity in hTERT-immortalized primary prostate cancer cells. Mol Cancer 10:112
96. Wei C, Guomin W, Yujun L, Ruizhe Q (2007) Cancer stem-like cells in human prostate carcinoma cells DU145: the seeds of the cell line? Cancer Biol Ther 6:763–768
97. Dubrovska A, Kim S, Salamone RJ, Walker JR, Maira S-M, García-Echeverría C et al (2009) The role of PTEN/Akt/PI3K signaling in the maintenance and viability of prostate cancer stem-like cell populations. Proc Natl Acad Sci USA 106:268–273
98. Marian CO, Wright WE, Shay JW (2010) The effects of telomerase inhibition on prostate tumor-initiating cells. Int J Cancer 127:321–331
99. Pfeiffer MJ, Schalken JA (2010) Stem cell characteristics in prostate cancer cell lines. Eur Urol 57:246–254
100. Chunthapong J, Seftor EA, Khalkhali-Ellis Z, Seftor REB, Amir S, Lubaroff DM et al (2004) Dual roles of E-cadherin in prostate cancer invasion. J Cell Biochem 91:649–661
101. Vander Griend DJ, Karthaus WL, Dalrymple S, Meeker A, De Marzo AM, Isaacs JT (2008) The role of CD133 in normal human prostate stem cells and malignant cancer-initiating cells. Cancer Res 68:9703–9711
102. Lang SH, Anderson E, Fordham R, Collins AT (2010) Modeling the prostate stem cell niche: an evaluation of stem cell survival and expansion in vitro. Stem Cells Dev 19:537–546
103. Taylor RA, Toivanen R, Frydenberg M, Pedersen J, Harewood L, Australian Prostate Cancer Bioresource et al (2012) Human epithelial basal cells are cells of origin of prostate cancer, independent of CD133 status. Stem Cells 30:1087–1096
104. Corbeil D, Röper K, Fargeas CA, Joester A, Huttner WB (2001) Prominin: a story of cholesterol, plasma membrane protrusions and human pathology. Traffic 2:82–91
105. Bauer N, Fonseca A-V, Florek M, Freund D, Jászai J, Bornhäuser M et al (2008) New insights into the cell biology of hematopoietic progenitors by studying prominin-1 (CD133). Cells Tissues Organs 188:127–138
106. Pine SR, Ryan BM, Varticovski L, Robles AI, Harris CC (2010) Microenvironmental modulation of asymmetric cell division in human lung cancer cells. Proc Natl Acad Sci USA 107:2195–2200
107. Takenobu H, Shimozato O, Nakamura T, Ochiai H, Yamaguchi Y, Ohira M et al (2010) CD133 suppresses neuroblastoma cell differentiation via signal pathway modification. Oncogene 30:95–105
108. Tost J, Gut IG (2007) DNA methylation analysis by pyrosequencing. Nat Protoc 2:2265–2275

Chapter 12
Prominin-1 (CD133) Reveals New Faces of Pancreatic Progenitor Cells and Cancer Stem Cells: Current Knowledge and Therapeutic Perspectives

Yuichi Hori

Abstract Islet transplantation-based therapies were proven successful for type 1 diabetes mellitus, but an extreme shortage of pancreatic islets has motivated recent efforts to develop renewable sources of islet-replacement tissue. Pancreatic progenitor cells hold a promising potential, yet attempts at their prospective isolation are scarce due to the lack of specific marker. We found that prominin-1 (often referred to as CD133 in humans) is expressed by the undifferentiated epithelial cells in the mouse embryonic pancreas. Putative pancreatic epithelial stem and progenitor cells were prospectively enriched in prominin-1^+ cell population by cell sorting and characterized. CD133 is also a cell surface marker of human pancreatic cancer stem cells (CSC), which are resistant to conventional treatments such as chemotherapy and radiotherapy. Therefore, a considerable interest in the specific targeting and eradication of CSC is emerging for the cancer therapy, and CD133 may represent a good molecular target. In this chapter, I will summarize our current knowledge about prominin-1/CD133 in mouse and human pancreas.

Keywords CD133 • Cancer stem cell • Diabetes • Pancreas • Prominin-1 • Tissue-specific stem cell

Y. Hori, M.D., Ph.D. (✉)
Division of Medical Chemistry, Department of Biophysics, Kobe University Graduate School of Health Science, 7-10-2 Tomogaoka, Suma-ku, Kobe 654-0142, Japan
e-mail: horiy@people.kobe-u.ac.jp

D. Corbeil (ed.), *Prominin-1 (CD133): New Insights on Stem & Cancer Stem Cell Biology*, Advances in Experimental Medicine and Biology 777, DOI 10.1007/978-1-4614-5894-4_12,

12.1 Introduction

The pancreas is a glandular organ with dual role in the digestive and endocrine (hormonal) systems. Because of its location and complexity within the body, it is difficult to diagnose the numerous disorders this crucial organ is subjected to, including acute and chronic pancreatitis, diabetes, and cancers. Physiologically, the pancreas secretes, as an exocrine gland, digestive enzymes into a network of ducts that finally join the main pancreatic duct. These enzymes pass through the pancreatic ducts and reach the common bile duct in an inactive form. When they enter into the duodenum via the ampulla of Vater and mix with bile, they are activated. These enzymes degrade proteins, lipids, carbohydrates, and nucleic acids by the process of luminal digestion. The exocrine tissue also releases bicarbonate ions that neutralize the acidic chyme as it enters the duodenum from the stomach. As an endocrine gland, which consists of the islets of Langerhans, pancreas secretes into the bloodstream polypeptide hormones such as insulin and glucagon that play an essential role in the carbohydrate metabolism. For instance, insulin promotes the uptake of glucose by most cells and, hence, lowers plasma glucose concentration. Glucagon has metabolic effects that oppose the action of insulin. These hormones have also other effects on energy metabolism, growth, and development. The insulin is generated from the proinsulin precursor molecule by the action of proteolytic enzymes, which remove the central portion of the molecule (i.e., C-peptide). The endocrine tissue secretes also somatostatin, which has a wide variety of effects on the gastrointestinal function and may also inhibit insulin and glucagon secretion.

From a therapeutic point of view, the identification of specific cell surface antigen(s) used either for the isolation or the targeting of cells endowed with stem cell properties has recently brought new perspectives in regenerative medicine particularly regarding type 1 diabetes mellitus as well as cancer treatment. The isolation of pancreatic tissue-specific stem and progenitor cells that give rise upon differentiation to insulin-producing cells (IPC) and the identification of cancer stem cells (CSC) harboring prominin-1 as a potential molecular target are good examples.

Prominin-1 (alias PROML1, AC133 antigen, or CD133 in humans) is now worldwide recognized as a marker of tissue-specific stem cells or CSC. For further details of the molecular and cellular features of this pentaspan membrane glycoprotein, I invite the readers to look at the first chapter of this book.

In this chapter, I will describe in the first part the enrichment of pancreatic stem and progenitor cells from mice by means of flow cytometry fluorescence-activated cell sorting using prominin-1 as a cell surface antigen and, in the second part, the identification of human CD133^{+} pancreatic CSC, which constitute potential targets in cancer therapy.

12.2 Pancreatic Stem and Progenitor Cells for Regenerative Medicine

12.2.1 Stem Cells for Replacement of Pancreatic β-cells

Diabetes mellitus is a metabolic disorder characterized by chronic hyperglycemia. One of its forms is characterized in terms of etiology and pathogenesis. Type 1 diabetes is the result of the autoimmune destruction of pancreatic endocrine β-cells. This category of diabetes is fatal unless treated with insulin. This means that multiple injections daily or insulin delivery through an insulin pump is necessary for survival. Moreover, patients must test their blood glucose levels several times per day. Nowadays, it is possible to transplant pancreas or islet cells with the result that the patients can live without insulin injections. Advances in the cell-replacement strategy for type 1 diabetes mellitus and the shortage of transplantable tissues have led the community to focus on renewable sources of IPC [1]. Although embryonic stem (ES) cells or induced pluripotent stem (iPS) cells hold a promising potential as a source of IPC, IPC clusters derived from them still have a high degree of cellular heterogeneity, tumor-forming potential, and low insulin levels compared with pancreatic islets [2–5].

Recent studies in the field of regenerative medicine have focused on the isolation of tissue-specific stem and progenitor cells as an alternative source of IPC. Unfortunately, only a few attempts have been made at their prospective isolation from the pancreas, due to the lack of suitable and specific marker and the failure to develop appropriate cell culture strategy to determine their capacity for self-renewal and multilineage differentiation potential. Nevertheless, two interesting studies have reported putative adult pancreatic stem and progenitor cells derived from mice that clonally expanded while expressing a low level of insulin and other pancreatic markers [6, 7]. Although these pioneering works have suggested the existence of stem cells in adult animals, the ability of isolated cells to self-renew and to differentiate into functional islets remains to be fully evaluated. On the other hand, Dor and colleagues have reported that after birth, new β-cells could be generated by the replication of existing ones rather than by putative pancreatic stem cells [8], thus challenging the existence of adult pancreatic stem and progenitor cells.

In contrast, the existence of pancreatic stem and progenitor cells in the developing pancreas is not controversial. The pancreas develops from the posterior foregut, emerging as buds from the ventral and dorsal area of the gut tube. Although it is thought that pancreatic and duodenal homeobox 1 (*Pdx1*)-expressing epithelial progenitor cells give rise to endocrine and exocrine cells and duct [9], clear evidence that isolated clonogenic pancreatic cells possess all stem cell characteristics remains to be confirmed. Transcription factor Pdx1 (also known as insulin-promoter factor 1) is a master regulator of pancreas development. To get more insights into these issues, we have recently investigated the possibility to use prominin-1 as a prospective marker for the isolation and characterization of potential pancreatic stem cells.

12.2.2 Prominin-1 Labels Putative Pancreatic Stem and Progenitor Cells Within the Pancreas

Prominin-1 has been recognized as a cell surface antigen of hematopoietic [10] and neural stem cells [11] as well as ES cell-derived progenitors [12], but its specific ligand and function are still a matter of speculation (see Chap. 1). Corbeil and colleagues have previously shown that human prominin-1 (hereafter named CD133 when referring to the human protein) is expressed at the apical plasma membrane of the embryonic gut tube and the neural tube just like previously reported for its murine ortholog [13]. Moreover, an accumulative set of data supports that the general expression of prominin-1/CD133 is far beyond stem cells originating from hematopoietic and neural systems but extends to other tissues/organs (e.g., see Chaps. 8 [kidney], 10 [colon], 11 [prostate]). Furthermore, the cell surface localization of prominin-1/CD133 makes this antigen a promising candidate for the prospective identification and isolation of pancreatic stem and progenitor cells.

In embryonic pancreas, the majority of the epithelium consists of undifferentiated stem and progenitor cells, *Pdx1*-expressing epithelial progenitor cells, which are surrounded by mesenchyme. Together with my colleagues, we could show by immunohistochemistry that murine prominin-1 is co-expressed specifically on the *Pdx1*-expressing epithelial cells, while platelet-derived growth factor receptor-β (PDGFRβ, CD140b) is expressed on the surrounding mesenchymal cells within the developing pancreas (Fig. 12.1a–d) [14].

Using a combination of prominin-1 (as a positive marker) and PDGFRβ (as a negative marker), we demonstrated by flow cytometry the existence of four distinct subpopulations of cells (here named a–d) derived from the embryonic pancreas [14]. Their phenotypes were fractionated as follows: (a) prominin-1^{high}PDGFRβ^{-}, (b) prominin-1^{dim}PDGFRβ^{-}, (c) prominin-1^{neg}PDGFRβ^{-}, and (d) PDGFRβ^{+} (see Fig. 12.1e; pink, violet, yellow, and blue, respectively). The gene expression analyses by means of reverse transcription polymerase chain reaction (RT-PCR) and microarray of the sorted cells demonstrated that putative markers normally associated with pancreatic stem and progenitor cells such as *Foxa2, HNF4a, HNF6, Pdx1, Hlbx9, Ptfa1,* and *neurogenin3* are enriched in prominin-1^{high}PDGFRβ^{-} fraction (i.e., subpopulation a), whereas markers of terminally differentiated cells including *insulin1, preproinsulin1* (β-cell marker), *glucagon, preproglucagon* (α-cell marker) are enriched in prominin-1^{neg}PDGFRβ^{-} fraction (subpopulation c) [14]. In the same line, Oshima and colleagues performed prospective isolation of pancreatic ductal progenitor cells by flow cytometry using prominin-1 [15]. Likewise, Sugiyama and colleagues demonstrated that prominin-1/CD133 labeled fetal mouse and human islet progenitor cells [16], and Koblas and colleagues reported that CD133 labeled human insulin-producing progenitor cells [17]. Collectively, these independent investigations suggest that prominin-1 as a cell surface antigen could be an ideal marker for the prospective isolation of stem and progenitor cells in developing pancreas and that its expression is decreased through differentiation into mature pancreatic cells notably islet cells. However, since its expression seems to persist at the

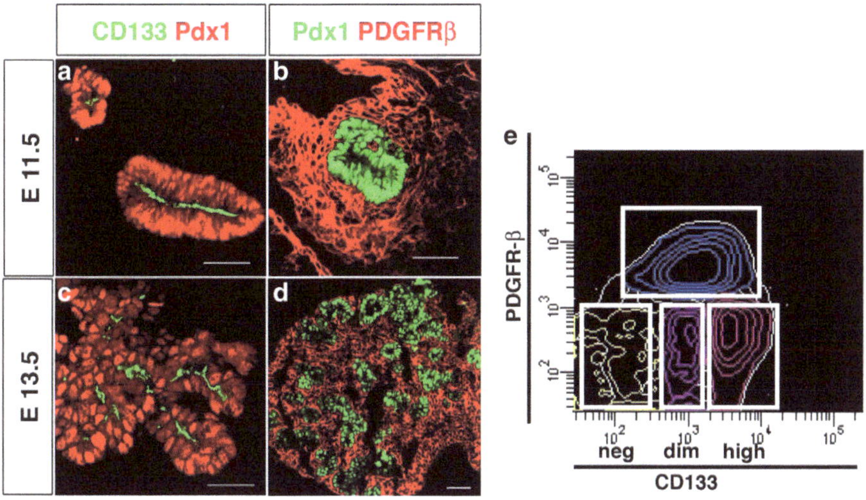

Fig. 12.1 Prominin-1 is expressed on the pancreatic epithelium of murine embryo. (**a**–**d**) Prominin-1 is detected at the apical membrane of the *Pdx1*-expressing pancreatic epithelial cells on embryonic (**e**) day 11.5 and 13.5, when the majority of the epithelium consist of undifferentiated progenitor cells (**a** and **c**), whereas PDGFRβ is expressed by surrounding pancreatic mesenchymal cells (**b** and **d**). Both proteins were observed using phycoerythrin-conjugated anti-prominin-1 mAb and biotin-conjugated anti-PDGFRβ mAb (eBioscience), respectively. Scale bars, 50 μm. (**e**) Flow-cytometric analysis of cells derived from embryonic pancreas using mAbs against prominin-1 and PDGFRβ. Four distinct gated cell subpopulations were observed, and the corresponding percentage is indicated as following phenotypes: prominin-1^{high}PDGFRβ^{-} (pink, 15.3±2.8%), prominin-1^{dim}PDGFRβ^{-} (violet, 11.4±2.2%), prominin-1^{neg}PDGFRβ^{-} (yellow, 6.5±1.2%), and PDGFRβ^{+} (blue, 66.8±9.8%). These panels are derived from our work published in stem cells in 2008 [14]

apical membrane of the peripheral exocrine acini in neonatal and adult pancreas, the use of prominin-1 as a marker of pancreatic stem cells might be more restricted to the embryonic stage [14, 15].

12.2.3 Differentiation Potential of Prominin-1^{high} Pancreatic Stem and Progenitor Cells In Vitro

To determine whether a cell of interest harbors stem and progenitor cell properties, it is imperative to show that it contains a multilineage differentiation potential. In pancreas, for instance, we would expect that the putative stem cells could give rise to endocrine and exocrine cells and duct. Although the culture strategy (including culture conditions and growth factors) for IPC has been intensively investigated especially in regenerative medicine for diabetes mellitus, the experimental conditions for other pancreatic cells have not been established yet.

Together with my colleagues, we demonstrated that isolated prominin-1^{high} cells plated on stromal cells during 7 days were generating IPC clusters, but no other pancreatic cell types [14]. Indeed, other markers characteristic of endocrine tissue (e.g., *glucagon*, *somatostatin*, and *pancreatic polypeptide*) were detected as mRNA using a sensitive RT-PCR analysis, but contrary to insulin, not as protein by immunohistochemistry, suggesting that both (i) their expression levels are very low, and (ii) insulin-producing lineage is a default pathway under this culture condition [14].

12.2.4 Prominin-1^{high} Pancreatic Stem and Progenitor Cells Have the Ability to Differentiate into Pancreatic Tissues in Mice

To evaluate the potential of prominin-1^{+} cells to differentiate and produce in vivo mature pancreatic cells, we have performed a transplantation assay. It is known that the microenvironment (often termed stem cell niche), which is composed of stromal cells and extracellular components, is important to maintain stem and progenitor cell properties. In this context, we engrafted prominin-1^{high} cells isolated from enhanced green fluorescent protein (eGFP) transgenic mice into the *nude* mice to trace them as well as their progenies [14]. One week after engraftment of prominin-1^{high} cells, but not prominin-1^{neg} cells, tubular structures expressing Pdx1 in the nucleus and prominin-1 at the apical membrane were identified in the graft, which was similar to that of the embryonic pancreatic epithelium [14]. In addition, insulin and glucagon were detected within GFP^{+} cell clusters (Fig. 12.2a). Specific markers of exocrine ($amylase^{+}$ or carboxypeptidase A^{+}) or ductal (Dolichos biflorus agglutinin^{+}) cells were also observed (Fig. 12.2b–d). Notably, 1 month after engraftment, the intracellular C-peptide content in IPC increased and showed levels comparable to that in pancreatic β-cells. Thus, most islet-like cell clusters consisted of IPC in the center and other pancreatic hormone positive cells in the periphery, including glucagon, somatostatin, and pancreatic polypeptide, which is characteristic of mature pancreatic islets [14].

12.2.5 Is Prominin-1 a Specific Marker of Pancreatic Stem and Progenitor Cells?

Engraftment of prominin-1^{+} cells resulted in the differentiation in endocrine-, ductal-, and exocrine-cell characteristics of pancreas. However, since the issue of the heterogeneity of the prominin1^{+} cell population remains to be elucidated, we are not able at the moment to rule out the possibility that the different cell types detected in the graft arose from distinct progenitor cells. Even in the in vitro experiment, we did not detect colonies with evidence of progeny with mixed lineage derived from a single cell [14]. Although the general concept of stem cells has been extended from

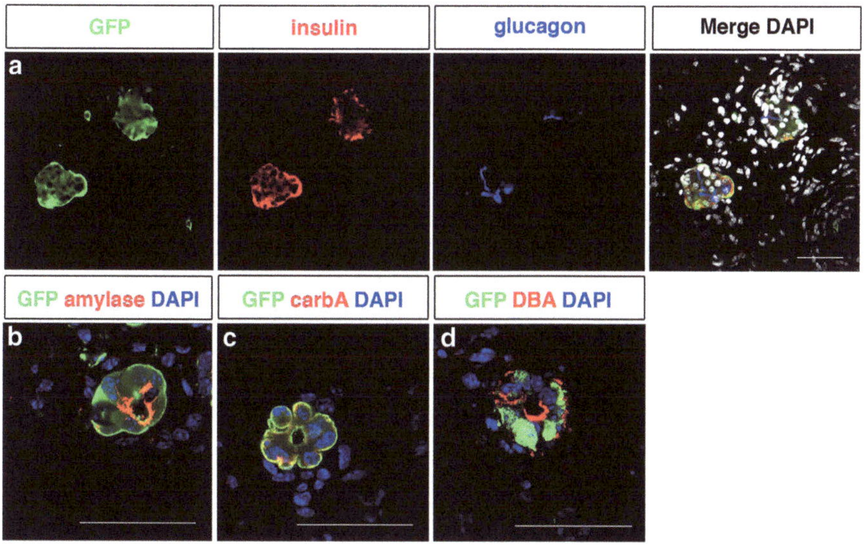

Fig. 12.2 Multilineage potential of pancreatic stem and progenitor cells. (**a–d**) 5×10^4 prominin-1$^+$ cells derived from enhanced green fluorescent protein (eGFP) transgenic mice were transplanted in the subcapsular renal space of nude mice, and after 7 days, the engrafted cells (visualized with GFP) were processed for immunohistochemistry for the following markers: insulin and glucagon (**a**), amylase (**b**) and carboxypeptidase A (**c**, carbA) as exocrine markers, and *Dolichos biflorus agglutinin* (**d**, DBA) as a ductal marker. Nuclei were labeled with 4',6-diamidino-2-phenylindole (DAPI). Scale bars, 50 μm. These panels are derived from our work published in stem cells in 2008 [14]

hematopoietic stem cells to many other tissues, only rarely stem cells have been identified as clonogenic precursors that include in their progeny both self-renewing and differentiation potential. Based strictly on this definition, stem cells reported in other systems are not clonogenic. To prove definitively that prominin1$^+$ cells isolated from pancreas have stem cell characteristics, it is now imperative to transplant single prominin-1$^+$ cell and reconstitute a pancreas as elegantly shown for hematopoietic [18] or mammary gland stem cells [19, 20]. Alternatively, we need to employ genetic lineage tracing in mice as established by Dor and colleagues [8] to demonstrate that a single prominin-1$^+$ cell has a multilineage differentiation potential.

There is a growing body of evidence that pancreatic stem and progenitor cells might be located in the terminal duct. In addition, the phenomenon of "acinar-to-ductal transdifferentiation" should be taken into account. The evidence of adult rodent acinar-to-ductal transdifferentiation and conversion into IPC seems to be promising for cell-replacement therapy in diabetes [21, 22]. More recently, Houbracken and colleagues demonstrated that the human acinar cell had plasticity similar to that described in rodent cells, which might be used to convert them into human IPC [23]. However, the plasticity of acinar cell to pancreatic stem and progenitor cells requires additional investigation, and it is obviously an exiting field to explore further.

12.3 Pancreatic Cancer Stem Cells

12.3.1 Human Pancreatic Cancer Stem/Tumor-Initiating/ Tumorigenic Cells

Pancreatic cancer is currently the fourth leading cause of cancer-related mortality. The ductal adenocarcinoma that arises within the exocrine component of the pancreas is the most common type accounting for 95% of these tumors. Morphologically, this cancer is typically characterized by moderately to poorly differentiated glandular structures. Other cancers might develop from islet cells and are classified as neuroendocrine tumors. Unfortunately, pancreatic cancer has a poor prognosis, and a high percentage of patients (>80%) will succumb from the disease over the year. The lack of symptoms and the high propensity of early metastasis normally lead to advanced disease at time of diagnosis. Despite our increasing knowledge in tumor biology, the efficacy of treatment of pancreatic cancer has not significantly improved over the last decade.

The current definition of CSC describes them as rare cells within a tumor that are able to self-renew and to produce the heterogeneous lineages of cancer cells [24]. The implementation of this concept explains the use of alternative terms in literature, such as "tumor-initiating cell" and "tumorigenic cell" to describe putative CSC. Moreover, CSC are often considered to be resistant to conventional treatment such as chemotherapy and radiotherapy, and treatments that fail to eliminate CSC may allow the tumor to relapse. For more details of CSC including their dynamics and the metastasis-initiating cell concept, I refer the readers to an excellent review [25]. Therefore, a considerable interest in the molecular targeting and eradication of CSC is emerging for cancer therapy.

As other tumors, recent reports have demonstrated that pancreatic cancer also contains a minute subpopulation of potential CSC. Li and colleagues isolated CSC from pancreatic cancer using cell surface markers such as CD44, CD24, and epithelial-specific antigen (ESA) [26]. Remarkably, cells harboring $CD44^+CD24^+ESA^+$ phenotype (i.e., only 0.2–0.8% of pancreatic cancer cells) had a 100-fold increased tumorigenic potential compared to non-tumorigenic cancer cells. Furthermore, the $CD44^+CD24^+ESA^+$ pancreatic cancer cells showed the stem cell properties of self-renewal and the ability to produce differentiated progeny. More recently, it was reported that CD44 could be co-expressed with CD133 in pancreatic ductal adenocarcinomas raising the possibility that CD133 might be an alternative marker of human pancreatic CSC [27].

12.3.2 CD133 Is an Alternative Marker of Human Pancreatic Cancer Stem Cells

The expression of CD133 on healthy pancreas was investigated by numerous independent research groups. As reported earlier in the mouse system (see above), Shimizu and colleagues found by means of immunohistochemistry that CD133 is expressed at the apical membrane of ductal cells in human adult pancreas (Fig. 12.3a, b) [28], consistent with a previous report from Karbanová and colleagues [29]. Lardon and colleagues have also shown that CD133 including its widely used stem cell AC133 epitope is expressed on all duct-lining cells of human pancreas [30]. However, Shimizu and Karbanová studies did not detect CD133 expression on the epithelium of larger interlobular and main ducts [28, 29]. Indeed, CD133 expression appeared more pronounced toward the acini than larger ducts [31]. Given the expression of CD133 within the pancreas [28–31], and its detection in CSC originating from other tissues/organs (see Chap. 1), it is not excluded that particular subpopulation of $CD133^+$ cells on the ductal epithelium might be the source of pancreatic CSC.

In diseased tissues, CD133 was concentrated in ductal metaplasia (the widespread interconversion of one cell type into another) of the acinar cells located in the border zone of the tumor (Fig. 12.3c, d) [28]. Metaplasia has been associated with pancreatic cancer in both humans and animal models, and a metaplasia-ductal adenocarcinoma sequence has been proposed for carcinogenesis in pancreas [32, 33]. As mentioned above, CD133 expression was also observed in ductal adenocarcinoma cells (Fig. 12.3e, f). Remarkably, Hermann and colleagues isolated human pancreatic CSC using CD133 as a cell surface marker and found that pancreatic CSC defined by its expression were exclusively tumorigenic and highly resistant to standard chemotherapy [34]. Moreover, in the invasive front of pancreatic cancer, double-positive CD133 and chemokine-related receptor-4 (CXCR4) cells were identified as the major metastatic CSC phenotype. Since depletion of this cell population abrogated the metastatic phenotype without affecting their tumorigenic potential, they concluded that a subpopulation of migrating $CD133^+CXCR4^+$ CSC is essential for tumor metastasis. It is known that CXCR4 and its ligand the stromal cell-derived factor-1α (SDF-1α, CXCL12) are involved in directional cell migration. A recent study has also proposed that CXCR4/SDF-1α axis plays a critical regulatory role in the genesis of human islets [35].

In analogy to the stem cell niche, a growing body of evidence is accumulating to show that the tumor microenvironment is essential to maintain CSC. For instance, Moriyama and colleagues demonstrated that $CD133^+$ pancreatic cancer cells exhibited a more aggressive behavior, such as migration and invasion, especially in the presence of pancreatic stromal cells [36]. They also reported that *CXCR4* transcript is overexpressed in $CD133^+$ pancreatic cancer cells. Likewise, Hashimoto and colleague have recently found that hypoxia induced the expansion of $CD133^+$ pancreatic cancer cells, subsequent tumor aggressiveness, and the expression of CXCR4 [37]. In these contexts, CD133 might play a role in the organization and/or polarization of migrating pancreatic cancer cells as it was suggested for $CD133^+$ hematopoietic stem cells growing on multipotent mesenchymal stromal cells as feeder cell layer [38]. Finally, to prove that $CD133^+$ and/or $CD133^+CXCR4^+$ cancer cells have all stem cell-like characteristics, it will be necessary to engraft a single cell and demonstrate that it reconstitutes the complex heterogeneity of a pancreatic cancer.

In conclusion, the recent effort in the identification and characterization of pancreatic $CD133^+$ CSC has brought some information concerning the development

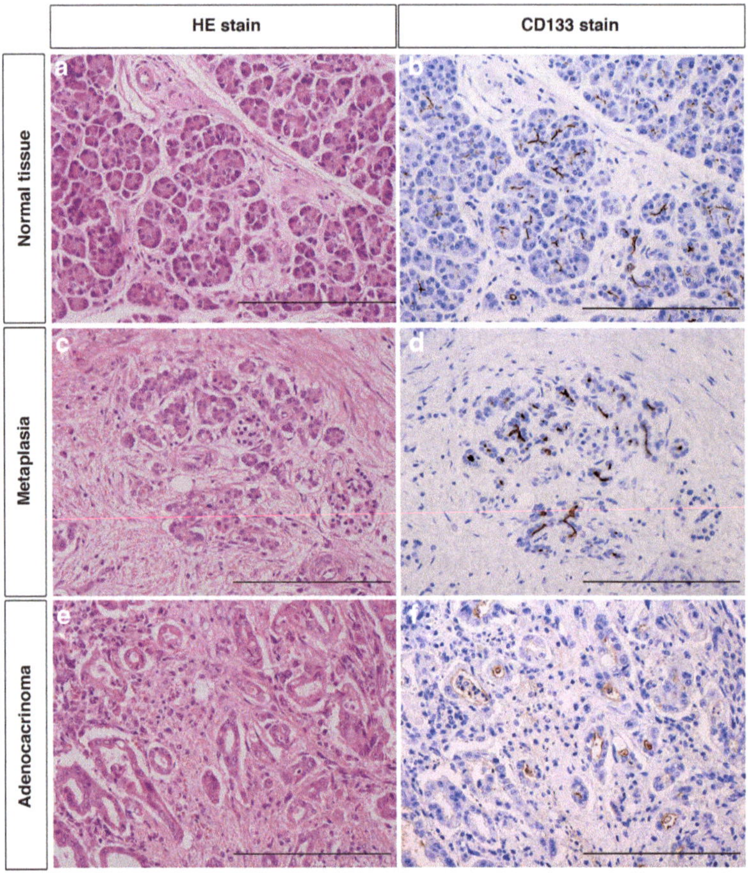

Fig. 12.3 CD133 expression in normal human pancreas, metaplasia, and ductal adenocarcinomas. (**a–f**) The expression of CD133 in normal human pancreas (**a**, **b**), metaplasia lesion (**c**, **d**), and in cancer cells of the ductal adenocarcinoma (**e**, **f**) were revealed by immunostaining using an antihuman CD133/1 mAb (Miltenyi Biotec) (**b**, **d**, **f**). Adjacent sections were stained with hematoxylin and eosin (HE stain) (**a**, **c**, **e**). Note that in the metaplasia lesion, a lobule undergoing a metaplastic transformation is shown adjacent to the normal acinar tissue (**c**, **d**). Scale bars, 200 μm. These panels are derived from our work published in pancreas in 2009 [28]

and spreading of the cancers as well as their resistance to treatment. However, some crucial questions are still opened such as the origin of CSC within pancreas. Further investigation on the molecular and cellular features of CD133 per se could indirectly provide new mechanisms underlying the proliferation of CSC as well as their migration. Likewise, CD133 as a cell surface antigen may become a potential target in the eradication of CSC.

Acknowledgements I would like to thank Drs. Kazuya Shimizu and Okito Hashimoto for valuable comments on the manuscript. I am also grateful to Norito Fujiwara, Rina Hirai, Yuki Kamino, Nancy Katayama, and Takuma Miura for their technical help. This research was supported by

Grants-in-Aid (21591773, 23592007, 24592025) for Scientific Research from the Ministry of Education, Culture, Sports, Science, and Technology of Japan.

References

1. Bonner-Weir S, Weir GC (2005) New sources of pancreatic beta-cells. Nat Biotechnol 23:857–861
2. Lumelsky N, Blondel O, Laeng P et al (2001) Differentiation of embryonic stem cells to insulin-secreting structures similar to pancreatic islets. Science 292:1389–1394
3. Hori Y, Rulifson IC, Tsai BC et al (2002) Growth inhibitors promote differentiation of insulin-producing tissue from embryonic stem cells. Proc Natl Acad Sci U S A 99:16105–16110
4. D'Amour KA, Bang AG, Eliazer S et al (2006) Production of pancreatic hormone-expressing endocrine cells from human embryonic stem cells. Nat Biotechnol 24:1392–1401
5. Tateishi K, He J, Taranova O et al (2008) Generation of insulin-secreting islet-like clusters from human skin fibroblasts. J Biol Chem 283:31601–31607
6. Seaberg RM, Smukler SR, Kieffer TJ et al (2004) Clonal identification of multipotent precursors from adult mouse pancreas that generate neural and pancreatic lineages. Nat Biotechnol 22:1115–1124
7. Suzuki A, Nakauchi H, Taniguchi H (2004) Prospective isolation of multipotent pancreatic progenitors using flow-cytometric cell sorting. Diabetes 53:2143–2152
8. Dor Y, Brown J, Martinez OI et al (2004) Adult pancreatic beta-cells are formed by self-duplication rather than stem-cell differentiation. Nature 429:41–46
9. Gu G, Dubauskaite J, Melton DA (2002) Direct evidence for the pancreatic lineage: NGN3+ cells are islet progenitors and are distinct from duct progenitors. Development 129:2447–2457
10. Yin AH, Miraglia S, Zanjani ED et al (1997) AC133, a novel marker for human hematopoietic stem and progenitor cells. Blood 90:5002–5012
11. Uchida N, Buck DW, He D et al (2000) Direct isolation of human central nervous system stem cells. Proc Natl Acad Sci USA 97:14720–14725
12. Kania G, Corbeil D, Fuchs J et al (2005) Somatic stem cell marker prominin-1/CD133 is expressed in embryonic stem cell-derived progenitors. Stem Cells 23:791–804
13. Corbeil D, Röper K, Hellwig A et al (2000) The human AC133 hematopoietic stem cell antigen is also expressed in epithelial cells and targeted to plasma membrane protrusions. J Biol Chem 275:5512–5520
14. Hori Y, Fukumoto M, Kuroda Y (2008) Enrichment of putative pancreatic progenitor cells from mice by sorting for prominin1 (CD133) and platelet-derived growth factor receptor ß. Stem Cells 26:2912–2920
15. Oshima Y, Suzuki A, Kawashimo K et al (2007) Isolation of mouse pancreatic ductal progenitor cells expressing CD133 and c-Met by flow cytometric cell sorting. Gastroenterology 132:720–732
16. Sugiyama T, Rodriguez RT, McLean GW et al (2007) Conserved markers of fetal pancreatic epithelium permit prospective isolation of islet progenitor cells by FACS. Proc Natl Acad Sci U S A 104:175–180
17. Koblas T, Pektorova L, Zacharovova K et al (2008) Differentiation of CD133-positive pancreatic cells into insulin-producing islet-like cell clusters. Transplant Proc 40:415–418
18. Weissman IL, Anderson DJ, Gage F (2001) Stem and progenitor cells: origins, phenotypes, lineage commitments, and transdifferentiations. Annu Rev Cell Dev Biol 17:387–403
19. Stingl J, Eirew P, Ricketson I et al (2006) Purification and unique properties of mammary epithelial stem cells. Nature 439:993–997
20. Shackleton M, Vaillant F, Simpson KJ et al (2006) Generation of a functional mammary gland from a single stem cell. Nature 439:84–88

21. Minami K, Okuno M, Miyawaki K et al (2005) Lineage tracing and characterization of insulin-secreting cells generated from adult pancreatic acinar cells. Proc Natl Acad Sci USA 102:15116–15121
22. Zhou Q, Brown J, Kanarek A et al (2008) In vivo reprogramming of adult pancreatic exocrine cells to beta-cells. Nature 455:627–632
23. Houbracken I, de Waele E, Larden J et al (2011) Lineage tracing evidence for transdifferentiation of acinar to duct cells and plasticity of human pancreas. Gastroenterology 141:731–741
24. Clarke MF, Dick JE, Dirks PB et al (2006) Cancer stem cells-perspectives on current status and future directions: AACR workshop on cancer stem cells. Cancer Res 66:9339–9344
25. Baccelli I, Trumpp A (2012) The evolving concept of cancer and metastasis stem cells. J Cell Biol 198:281–293
26. Li C, Heidt DG, Dalerba P et al (2007) Identification of pancreatic cancer stem cells. Cancer Res 67:1030–1037
27. Immervoll H, Hoem D, Steffensen OJ et al (2011) Visualization of CD44 and CD133 in normal pancreas and pancreatic ductal adenocarcinomas: non-overlapping membrane expression in cell populations positive for both markers. J Histochem Cytochem 59:441–455
28. Shimizu K, Itoh T, Shimizu M et al (2009) CD133 expression pattern distinguishes intraductal papillary mucinous neoplasms from ductal adenocarcinomas of the pancreas. Pancreas Nov 38(8):e207–e214
29. Karbanová J, Missol-Kolka E, Fonseca AV et al (2008) The stem cell marker CD133 (Prominin-1) is expressed in various human glandular epithelia. J Histochem Cytochem 56:977–993
30. Lardon J, Corbeil D, Huttner WB et al (2008) Stem cell marker prominin-1/AC133 is expressed in duct cells of the adult human pancreas. Pancreas 36:e1–e6
31. Immervoll H, Hoem D, Sakariassen P et al (2008) Expression of the "stem cell marker" CD133 in pancreas and pancreatic ductal adenocarcinomas. BMC Cancer 8:48
32. Miyamoto Y, Maitra A, Ghosh B et al (2003) Notch mediates TGF alpha-induced changes in epithelial differentiation during pancreatic tumorigenesis. Cancer Cell 3:565–576
33. Stanger BZ, Stiles B, Lauwers GY et al (2005) Pten constrains centroacinar cell expansion and malignant transformation in the pancreas. Cancer Cell 8:185–195
34. Hermann PC, Huber SL, Herrler T et al (2007) Distinct populations of cancer stem cells determine tumor growth and metastatic activity in human pancreatic cancer. Cell Stem Cell 1:313–323
35. Kayali AG, Lopez AD, Hao E et al (2012) The SDF-1α/CXCR4 Axis is required for proliferation and maturation of human fetal pancreatic endocrine progenitor cells. PLoS One 7:e38721
36. Moriyama T, Ohuchida K, Mizumoto K et al (2010) Enhanced cell migration and invasion of CD133+ pancreatic cancer cells cocultured with pancreatic stromal cells. Cancer 116:3357–3368
37. Hashimoto O, Shimizu K, Semba S et al (2011) Hypoxia induces tumor aggressiveness and the expansion of CD133-positive cells in a hypoxia-inducible factor-1α-dependent manner in pancreatic cancer cells. Pathobiology 78:181–192
38. Freund D, Bauer N, Boxberger S et al (2006) Polarization of human hematopoietic progenitors during contact with multipotent mesenchymal stromal cells: effects on proliferation and clonogenicity. Stem Cells Dev 15:815–829

Chapter 13
Prominin-1 (CD133) and Metastatic Melanoma: Current Knowledge and Therapeutic Perspectives

Aurelio Lorico, Javier Mercapide, and Germana Rappa

Abstract Innovative approaches to specifically target the melanoma subpopulation responsible for local invasion and metastatic dissemination are needed. Prominin-1 (CD133) expression has been observed in many melanoma cell lines, as well as in primary and metastatic melanomas from patients. Although its function(s) in melanoma is presently unknown, prominin-1 may represent a molecular target, due to its association with melanoma stem cells and with the metastatic phenotype.

Keywords CD133 • Melanoma • Immunotherapy • Stem cells • AC133 • Wnt

13.1 Introduction

The incidence of melanoma is steadily increasing in Caucasian populations worldwide, with approximately 70,000 new cases per year in the United States alone [1]. Of all deaths associated with any type of skin cancer, about 85% are caused by melanoma, and the development of metastasis is by far the major cause of melanoma deaths [1]. Metastatic melanoma is largely refractory to all currently available systemic therapies, with infrequent durable responses to conventional chemotherapy,

A. Lorico (✉) • G. Rappa
Cancer Research Program, Roseman University of Health Sciences, 89104 Henderson, NV, USA

College of Pharmacy, Roseman University of Health Sciences, 89104 Henderson, NV, USA
e-mail: alorico@roseman.edu

J. Mercapide
Cancer Research Program, Roseman University of Health Sciences, 89104 Henderson, NV, USA

D. Corbeil (ed.), *Prominin-1 (CD133): New Insights on Stem & Cancer Stem Cell Biology*, Advances in Experimental Medicine and Biology 777, DOI 10.1007/978-1-4614-5894-4_13,

immunotherapy, or to the recent BRAF (v-raf homolog B1) inhibitors [2]. Potentially, metastatic cells represent a minority of the bulk population of melanoma cells that must be specifically targeted to prevent metastatic dissemination. Many studies have identified expression of prominin-1 (CD133) in both established melanoma cell lines and clinical specimens derived from melanoma patients [3–5], generally by immunoreactivity for the glycosylation-dependent monoclonal antibody (mAb) CD133/1 (clone AC133; Miltenyi Biotec, Bergisch Gladbach, Germany). Our laboratory showed that prominin-1 knockdown in human FEMX-I metastatic melanoma cells resulted in decreased metastatic potential [6]. Prominin-1, normally expressed on undifferentiated cells, including hematopoietic stem cells, endothelial progenitor cells, fetal brain stem cells, and prostate epithelial cells, has also been exploited to identify and purify cancer stem cells (CSC) from various solid tumors including brain, colon, prostate, and pancreatic cancers [7]. Thus, despite its unknown function in stem cell biology, prominin-1 appears to mark both normal and CSC. In this chapter, we will summarize the existing literature data on the expression of prominin-1 in melanoma and in melanoma stem cells (MSC) and its association with the metastatic phenotype, and we will examine its potential role as a therapeutic target for melanoma.

13.2 Melanoma Development

Two, probably coexisting, models of melanoma development have been proposed:

1. Normal mature melanocytes acquire progressive mutations, which lead the cells through the classical phases of benign nevus, dysplastic nevus, and radial and vertical growth phases, ultimately resulting in metastatic melanoma.
2. Melanomas arise from the transformation of neural crest-derived melanocytic stem or immature progenitor cells [8], resulting directly in local growth, followed by invasion and metastatic colonization. Interestingly, prominin-1 is expressed on the surface of dermal-derived stem cells that are capable of differentiating into neural cells [9].

In both developmental models, an important role of the tumor microenvironment, including inflammatory, mesenchymal, and endothelial cells in the invasive/metastatic process, has been postulated. The observation that the majority of melanomas do not evolve from melanocytic proliferations, such as nevi [10], is in favor of the clinical relevance of the stem cell developmental model [11].

13.3 Melanoma Stem Cells

The definition of MSC, as for all CSC, is at present only functional, encompassing the capacities to initiate melanomas when implanted into immunodeficient animals, to reconstitute the heterogeneity of the original tumor, and to successfully establish tumors upon serial transplantation. While self-renewing MSC maintain the stem cell compartment, they also give rise to progenitor cells with limited self-renewing capacity and ultimately to differentiating clones of varying dominance, which contribute to the cell heterogeneity typically found in melanomas. The MSC model does not address the question of whether melanomas arise from normal melanocyte stem cells. Rather, it suggests that, irrespective of the cell of origin, melanomas are hierarchically organized, with MSC undergoing epigenetic changes analogous to the differentiation of normal cells, forming phenotypically diverse non-tumorigenic cancer cells that make up the bulk of the tumor.

For clinical applications, molecules present on the surface of MSC would allow their diagnostic identification in melanoma biopsies and might be used as prognostic indicators and therapeutic targets. In fact, while most treatment strategies are currently aimed at the bulk tumor population, MSC surface molecules may constitute targets for antibody-based or other types of therapies that would selectively eliminate these cells.

The notorious resistance of melanoma cells to all major chemotherapeutic drugs strongly suggests that this tumor type possesses major properties for drug resistance, presumably intrinsic to the MSC subpopulation. Without targeting the minor hidden fraction of partially quiescent or drug-resistant MSC, the goal of curing this disease may be unattainable, with anti-melanoma regimens, resulting in melanoma regrowth.

Cells with stem cell markers and features have recently been identified in melanoma tissues and cell lines. While several of these proteins, including prominin-1, CD271, CD20, nestin, ABCB5, and Bmi-1 have been proposed as MSC markers [3–5, 12–14], none of them have so far been shown to conclusively identify the MSC subpopulation [15]. It has been suggested that the use of nonobese diabetic mice with severe combined immunodeficiency disease (NOD/SCID mice) in xenotransplantation experiments can underestimate the frequency of human cancer cells with tumorigenic potential due to the xenogeneic immune response in these mice [16]. Thus, although only ~1 in a million melanoma cells reportedly formed tumors in NOD/SCID mice, 1 in 4 could form tumors in NOD/SCID IL2Rγ^{null} mice when coinjected with Matrigel, a substance mimicking the basement membrane. The possibility has been raised that some melanomas follow a cancer stem cell model, whereas others do not [16].

Melanoma side population (SP), a functional subpopulation distinct by the capacity to efflux a Hoechst dye with great efficiency, has also been proposed to contain the putative MSC. Interestingly, the major representative of the SP phenotype, the ABC transporter ABCG2 [17], has also been shown to be co-expressed on a subpopulation of prominin-1$^+$ melanoma cells [5]. In another study, melanoma cells expressing ABCB5, another energy-dependent drug efflux transporter highly similar to ABCB1, highly co-expressed prominin-1 and could enrich for prominin-1$^+$ cells, indicating that ABCB5$^+$/prominin-1$^+$ may mark MSC [4].

Whether prominin-1$^+$, CD271$^+$, and any other putative MSC subpopulation isolated from cultures or from clinical specimens, based on surface markers, overlap or are indeed the true tumor-initiating cells requires further characterization, and how these correlate with clinical outcome remains to be determined.

13.4 Expression of Prominin-1 in Established Human Melanoma Cell Lines

Several established melanoma cell lines express prominin-1, generally at relatively high levels and in the great majority of cells. Gil-Benso and colleagues [18] recently established the MEL-RC08 metastatic melanoma cell line with mutations in both BRAF and TP53 genes, derived from a pericranial metastasis of a malignant melanoma of the skin. A high percentage (84.1%) of MEL-RC08 cells expressed prominin-1. Four clonal cell lines obtained from the parental MEL-RC08 cell line also showed a high percentage of prominin-1^+ cells, ranging from 87% to 99%.

In a separate study, Monzani and colleagues investigated [5] prominin-1 expression in four independently derived human melanoma cell lines, namely, WM115, A375, IGR 32, and IGR 39. In all cases, practically all cells expressed prominin-1. In particular, for WM115 cells, prominin-1 expression was observed in all cells by means of fluorescent-activated cell-sorting (FACS) analysis, immunofluorescence, and reverse transcriptase-polymerase chain reaction (RT-PCR). WM115 expressed both prominin-1 and ABCG2, another putative MSC marker (see above). This cell line grows as floating spheroids, expresses typical progenitor and mature neuronal/oligodendrocyte markers, and is able to transdifferentiate into astrocytes or mesenchymal lineages under specific growth conditions. In WM115 melanoma xenografts, prominin-1 levels significantly decreased in tumors propagated in mice; yet on in vitro reculturing, cells readily regained high levels of prominin-1, suggesting that in vivo tumor growth conditions can significantly differ from in vitro culture conditions and can influence observed expression patterns [5].

Pietra and colleagues reported that five out of eight melanoma cell lines were negative for prominin-1, while in two (i.e., MeTA and Me1386), essentially all cells were prominin-1^+, and one (i.e., FO-1) displayed a bimodal distribution of prominin-1 on the cell surface [19].

Our laboratory, employing the AC133 mAb, found that essentially all human FEMX-I metastatic melanoma cells in culture expressed prominin-1 on their surface, with an average 400,000 molecules/cell [6]. As described later, while in vitro, a high percentage of cells expressing prominin-1 is generally observed, in freshly isolated melanoma cells from patients, only a minor subpopulation of prominin-1^+ cells (<1%) is present.

Combined, these studies establish the expression of prominin-1 in melanomas, supporting the notion that melanomas contain a stem cell component. However, comprehensive studies to correlate prominin-1 expression to self-renewal, differentiation, and tumorigenicity remain to be conducted.

13.5 Prominin-1 in Experimental Metastasis Models

Kim and colleagues [20] showed that administration of dacarbazine enriched a distinct prominin-1^+ subpopulation of B16F10 murine melanoma, with enhanced metastatic potential in vivo. Prominin-1^+ tumor cells were located close to tumor-associated lymphatic vessels in metastatic organs. Lymphatic vessels in metastatic tissues stimulated chemokine receptor 4 $(CXCR4)^+$/prominin-1^+ cell metastasis to target organs by secretion of stromal cell-derived factor-1 (SDF-1, CXCL12), a ligand of CXCR4. The authors suggested that targeting the SDF-1/CXCR4 axis in addition to dacarbazine treatment could therapeutically block chemoresistant prominin-1^+ cell metastasis toward a lymphatic metastatic niche.

In a study published by our laboratory [6], downregulation of prominin-1 had profound effects on human FEMX-I metastatic melanoma cells, including slower cell growth, reduced cell motility, and decreased capacity to form spheroids under stem cell-like growth conditions. Also, prominin-1 downregulation severely reduced the metastatic capacity of melanoma tumor xenografts, particularly to lung and spinal cord. These data strongly suggest that prominin-1 plays an important role in the melanoma cells' capacity to seed to distant sites. By microarray analysis, we also observed that many genes upregulated in prominin-1-knockdown cells coded for Wnt inhibitors, suggesting an interaction between prominin-1 and the canonical Wnt pathway [6]. In a subsequent study, our laboratory reported that three distinct pools of prominin-1 coexisted in cultures of FEMX-I cells. Morphologically, in addition to the plasma membrane localization, prominin-1 was found within the intracellular compartments (e.g., Golgi apparatus) and in association with extracellular membrane vesicles, most likely exosomes (unpublished observations). Since the function of microvesicles is dependent on the molecular cargo that they carry (e.g., protein, RNAs), the observation that microvesicles released by metastatic melanoma cells contain prominin-1 suggests that this molecule may have a certain role in extracellular communication.

Degradation of the basement membrane is a critical step in promoting tumor metastasis. We found that prominin-1 knockdown results in reduced cell motility and misalignment, presumably because cells could not organize properly due to impaired migration. We also observed loss of their capacity to invade through Matrigel, a basement membrane-mimicking substance [6]. Interestingly, it was previously shown that tumor-derived microvesicles often carry protein-degrading enzymes, including matrix metalloproteinases. By releasing them, tumor cells can degrade the extracellular matrix and invade surrounding tissues. Functionally, the downregulation of prominin-1 in FEMX-I cells resulted in upregulation of Wnt inhibitor genes [6] and decrease in the nuclear localization of β-catenin, a surrogate marker of Wnt activation. Moreover, the T cell factor/lymphoid enhancer factor (TCF/LEF) promoter activity was higher in parental than in prominin-1-knockdown cells (unpublished observations). A constitutive complex of β-catenin and LEF-1 was detected in melanoma cell lines, expressing either mutant β-catenin or mutant adenomatous polyposis coli tumor suppressor protein (APC) [21]; however, β-catenin mutations are rare in primary malignant melanoma, while its nuclear localization, a potential indicator of Wnt/β-catenin pathway activation, is frequently observed in melanoma [22]. The relevance of Wnt signaling to the metastatic process has only recently been unveiled by the observation that β-catenin is a central mediator of melanoma metastasis to the lymph nodes and lungs [23]. Interestingly, a clear association between prominin-1 and nuclear β-catenin exists for colon carcinoma [24]. It is conceivable, as recently proposed by Boivin and colleagues [25], that phosphorylation of prominin-1 in response to an unknown extracellular ligand might activate the Wnt pathway that would influence cell growth, motility, and metastatic potential. Also, the presence of conserved tandem TCF/LEF binding sites in the *prominin-1* (*Prom1*) gene suggests a feedback regulation of prominin-1 expression by the Wnt pathway [26]. Prominin-1-mediated Wnt activation, probably linked to the dynamic organization of cholesterol-rich membrane microdomains, may be responsible for the precise coordination and integration of the expression of multiple genes, resulting in dynamic, pro-metastatic changes of cell adhesion, and motility. Collectively, our recent results pointed to Wnt signaling and/or release of prominin-1-containing membrane vesicles as mediators of the pro-metastatic activity of prominin-1 in FEMX-I melanoma.

13.6 Expression of Prominin-1 in Patient-Derived Primary and Metastatic Melanoma Specimens

Surprisingly, little data are present in the literature on prominin-1 expression in primary melanomas, sentinel lymph nodes, and distant metastasis. Studies on the presence of distinct intracellular pools of prominin-1 have neither been performed nor has the reactivity of clinical specimens with different anti-prominin-1 mAbs been investigated. Monzani and colleagues determined that less than 1% of cells within metastatic melanomas expressed prominin-1 [5]. Only prominin-1$^+$ cells collected from these biopsies induced tumors in NOD/SCID mice, whereas the prominin-1$^-$ fraction failed to regenerate tumors, indicating that only a small percentage of cells was capable of recapitulating tumor growth [5]. Klein and colleagues [3] observed by immunohistochemistry significant increases in the expression of prominin-1 as well as of CD166 and nestin in primary and metastatic melanomas compared with banal nevi. Increased melanoma aggressiveness corresponded with greater expression of these markers, suggesting that during melanoma progression, stem cell markers become more evident, probably due to an increase of the dysregulation of stem/progenitor cell function and proliferation. However, in that study, only nestin showed a statistical difference when comparing primary and metastatic melanoma [3]. In 30 cases of lentigo maligna melanoma, analyzed by immunohistochemical staining, Bongiorno and colleagues [27] found in the vast majority of cases the presence of prominin-1$^+$ cells in the outer root sheath of the mid-lower hair follicles, intermixed with atypical melanocytes extending along layers of the hair follicles. While the authors concluded that this finding supports the origin of melanoma from melanocyte stem cells, no conclusion on the expression of prominin-1 in lentigo maligna melanoma cells was possible. Piras and colleagues [28] also analyzed prominin-1 expression by immunohistochemistry in 130 primary melanomas and 32 nodal metastasis biopsy specimens. Therein, prominin-1 staining was neither associated with survival, nor significant differences in prominin-1 expression were observed between primary tumors and metastases. Employing RT-PCR, Gazzaniga and colleagues [29] detected prominin-1$^+$ cells in the sentinel lymph nodes from 18 out of 45 total patients with stage I, II, and III. However, presumably for the low number of patients analyzed, no statistically significant correlations between prominin-1 expression and overall survival were found.

In a recent study [30], patient-derived melanoma cells could be maintained in cell culture for more than 16 months as melanospheres in serum-free cultures. The transition from melanospheres to monolayers was accompanied by an apparent loss of clonogenic potential and reversible changes in expression of cell surface markers, including prominin-1. Compared with adherent monolayer cultures, melanospheres were enriched in cells with clonogenic potential, reflecting the self-renewing capacity of cancer stem-like cells. The authors concluded that melanoma cells easily change their function upon exposure to external stimuli and suggested that the frequency of melanoma stem-like cells strongly depends on the microenvironment.

In a report by Fusi and colleagues [31], in about 50% of patients affected by metastatic melanoma, a small fraction of circulating melanoma cells isolated from peripheral blood expressed prominin-1. Its expression was not associated with a shorter overall survival. However, tissue positivity for prominin-1 was detected by immunohistochemistry in all patients with accessible metastases [31]. In support of a role of prominin-1 in the metastatic process, Thill and colleagues [32] found by immunohistochemistry the presence of prominin-1$^+$ cells in uveal melanomas, predominantly at the invading tumor front. Further evidence of association of prominin-1 with the metastatic potential stems from an analysis by Kupas and colleagues [33], showing that melanoma cells expressing the receptor activator of NF-κB (RANK) co-expressed prominin-1 and were able to induce tumor growth in immunodeficient mice. The RANK-RANK ligand pathway, involved in the migration and metastasis of epithelial tumor cells [33], increased significantly in peripheral circulating melanoma cells, primary melanomas, and metastases from stage IV melanoma patients compared with tumor cells from stage I melanoma patients. Also, a statistically significant overexpression of prominin-1 in 15 patients with lymph node metastasis and 19 patients with distant metastases compared to benign nevi was observed by Sharma and colleagues [34] (Table 13.1).

13.7 Therapeutic Perspectives

Since CSC in general have been reported to be both drug and radioresistant [35, 36], it is generally accepted that novel therapeutic approaches are urged to eradicate the malignant melanoma clone(s). One attractive approach is the selective targeting of MSC markers. Although the physiological function of prominin-1 in stem and CSC has not been elucidated, it is conceivable, based on data from our laboratory and others [6, 37], that prominin-1 represents a potential molecular target as well as cells expressing it. Efficacy of anti-prominin-1 immunotoxins was elegantly shown in hepatocellular and gastric cancer: the AC133 mAb, conjugated to monomethyl auristatin F, effectively inhibited the growth of Hep3B hepatocellular and KATO III

Table 13.1 Expression of prominin-1 in melanoma specimens

Reference	Specimen	Prominin-1⁺ cases/total	Detection method	Comment
Monzani, 2007 [5]	Distant metastasis	7/7	Flow cytometry	0.2% to 0.8% of cells positive
Klein, 2007 [3]	Primary melanoma	28/71	IHC	
	Distant metastasis	39/84		
Bongiorno, 2008 [27]	Lentigo maligna	Vast majority/30	IHC	Staining in the outer root sheath of mid-lower hair follicles
Piras, 2010 [28]	Primary melanoma	54/106	IHC	Staining at the invading front
	Nodal metastasis	17/32		
Gazzaniga, 2010 [29]	Sentinel lymph node	18/45	RT-PCR	
Fusi, 2011 [31]	Circulating melanoma cells	13/28	Flow cytometry	
	Distant metastasis	6/6	IHC	
Thill, 2011 [32]	Uveal melanoma	8/8	IHC	Staining at the invading front
Sharma, 2010 [34]	Primary melanoma	18/40	IHC	Statistically significant overexpression in lymph node and distant metastases compared to benign nevi
	Lymph node metastasis	12/15		
	Distant metastasis	13/19		

IHC immunohistochemistry, *RT-PCR* reverse transcriptase-polymerase chain reaction

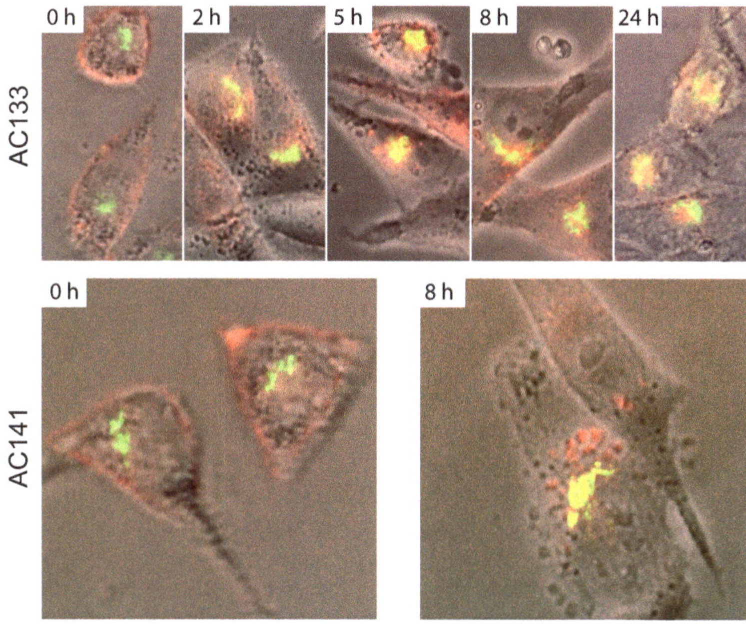

Fig. 13.1 Co-localization of anti-prominin-1-phycoerythrin (PE) antibodies with Golgi apparatus in human FEMX-I melanoma cells. The Golgi apparatus was labeled by a baculovirus expressing a signal peptide for N-acetylgalactosaminyltransferase fused to green fluorescent protein (GFP) under the control of a cytomegalovirus (CMV) promoter. *Left panel*: After 30 min, binding at 4°C with anti-prominin-1-PE antibody (AC133, *top panels*; AC141, *bottom panels*), cells were washed and incubated at 37°C for a given time (as indicated); *red*, anti-prominin-1-PE; *green*, GFP

gastric cancer cells in vitro and resulted in significant delay of Hep3B tumor growth in NOD-SCID mice [37].

Our laboratory performed preliminary studies to investigate the feasibility of an immunotoxin approach to malignant melanoma. First, we investigated whether anti-prominin-1 mAbs were internalized upon binding to prominin-1 on the surface of FEMX-I cells. To this aim, we incubated FEMX-I cells with phycoerythrin (PE)-conjugated anti-prominin-1 AC133 or AC141 mAbs for 30 min at 4°C (Fig. 13.1). The Golgi apparatus was highlighted by a baculovirus expressing a signal peptide for N-acetylgalactosaminyltransferase fused to green fluorescent protein (GFP) under the control of a cytomegalovirus (CMV) promoter. After extensive washing of the unbound antibody, cells were incubated at 37°C in an antibody-free medium. Progressively, the anti-prominin-1 antibodies were internalized and transported to the Golgi apparatus, with an apparent accumulation therein between 12 and 24 h.

We employed saporin, the ribosome-inactivating toxin derived from the seeds of *Saponaria officinalis*, for conjugation with the AC133 mAb, because of its previously reported efficacy in several xenograft models [38, 39], and it is simple to conjugate. The conjugated antibody had a 1.9 saporin/antibody molar ratio. For nontargeted saporin control, we used pre-immune mouse IgG antibody (with no known specificity) conjugated to saporin using the same protocol as the targeted immunotoxin.

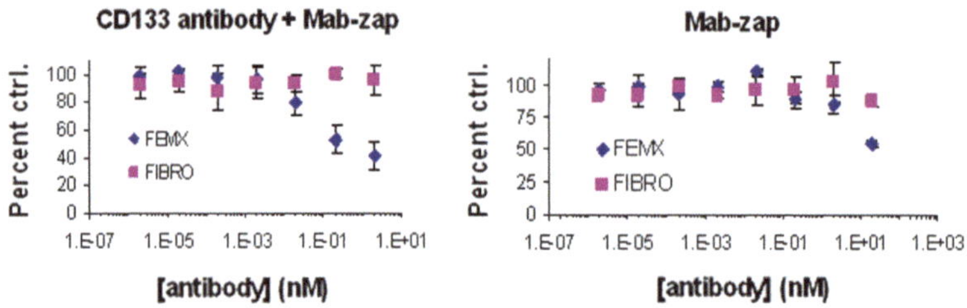

Fig. 13.2 In the presence of anti-prominin-1 monoclonal antibodies, saporin-conjugated secondary antibodies (Mab-zap) are toxic to prominin-1-expressing FEMX-I (FEMX) cells, but not to primary human fibroblasts (FIBRO)

To determine AC133/saporin conjugated as an immunotoxin, we evaluated by indirect and direct methods its cytotoxicity. For indirect immunotoxin assay, FEMX-I and normal adult human primary dermal fibroblasts (from the American Tissue Culture Collection) were plated at 1,000 per well (10,000/mL) onto 96-well plates in serum-supplemented tissue culture medium and incubated overnight at 37°C under 5% CO_2 atmosphere. 24 h later, cells were incubated with saporin-conjugated secondary antibody alone (Fig. 13.2, Mab-zap; right panel) or with AC133 mAb + Mab-zap (left panel). AC133 mAb and Mab-zap were mixed at 1:1 molar ratio (20 nM each) in serum-supplemented culture medium; as control, Mab-zap was incubated at the same concentration (20 nM) with serum-supplemented tissue culture medium. After 2–3 h at room temperature, mixing on a rotating platform, the antibody mixtures were added to the cells. Three days later, cytotoxicity was assessed using the 3-(4,5-dimethylthiazol-2-yl)-2, 5-diphenyltetrazolium bromide (MTT) assay. We observed that, in the presence of anti-prominin-1 monoclonal antibody, saporin-conjugated secondary antibody was toxic to prominin-1-expressing FEMX-I cells, but not to control human fibroblasts (Fig. 13.2).

In a direct immunotoxin experiments, we compared the effects of saporin-conjugated AC133 mAb with saporin-conjugated pre-immune mouse IgG antibody on both FEMX-I and prominin-1-knockdown FEMX-I, the latter expressing only 15,000 prominin-1 molecules/cell. The saporin-conjugated AC133 antibody was much more effective on FEMX-I cells than on the prominin-1 knockdown, while the pre-immune conjugate showed modest toxicity only at 2-log higher concentrations (Fig. 13.3). These data indicate that, as expected, the level of expression of prominin-1 is an important determinant of the immunotoxin activity.

In summary, the rationale for development of anti-prominin-1 immunotoxins lies in (i) our previous findings that prominin-1 downregulation prevents the formation of metastasis [6]; (ii) our preliminary data showing that saporin-conjugated anti-prominin-1 antibodies are toxic to prominin-1-expressing FEMX-I cells, but not to control human fibroblasts, and that anti-prominin-1 antibodies are effectively internalized and localized not in the lysosomes, where immunotoxins and antibodies are rapidly degraded [40], but in the Golgi apparatus; and (iii) the recent report of in vivo antitumor efficacy of an anti-prominin-1-drug conjugate in prominin-1$^+$ hepatocellular and gastric cancer cells in vivo [37]. However, before clinical translation,

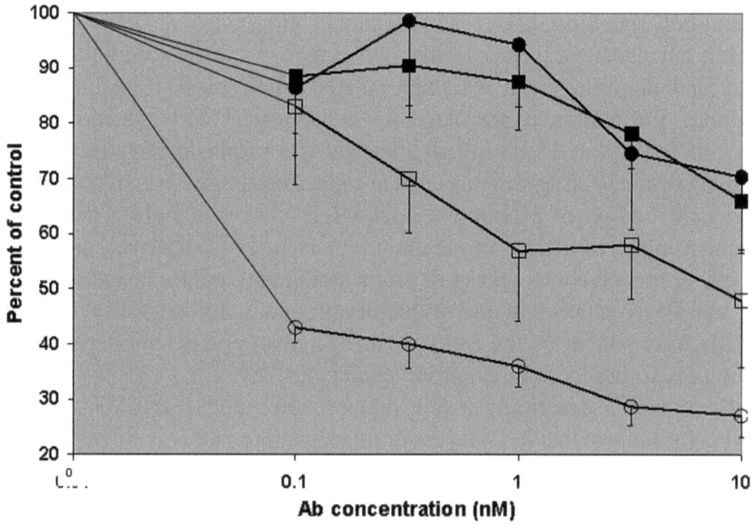

Fig. 13.3 Effects of saporin-conjugated AC133 antibody (*open symbols*) and pre-immune saporin conjugate (*solid symbols*) on FEMX-I (*circles*) and FEMX-I prominin-1-knockdown cells (*squares*). *Points* represent the mean of three separate experiments. *Bars*, standard deviation

the potential toxicity of prominin-1 targeting to nonmalignant prominin-1-expressing stem cell compartments should be investigated. The issue of potential toxicity to host tissues is crucial and difficult to tackle in mouse models because it is highly difficult to have a good handle on comparative toxicities. Furthermore, anti-prominin-1 immunotoxins could target and subsequently deplete a pool of long-term repopulating hematopoietic stem cells given the expression of prominin-1 in a subset of human CD34$^+$ stem and progenitor cells [41–44]. The restricted localization of prominin-1 at the apical plasma membrane in polarized epithelial cells found in healthy human tissues [45] may limit however the accessibility of the antibody and thereby lower the risk of antigen-dependent toxicities. Often, cell polarization is lost, at least in poorly differentiated tumors, potentially enhancing the accessibility to antibody targeting [46].

Other prominin-1-based immunotherapeutic approaches have been proposed. In a paper by Pietra and colleagues [19], natural killer (NK) cells eradicated human melanoma cells with characteristics of CSC. While prominin-1$^+$ melanoma cells preferentially survived radiations as compared with prominin-1$^-$ cells, prominin-1$^+$ melanoma cells displayed the same susceptibility to NK cell-mediated lysis than their prominin-1$^-$ counterpart, thus suggesting that NK cell-based immunotherapy might be considered as a possible approach in the cure of patients with metastatic melanoma. Miyabayashi and colleagues [47] demonstrated that a prominin-1$^+$ subpopulation in murine melanoma was immunogenic and that effector T cells specific for the prominin-1$^+$ melanoma cells mediated potent antitumor reactivity, curing the mice of the parental melanoma. Prominin-1$^+$ melanoma antigens preferentially induced type 17 T helper (Th17) cells and Th1 cells but not Th2 cells. Prominin-1$^+$

melanoma cell-specific $CD4^+$ T cell treatment eliminated not only prominin-1^+ tumor cells but also negative ones while inducing long-lasting accumulation of lymphocytes and dendritic cells with upregulated MHC class II in tumor tissues. Furthermore, the treatment prevented the regulatory T cell induction [47]. The authors concluded that T cell immunotherapy is a promising treatment option to eradicate prominin-1^+ drug-tolerant cells in cancerous tissues. It is of note that based on our recent finding of prominin-1-containing exosomes and on their potential involvement in the formation of melanoma metastases (see above), an alternative therapeutic approach could consist of drugs modulating either the exosome release or their uptake by supporting and/or neighboring cells. In line with this objective, recent data have suggested the endocytosis of hematopoietic cell-derived CD133-containing exosomes by microenvironmental cells [48].

Finally, our data describing a link between prominin-1 and Wnt pathway in FEMX-I cells suggest that key players of this signaling pathway might be potential targets to inhibit melanoma progression in prominin-1-expressing melanomas [6]. The relevance of Wnt signaling to the metastatic process has recently been unveiled by the observation that β-catenin is a central mediator of melanoma metastasis to the lymph nodes and lungs [23]. Prominin-1-mediated Wnt activation, which is probably linked to a dynamic organization of prominin-1-containing membrane microdomains [49], may be responsible for the precise coordination and integration of the expression of multiple genes, resulting in dynamic, pro-metastatic changes of cell adhesion and motility.

13.8 Conclusions

The refractoriness of metastatic melanoma to conventional therapy requires innovative approaches to prevention of metastatic dissemination or to specific targeting of the malignant population. Experimental data from our laboratory and others indicate a possible role of prominin-1 in the metastatic process. The frequent expression of prominin-1 in both melanoma cell lines and clinical melanoma specimens suggests that this molecule [and its underlying signaling pathway(s)] is a potential therapeutic target. A better understanding of the function of prominin-1 in melanoma cells as well as in healthy tissues and careful consideration of potential toxicities associated with its direct (or indirect) targeting is nevertheless required before clinical trials.

Acknowledgements This work was supported by US NIH R01CA133797 (G. R.). The content is solely the responsibility of the authors and does not necessarily represent the official views of the National Cancer Institute or the National Institutes of Health. The authors declare to have no conflict of interests. The authors thank H. Rosenberg and R. R. Fiscus for their support.

References

1. Jemal A, Siegel R, Xu J, Ward E (2010) Cancer statistics. CA Cancer J Clin 60:277–300
2. Tsai J, Lee JT, Wang W et al (2008) Discovery of a selective inhibitor of oncogenic B-Raf kinase with potent antimelanoma activity. Proc Natl Acad Sci USA 105:3041–3046
3. Klein WM, Wu BP, Zhao S, Wu H, Klein-Szanto AJ, Tahan SR (2007) Increased expression of stem cell markers in malignant melanoma. Mod Pathol 20:102–107
4. Frank NY, Margaryan A, Huang Y et al (2005) ABCB5-mediated doxorubicin transport and chemoresistance in human malignant melanoma. Cancer Res 65:4320–4333
5. Monzani E, Facchetti F, Galmozzi E et al (2007) Melanoma contains CD133 and ABCG2 positive cells with enhanced tumourigenic potential. Eur J Cancer 43:935–946
6. Rappa G, Fodstad O, Lorico A (2008) The stem cell-associated antigen CD133 (Prominin-1) is a molecular therapeutic target for metastatic melanoma. Stem Cells 26:3008–3017
7. Mizrak D, Brittan M, Alison MR (2008) CD133: molecule of the moment. J Pathol 214:3–9
8. Grichnik JM, Burch JA, Schulteis RD et al (2006) Melanoma, a tumor based on a mutant stem cell? J Invest Dermatol 126:142–153
9. Belicchi M, Pisati F, Lopa R et al (2004) Human skin-derived stem cells migrate throughout forebrain and differentiate into astrocytes after injection into adult mouse brain. J Neurosci Res 77:475–486
10. Bevona C, Goggins W, Quinn T, Fullerton J, Tsao H (2003) Cutaneous melanomas associated with nevi. Arch Dermatol 139:1620–1624, discussion 4
11. Zabierowski SE, Herlyn M (2008) Melanoma stem cells: the dark seed of melanoma. J Clin Oncol 26:2890–2894
12. Boiko AD, Razorenova OV, van de Rijn M et al (2011) Human melanoma-initiating cells express neural crest nerve growth factor receptor CD271. Nature 466:133–137
13. Fang D, Nguyen TK, Leishear K et al (2005) A tumorigenic subpopulation with stem cell properties in melanomas. Cancer Res 65:9328–9337
14. Mihic-Probst D, Kuster A, Kilgus S et al (2007) Consistent expression of the stem cell renewal factor BMI-1 in primary and metastatic melanoma. Int J Cancer 121:1764–1770
15. Quintana E, Shackleton M, Foster HR et al (2010) Phenotypic heterogeneity among tumorigenic melanoma cells from patients that is reversible and not hierarchically organized. Cancer Cell 18:510–523
16. Quintana E, Shackleton M, Sabel MS, Fullen DR, Johnson TM, Morrison SJ (2008) Efficient tumour formation by single human melanoma cells. Nature 456:593–598
17. Scharenberg CW, Harkey MA, Torok-Storb B (2002) The ABCG2 transporter is an efficient Hoechst 33342 efflux pump and is preferentially expressed by immature human hematopoietic progenitors. Blood 99:507–512
18. Gil-Benso R, Monteagudo C, Cerda-Nicolas M et al (2012) Characterization of a new human melanoma cell line with CD133 expression. Hum Cell 25:61–67
19. Pietra G, Manzini C, Vitale M et al (2009) Natural killer cells kill human melanoma cells with characteristics of cancer stem cells. Int Immunol 21:793–801
20. Kim M, Koh YJ, Kim KE et al (2010) CXCR4 signaling regulates metastasis of chemoresistant melanoma cells by a lymphatic metastatic niche. Cancer Res 70:10411–10421
21. Rubinfeld B, Robbins P, El-Gamil M, Albert I, Porfiri E, Polakis P (1997) Stabilization of beta-catenin by genetic defects in melanoma cell lines. Science 275:1790–1792
22. Rimm DL, Caca K, Hu G, Harrison FB, Fearon ER (1999) Frequent nuclear/cytoplasmic localization of beta-catenin without exon 3 mutations in malignant melanoma. Am J Pathol 154:325–329
23. Damsky WE, Curley DP, Santhanakrishnan M et al (2011) beta-catenin signaling controls metastasis in Braf-activated Pten-deficient melanomas. Cancer Cell 20:741–754
24. Horst D, Kriegl L, Engel J, Jung A, Kirchner T (2009) CD133 and nuclear beta-catenin: the marker combination to detect high risk cases of low stage colorectal cancer. Eur J Cancer 45:2034–2040

25. Boivin D, Labbe D, Fontaine N et al (2009) The stem cell marker CD133 (prominin-1) is phosphorylated on cytoplasmic tyrosine-828 and tyrosine-852 by Src and Fyn tyrosine kinases. Biochemistry 48:3998–4007
26. Katoh Y, Katoh M (2007) Comparative genomics on PROM1 gene encoding stem cell marker CD133. Int J Mol Med 19:967–970
27. Bongiorno MR, Doukaki S, Malleo F, Arico M (2008) Identification of progenitor cancer stem cell in lentigo maligna melanoma. Dermatol Ther 21(Suppl 1):S1–S5
28. Piras F, Perra MT, Murtas D et al (2010) The stem cell marker nestin predicts poor prognosis in human melanoma. Oncol Rep 23:17–24
29. Gazzaniga P, Cigna E, Panasiti V et al (2010) CD133 and ABCB5 as stem cell markers on sentinel lymph node from melanoma patients. Eur J Surg Oncol 36:1211–1214
30. Sztiller-Sikorska M, Koprowska K, Jakubowska J et al (2012) Sphere formation and self-renewal capacity of melanoma cells is affected by the microenvironment. Melanoma Res 22:215–224
31. Fusi A, Reichelt U, Busse A et al (2011) Expression of the stem cell markers nestin and CD133 on circulating melanoma cells. J Invest Dermatol 131:487–494
32. Thill M, Berna MJ, Grierson R et al (2011) Expression of CD133 and other putative stem cell markers in uveal melanoma. Melanoma Res 21:405–416
33. Kupas V, Weishaupt C, Siepmann D et al (2011) RANK is expressed in metastatic melanoma and highly upregulated on melanoma-initiating cells. J Invest Dermatol 131:944–955
34. Sharma BK, Manglik V, Elias EG (2010) Immuno-expression of human melanoma stem cell markers in tissues at different stages of the disease. J Surg Res 163:e11–e15
35. Singh A, Settleman J (2010) EMT, cancer stem cells and drug resistance: an emerging axis of evil in the war on cancer. Oncogene 29:4741–4751
36. Diehn M, Cho RW, Lobo NA et al (2009) Association of reactive oxygen species levels and radioresistance in cancer stem cells. Nature 458:780–783
37. Smith LM, Nesterova A, Ryan MC et al (2008) CD133/prominin-1 is a potential therapeutic target for antibody-drug conjugates in hepatocellular and gastric cancers. Br J Cancer 99:100–109
38. Siva AC, Wild MA, Kirkland RE et al (2008) Targeting CUB domain-containing protein 1 with a monoclonal antibody inhibits metastasis in a prostate cancer model. Cancer Res 68:3759–3766
39. Foehr ED, Lorente G, Kuo J, Ram R, Nikolich K, Urfer R (2006) Targeting of the receptor protein tyrosine phosphatase beta with a monoclonal antibody delays tumor growth in a glioblastoma model. Cancer Res 66:2271–2278
40. Wargalla UC, Reisfeld RA (1989) Rate of internalization of an immunotoxin correlates with cytotoxic activity against human tumor cells. Proc Natl Acad Sci USA 86:5146–5150
41. Yin AH, Miraglia S, Zanjani ED et al (1997) AC133, a novel marker for human hematopoietic stem and progenitor cells. Blood 90:5002–5012
42. Bühring HJ, Seiffert M, Marxer A et al (1999) AC133 antigen expression is not restricted to acute myeloid leukemia blasts but is also found on acute lymphoid leukemia blasts and on a subset of CD34+ B-cell precursors. Blood 94:832–833
43. Gallacher L, Murdoch B, Wu DM, Karanu FN, Keeney M, Bhatia M (2000) Isolation and characterization of human CD34(−)Lin(−) and CD34(+)Lin(−) hematopoietic stem cells using cell surface markers AC133 and CD7. Blood 95:2813–2820
44. de Wynter EA, Buck D, Hart C et al (1998) CD34+AC133+ cells isolated from cord blood are highly enriched in long-term culture-initiating cells, NOD/SCID-repopulating cells and dendritic cell progenitors. Stem Cells 16:387–396
45. Karbanová J, Missol-Kolka E, Fonseca A-V et al (2008) The stem cell marker CD133 (Prominin-1) is expressed in various human glandular epithelia. J Histochem Cytochem 56:977–993
46. Christiansen J, Rajasekaran AK (2004) Biological impediments to monoclonal antibody-based cancer immunotherapy. Mol Cancer Ther 3:1493–1501

47. Miyabayashi T, Kagamu H, Koshio J et al (2011) Vaccination with CD133(+) melanoma induces specific Th17 and Th1 cell-mediated antitumor reactivity against parental tumor. Cancer Immunol Immunother 60:1597–1608
48. Bauer N, Wilsch-Bräuninger M, Karbanová J et al (2011) Haematopoietic stem cell differentiation promotes the release of prominin-1/CD133-containing membrane vesicles-a role of the endocytic-exocytic pathway. EMBO Mol Med 3:398–409
49. Fargeas CA, Karbanová J, Jászai J, Corbeil D (2011) CD133 and membrane microdomains – Old facets for future hypotheses. World J Gastroenterol 17:4149–4152

Part IV
Future Perspectives: From Basic Research to Medical Application

Chapter 14
CD133-Positive Cells for Cardiac Stem Cell Therapy: Current Status and Outlook

Peter Donndorf and Gustav Steinhoff

Abstract Ischemic heart disease represents one major cause of death in developed countries. Ten years ago, cardiac application of bone marrow-derived progenitor cells was introduced as a new therapeutic strategy with the aim of restoring the function of ischemic myocardium. Among other cell populations, $CD133^{+}$ bone marrow stem cells form a major subpopulation of progenitor cells studied in this context. Following promising preclinical evidence, both cardiac surgeons and interventional cardiologists have applied $CD133^{+}$ cells in setting of chronic ischemic heart failure as well as acute myocardial infarction within phase I and II clinical trials. This chapter summarizes the rationale for the use of this stem cell subpopulation in the field of regenerative cardiac therapy strategies and gives an overview on the current clinical evidence as well as upcoming phase III trials.

Keywords CD133 • CABG • Coronary intervention • Intramyocardial • Surgery stem cell

14.1 Introduction

Chronic ischemic heart disease remains one of the most important causes of morbidity and mortality worldwide. Although revascularization procedures and conventional drug therapy may delay ventricular remodeling, there is no basic therapeutic regime available to prevent or even reverse this process. Chronic coronary artery

P. Donndorf, M.D. (✉) • G. Steinhoff
Department of Cardiac Surgery, University of Rostock,
Schillingallee 35, 18057 Rostock, Germany
e-mail: peter.donndorf@med.uni-rostock.de

D. Corbeil (ed.), *Prominin-1 (CD133): New Insights on Stem & Cancer Stem Cell Biology*, Advances in Experimental Medicine and Biology 777, DOI 10.1007/978-1-4614-5894-4_14,

disease and heart failure impair quality of life and are associated with subsequent worsening of the ventricular function. Recent experimental studies have demonstrated the capacity of bone marrow stem cells in restoring the function of ischemic myocardium [1, 2]. Since intrinsic myocardial regeneration has been shown to take place but to be reduced during a normal life span [3], it seems conceivable that – in addition to established revascularization procedures – the myocardium, damaged by chronic or acute ischemia, might profit from cardiac cell therapy aimed at stimulating the endogenous regenerative processes. After comparably quick translation of the solid preclinical results from "bench to bedside," respective phase I and II clinical trials were able to demonstrate feasibility and safety of this new therapeutic approach for different populations of progenitor cells, as well as different routes of application. The functional benefits achieved remained modest. However, compared with the functional results achieved by long-term pharmaceutical therapy or revascularization procedures and taking into account the fact that the functional effect of stem cell application adds on the effects gained with established therapeutic protocols, the functional efficacy of cardiac stem cell therapy can be considered relevant [4]. Several clinical trials showed the safety and efficacy of autologous bone marrow stem cell (BMSC) transplantation in the patients with acute myocardial infarction or chronic ischemic heart disease [5, 6] including an improved long-term survival. Today the therapeutic strategy of stem cell administration during cardiac surgery or coronary artery intervention is entering the clinical practice. In this chapter, we will focus on cardiac application of CD (cluster of differentiation) 133^{+} bone marrow stem cells in the setting of acute myocardial infarction and chronic ischemic heart disease.

14.2 Backgrounds

14.2.1 Cardiac Stem Cell Therapy for Regenerative Purposes

Stem cells exhibit the important properties of self-regeneration and differentiation plasticity [7–10]. Therefore, they form ideal candidates for regeneration of myocardial tissue damaged by acute or chronic ischemia [11].

By coronary revascularization procedures, e.g., percutaneous coronary intervention or coronary bypass surgery, myocardial perfusion may be normalized; however, myocardial function is restored only to a minor degree, as consecutive ventricular remodeling is not prevented. Prevention of the remodeling process, occurring in a majority of all postinfarction patients, however, might be achieved by cell transplantation, which leads to myocardial restitution with the beneficial aim of restoring or normalizing compromised heart function.

One possible way of myocardial repair is to transplant progenitor cells exhibiting angiogenic properties, such as human bone marrow-derived stem cells [12]. Myocardial regeneration by direct injection c-kit^{+} (CD117) bone marrow stem cells following myocardial infarction in a mouse model was reported the first time by Orlic and colleagues in 2001 [13]. This work, reporting of cardiac cell type regeneration by more than 50% in the infarcted area, initiated a controversial but very inspiring discussion. Although a straight reproduction of the reported results failed and subsequent experimental studies showed that bone marrow-derived stem cells of hematopoietic or endothelial lineage are most likely not able to transdifferentiate into functional cardiomyocytes [14], plausible confirmation of the principle of functional cardiac regeneration caused by the transplantation of bone marrow-derived stem cells in different animal models followed until today [15]. The isolation and the systemic delivery of bone marrow stem cells have been established before in the treatment of hematopoietic diseases [16]. Surface markers characterizing hematopoietic progenitor cells in the adult human bone marrow included primarily CD34 and CD117. Additionally, CD133 was introduced as an alternative marker to CD34 antigen. The CD133 antigen (prominin-1) is a 120-kDa 5-transmembrane domain glycoprotein that is expressed in immature hematopoietic stem and progenitor cells. Intramyocardial transplantation of bone marrow-derived hematopoietic stem cells is associated to myocardial regeneration most likely due to neovascularization and reduction of apoptosis [17, 18]. By intramyocardial injection, these cells are accumulated within the infarcted area and the border zone where they contribute to myocardial regeneration and scar size reduction, although contractile neotissue is not generated.

Vessel formation is shown as a result of both stem cell and resident cell action in the infarction area. Thereby, rescue of hibernating myocardium in the infarction border zone by improved myocardial oxygen supply leads to functional improvement due to improved regional ventricular wall function.

14.2.2 CD133$^+$ Bone Marrow Stem Cells for Myocardial Regeneration

In our own group, the capability of human CD133$^+$ stem cells to restore cardiac function after myocardial cryoinjury was tested utilizing immunodeficient mouse models. Compared to CD133$^+$ progenitor cells from the cord blood, CD133$^+$ cells derived from the bone marrow were found to have beneficial effects on both post-injury angiogenesis and particularly on myocardial functional recovery [19]. Moreover, by tracking of human DNA, the persistence of progenitor cell progeny in the heart after intramyocardial injection could be shown. As other groups, we did not see direct myocardial differentiation of these cells, although differentiation toward endothelial phenotypes was clearly shown. Therefore, it is questionable whether transdifferentiation of bone marrow stem cells from hematopoietic lineages to functional cardiomyocytes – the hypothesis on which regenerative strategies in the cardiovascular field were once based – truly takes place. The reported functional improvement is more likely due to the support of angiogenesis and arteriogenesis in the ischemic myocardium and the protection of cardiomyocytes from apoptotic cell death [17, 20]. Within the population of CD133$^+$ bone marrow stem cells especially a non-hematopoietic (CD34$^-$) subpopulation, it has been reported to exhibit a relevant potential inducing angiogenesis. The differentiation capacity of human CD133$^+$ bone marrow-derived stem cells toward endothelial lineages was recently described also by Ong and colleagues [21], demonstrating that the endothelial differentiation potential of CD133$^+$ cells is increased by hypoxic/normoxic preconditioning and thereby implicating that this cell population might be capable for distinct and specific pretreatment. In this way, "cell products" for cardiac regeneration might be optimized prior to transplantation to enhance their therapeutic capacity. Another regenerative mechanism, discussed in the context of CD133$^+$ stem cell-related improvement of cardiac function, includes paracrine effects of the applied progenitor cells in the injured myocardium, leading to higher levels of pro-angiogenic signaling molecules, inotropic peptides [22], as well as chemoattractants, like SDF-1α (stromal-derived factor-1 alpha) [23]. Although it is a reasonable question as to which extent results from immune-deficient animal models can be extrapolated to the clinical setting and despite the fact that chronic ischemic heart disease is not represented by an animal model of acute myocardial ischemia, the clear and reproducible beneficial effects of CD133$^+$ cell therapy regarding functional recovery of the ischemic myocardium and the absence of any relevant side effects of intramyocardial cell injection in the preclinical setting allowed the early translation from "bench to bedside."

The first intramyocardial injection of bone marrow-derived CD133$^+$ stem cells for treatment of chronic ischemic heart disease due to myocardial infarction applying this rationale was performed in 2001 by the Rostock group [24], utilizing CD133$^+$ stem cells isolated from the patient's bone marrow. Functional results gained in this phase I trial were highly promising.

Based on these results, several clinical trials followed investigating the safety and efficacy of cardiac CD133$^+$ bone marrow stem cell application for regenerative purposes.

14.3 Clinical Applications

14.3.1 Clinical Setting and Routes of Cell Delivery

Chronic ischemic heart failure and acute myocardial infarction represent the major clinical settings in which cardiac stem cell therapy has been investigated. These settings, however, exhibit significant differences regarding the favorable revascularization strategy and route of cell application as well as the progress of myocardial damage and the dynamics of myocardial remodeling. In contrast to acute infarction, with high levels of proinflammatory cytokines but yet no relevant remodeling, chronic ischemia, with the inflammatory reactions having subsided, leads to a loss of functional myocardial tissue, scar formation, and consequent ventricular remodeling [25]. Early injection after infarction could be beneficial to prevent a large fibrotic scare. On the other hand, since myocardial infarction leads to severe impairment of heart function associated with rhythmic instability and poorer tolerance of additional treatment, including further ischemia during cardiac surgery, it might be reasonable to wait for the acute phase to pass until the infarction zone is consolidated. For surgical reasons, myocardial consolidation is also preferable for any elective operative revascularization procedure. Furthermore, cell transplantation should be more effective after the postischemic inflammatory reaction has subsided, i.e., after day 8–12 following the acute attack [25, 26]. Stem cell transplantation within the immediate "hot" inflammatory phase following myocardial infarction might lead the cells to take part in the inflammation cascade rather than in the formation of functional myocardium and vessels [27].

In our clinic, we assign patients with impaired heart function after myocardial infarction and the presence of hibernating myocardium to stem cell treatment. The goal of intramyocardial stem cell transplantation in these patients is the improvement of myocardial function by augmentation of myocardial perfusion in the infarction border zone or regions of low contractility due to poor perfusion despite coronary revascularization. The treatment indication is additional to currently available revascularization strategies. Therefore, we select patients with the following characteristics [26]:

- Presence of ischemic heart disease with the need for extended revascularization; presence of akinetic/hypokinetic left ventricular wall and regional hibernating myocardium
- Reduced left ventricular ejection fraction (LVEF) below 50%

Regarding the revascularization strategies applied in the setting of chronic ischemic heart failure, coronary artery bypass grafting (CABG), not reasonably feasible in the most cases of acute myocardial infarction, represents one major therapeutic option besides elective interventional procedures. The route of cardiac cell application for regenerative purposes is preferentially influenced by the revascularization procedure performed simultaneously [4]. Stem cell injection without any simultaneous or staged revascularization has not been investigated in large-scale randomized trials and from the current point of view, this kind of approach might only be conceivable within a relatively small cohort of patients not suitable for any kind of revascularization or as an "ultima ratio" option in nonischemic heart failure. In the case of elective patients with a chronic ischemic disease, surgical epicardial or endocardial catheter-based cell injections have been performed successfully in addition to the concomitant revascularization procedure. In this context, it is of special interest that direct intramyocardial injection has been shown to allow for greater myocardial retention of applied stem cell compared to intracoronary or systemic application leading to superior functional improvements [28]. Therefore, in patients with chronic ischemic heart failure, the above mentioned routes of cell delivery should probably be preferred [29]. Surgical, epicardial injection is performed manually by the operating surgeon, with a high level of accuracy regarding the targeted region. For direct catheter-based injection, special tools, like the NOGA® system, have been developed, enabling for direct endocardial cell injection. A recent experimental study performed in pigs comparing surgical epicardial and endocardial myoblast injection using the NOGA® system reported similar functional benefits for both methods [30].

An alternative route of progenitor cell delivery to the damaged myocardium in combination with CABG was introduced recently by Hu and colleagues in a randomized safety and efficacy trial, by performing cell delivery through a graft vessel after completion of the respective distal anastomosis [31].

14.3.2 Current Clinical Evidence

After the initial phase I studies were able to proof the principle of myocardial functional regeneration by cardiac application of bone marrow-derived stem cells in humans, subsequent phase II studies were launched. The population of $CD133^+$ bone marrow stem cells represents one major progenitor cell population applied in this phase of clinical testing.

14.3.2.1 Intramyocardial Stem Cell Application During Cardiac Surgery

Until today, several studies investigating intramyocardial stem cell transplantation during cardiac surgery in patients with chronic ischemic heart disease have been launched. While some of these studies have already presented their results, others are still ongoing.

Most of the surgical trials designed for this setting performed intramyocardial stem cell transplantation in combination with "on-pump" coronary artery bypass surgery using the heart-lung machine. Surgical studies completed so far have been randomized [6, 32–34] as well as nonrandomized [35, 36].

Small cohort studies investigated also the safety and feasibility of surgical "stand-alone" bone marrow stem cell injection without concomitant coronary surgery [37].

As demonstrated in a recent meta-analysis on surgical stem cell application, most of these studies included a rather small number of patients weakening the validity of the results [38]. While some investigators used a specified population of bone marrow stem cells, others injected a semi-enriched population of bone marrow mononuclear cells. The major completed and ongoing studies using $CD133^+$ bone marrow stem cells are listed in Table 14.1. In this context, the study performed by Stamm and colleagues represents an essential analysis on clinical safety and efficacy of intramyocardial $CD133^+$ bone marrow stem cell transplantation in combination with CABG for treatment of chronic ischemic heart disease [6]. The study consisted of a dose-escalating phase I analogous safety trial including 15 patients and was followed by a phase I/II efficacy trial including 40 patients allocated either to a treatment group (20 patients) or a control group (20 patients). In line with the preclinical results and in order to work with a distinct, well-characterized cell population, bone marrow stem cells enriched for CD133 were chosen, whereby potential proinflammatory side effects of unselected mononuclear cell preparations were avoided. A high rhythmic stability after cell injection with no relevant ventricular

Table 14.1 Major clinical trials investigating surgical intramyocardial CD133+ bone marrow stem cell injection

Author year [ref.]	Sample size design	Primary intervention	Stem cell type and origin; mean stem cell dose (SD)	Route of injection	Follow-up duration (month)
Hendrikx 2006 [33]	20 RCT	CABG	CD133, bone marrow; $60.25\ (31) \times 10^6$	IM	4
Klein 2007 [37]	10 cohort	"Stand-alone" cell injection	CD133, bone marrow; $1.5–9.7 \times 10^6$ cells	IM	9
Ahmadi 2007 [36]	27 cohort	CABG	CD133, bone marrow; $1.89\ (0.03) \times 10^6$	IM	6
Pompilio 2008 [45]	5 cohort	"Stand-alone" cell injection	CD133, Bone marrow; 7.6×10^6	IM	12
Stamm 2007 [6]	40 RCT	CABG	CD133, bone marrow; 5.80×10^6	IM	6
Steinhoff ClinicalTrials. gov identifier: NCT00950274	142 RCT ongoing	CABG	CD133, bone marrow; n.a.	IM	6

CABG coronary artery bypass grafting, *CD* cluster of differentiation, *IM* intramyocardial, *n.a.* not announced, *RCT* randomized controlled trial arrhythmia was recorded at any point by online telemetric monitoring, and the average LVEF increased significantly, from 39.0% ± 8.7% preoperatively to 50.2% ± 8.5% at 6 months and 47.9% ± 6.0% at 18 months follow-up. In the subsequent efficacy trial, 40 patients were stratified to undergo CABG with intramyocardial $CD133^+$ cell injection or CABG alone, and the global LV systolic function at 6 months was found to be moderately but significantly better in cell-treated patients (37.4% ± 8% to 47.1% ± 7% vs. 37.9% ± 10.3% to 41.3% ± 9.1%). The results demonstrated that functional benefits could be achieved by combining $CD133^+$ stem cell injection and CABG. Patients treated with $CD133^+$ stem cells and CABG showed a significantly greater improvement of contractile function than those treated with CABG alone as measured by postoperative echocardiography. At the same time, the high level of procedural safety established in the phase I trial could be further confirmed. Interestingly, a subgroup analysis revealed that all patients with cell treatment in addition to CABG patients with poorer preoperative LVEF (<35% vs. ≥35%) benefit more. Patients with a preoperative LVEF less than 35% showed a mean increase of 15.3% (95% confidence interval 10.8–20.4%), significantly greater than the change in LVEF in patients with preoperative LVEF of at least 35% (increase of 7.8%, 95% confidence interval 4.1–11.5%, $F = 5.87$, $P = 0.02$).

Long-term follow-up analysis after 36 months confirmed the association of a significantly depressed preoperative ventricular function with a higher chance of improvement in LVEF, following intramyocardial $CD133^+$ cell injection in combination with CABG. Meanwhile, a significant lasting clinical effect could not be detected beyond 6 months [39].

However, this study, like others performed in this period, was characterized by methodological limitations such as lack of adequate blinding and placebo groups. Therefore, in order to give a conclusive answer regarding the functional efficacy of intramyocardial $CD133^+$ stem cell application during CABG, subsequent phase III clinical trials were mandatory. Consequently, a randomized, double-blinded, placebo-controlled, phase III multicenter clinical trial – PERFECT, Intramyocardial Trans**P**lantation of Bon**E** Ma**R**row Stem Cells **F**or Improv**E**ment of Post-Infarct Myo**C**ardial Regenera**T**ion in addition to CABG Surgery – was launched at our institution. This trial is ongoing and currently recruiting patients.

14.3.2.2 Intracoronary Stem Cell Application

Since interventional revascularization strategies represent the standard therapeutic regime in most cases of acute and subacute myocardial infarction, stem cell delivery in these settings is preferentially performed by applying the intravascular route via the coronary arteries. Similar to the surgical setting described above, phase II clinical testing has been undertaken for this approach in the recent past utilizing $CD133^+$ cells as well as other bone marrow stem cell subpopulations and unselected bone marrow mononuclear cells. The use of selected hematopoietic subpopulations in this setting is of special interest, as Hofmann and colleagues described a higher engraftment potential within infarcted myocardium compared with mononuclear cells after intracoronary injection [40]. Furthermore, intracoronary bone marrow stem cell application has also been investigated in the scenario of chronic ischemia, as an alternative to intramyocardial injection during coronary bypass surgery. Major clinical studies investigating the safety and functional efficacy of intracoronary $CD133^+$ bone marrow stem cell delivery for the treatment of acute and chronic myocardial ischemia are summarized in Table 14.2. The majority of clinical studies performed so far, however, used unselected bone marrow mononuclear cell preparations. Therefore – although recent meta-analyses suggest a beneficial functional effect of intracoronary stem cell infusion following ischemic myocardial damage [41, 42] – based on the current clinical evidence, it is difficult to judge the specific functional efficacy of $CD133^+$ bone marrow stem cells after intracoronary administration. In order to further address this issue, results from currently ongoing, elaborated phase II trials, like the COMPARE-AMI trial, are awaited with great interest [43]. Besides analyzing safety and functional efficacy of intracoronary $CD133^+$ bone marrow stem cell injection following acute myocardial infarction, this trial is assessing the potential effect on the underlying coronary disease, namely, the coronary atherosclerotic burden progression proximal and distal to the coronary stent in the infarct related artery.

In contrast to surgical intramyocardial $CD133^+$ cell injection, no trial reaching phase III level is currently ongoing.

14.4 Summary and Outlook

Based on solid preclinical evidence demonstrating the capability of $CD133^+$ bone marrow stem cells to partially restore cardiac function after ischemic damage in animal models, clinical phase I and II trials investigating feasibility and safety of intramyocardial $CD133^+$ cell injection as well as intracoronary $CD133^+$ cell application have been completed. Currently, the completion of elaborated phase III trials is mandatory to make a step forward in evaluating cardiac $CD133^+$ cell therapy as a relevant therapeutic tool. However, if the phase III trial results support cardiac $CD133^+$ cell therapy as an effective therapeutic option in chronic ischemic heart disease, further efforts will be needed in order answer questions regarding the exact mechanisms of stem cell migration, differentiation, and myocardial regeneration, in order to optimize cell therapy protocols. Additionally, the optimal time point of cell application, as well as the favorable cell number, is still unclear. Furthermore, despite the fact that recent evidence seems to favor stem cell subpopulations against unspecified bone marrow cell preparations, it is not clear whether $CD133^+$ cells form the optimal population. Only by a combined approach consisting of both preclinical research and consequent clinical translation, stem cell therapy can become a truly standardized, therapeutic "tool" in the field of regenerative medicine. Furthermore, the authors believe that exact, critical, and standardized pre-procedural evaluation of left ventricular function as well as examination of regional myocardial viability, preferably by cardiac MRI, form another key issue for reaching possible

Table 14.2 Major clinical trials investigating intracoronary CD133+ bone marrow stem cell application

Author year [ref.]	Sample size design	Primary intervention	Stem cell type and origin; mean stem cell dose (SD)	Route of injection	Follow-up duration (month)
Bartunek 2005 [46]	35 cohort	PCI (AMI)	CD133 SC, bone marrow; $12.6 \pm 2.2 \times 10^6$	IC	4
Manginas 2007 [47]	24 cohort	PTCA (CHF)	CD133 SC, bone marrow; $16.9 \pm 4.9\ 10^6$	IC	11
Colombo 2011 [48]	15 RCT	PCI (AMI)	CD133 SC, bone marrow + Peripheral blood; n.a.	IC	12
Bartunek ClinicalTrials.gov identifier: NCT00529932	Ongoing RCT	PCI (AMI)	CD133 SC, bone marrow; n.a.	IC	6
Mansour 2010 [43] ClinicalTrials.gov identifier: NCT00400959	Ongoing RCT	PCI (AMI)	CD133 SC, bone marrow; n.a.	IC	4

AMI acute myocardial infarction, *CD* cluster of differentiation, *CHF* chronic ischemic heart failure, *IC* intracoronary, *n.a.* not announced, *PCI* percutaneous coronary intervention, *PTCA* percutaneous transluminal coronary angioplasty, *RCT* randomized controlled trial, *SC* stem cell therapeutic efficacy by means of cardiac cell therapy. However, novel imaging modalities like speckle tracking by two-dimensional strain echocardiography might offer useful additional tools for planning and evaluation of cardiac cell therapy in the future [44]. On the other hand, alternative therapeutic goals for cardiac cell applications, besides global ventricular functional recovery, are already sought. For example, a potentially relevant rhythmic stabilization after myocardial infarction by regional gain in myocardial perfusion is hypothesized and respective phase I trials are under preparation.

References

1. Orlic D, Kajstura J, Chimenti S, Limana F, Jakoniuk I, Quaini F et al (2001) Mobilized bone marrow cells repair the infarcted heart, improving function and survival. Proc Natl Acad Sci U S A 98:10344–10349
2. Uemura R, Xu M, Ahmad N, Ashraf M (2006) Bone marrow stem cells prevent left ventricular remodeling of ischemic heart through paracrine signaling. Circ Res 98:1414–1421
3. Bergmann O, Bhardwaj RD, Bernard S, Zdunek S, Barnabe-Heider F, Walsh S et al (2009) Evidence for cardiomyocyte renewal in humans. Science 324:98–102
4. Donndorf P, Strauer BE, Steinhoff G (2012) Update on cardiac stem cell therapy in heart failure. Curr Opin Cardiol 27:154–160
5. Strauer BE, Yousef M, Schannwell CM (2012) The acute and long-term effects of intracoronary Stem cell Transplantation in 191 patients with chronic heARt failure: the STAR-heart study. Eur J Heart Fail 12:721–729
6. Stamm C, Kleine HD, Choi YH, Dunkelmann S, Lauffs JA, Lorenzen B et al (2007) Intramyocardial delivery of CD133+ bone marrow cells and coronary artery bypass grafting for chronic ischemic heart disease: safety and efficacy studies. J Thorac Cardiovasc Surg 133:717–725
7. Allgöwer M (1956) The cellular basis of wound repair. Charles C. Thomas. Oxford Blackwell, Springfield, III
8. Krause DS, Theise ND, Collector MI, Henegariu O, Hwang S, Gardner R et al (2001) Multi-organ, multi-lineage engraftment by a single bone marrow-derived stem cell. Cell 105:369–377
9. Jiang Y, Jahagirdar BN, Reinhardt RL, Schwartz RE, Keene CD, Ortiz-Gonzalez XR et al (2002) Pluripotency of mesenchymal stem cells derived from adult marrow. Nature 418:41–49
10. Quaini F, Urbanek K, Graiani G, Lagrasta C, Maestri R, Monica M et al (2004) The regenerative potential of the human heart. Int J Cardiol 95:S26–S28
11. Goodell MA, Jackson KA, Majka SM, Mi T, Wang H, Pocius J et al (2001) Stem cell plasticity in muscle and bone marrow. Ann N Y Acad Sci 938:208–218, discussion 18–20
12. Ferrari G, Cusella-De Angelis G, Coletta M, Paolucci E, Stornaiuolo A, Cossu G et al (1998) Muscle regeneration by bone marrow-derived myogenic progenitors. Science 279:1528–1530
13. Orlic D, Kajstura J, Chimenti S, Jakoniuk I, Anderson SM, Li B et al (2001) Bone marrow cells regenerate infarcted myocardium. Nature 410:701–705
14. Murry CE, Soonpaa MH, Reinecke H, Nakajima H, Nakajima HO, Rubart M et al (2004) Haematopoietic stem cells do not transdifferentiate into cardiac myocytes in myocardial infarcts. Nature 428:664–668
15. Segers VF, Lee RT (2008) Stem-cell therapy for cardiac disease. Nature 451:937–942
16. Hock H (2010) Some hematopoietic stem cells are more equal than others. J Exp Med 207:1127–1130

17. Tse HF, Yiu KH, Lau CP (2007) Bone marrow stem cell therapy for myocardial angiogenesis. Curr Vasc Pharmacol 5:103–112
18. Ma N, Ladilov Y, Kaminski A, Piechaczek C, Choi YH, Li W et al (2006) Umbilical cord blood cell transplantation for myocardial regeneration. Transplant Proc 38:771–773
19. Ma N, Ladilov Y, Moebius JM, Ong L, Piechaczek C, David A et al (2006) Intramyocardial delivery of human CD133+ cells in a SCID mouse cryoinjury model: bone marrow vs. cord blood-derived cells. Cardiovasc Res 71:158–169
20. Kocher AA, Schuster MD, Szabolcs MJ, Takuma S, Burkhoff D, Wang J et al (2001) Neovascularization of ischemic myocardium by human bone-marrow-derived angioblasts prevents cardiomyocyte apoptosis, reduces remodeling and improves cardiac function. Nature Med 7:430–436
21. Ong LL, Li W, Oldigs JK, Kaminski A, Gerstmayer B, Piechaczek C et al (2007) Hypoxic/normoxic preconditioning increases endothelial differentiation potential of human bone marrow CD133+ cells. Tissue Eng Part C Methods 16:1069–1081
22. Gao LR, Xu RY, Zhang NK, Chen Y, Wang ZG, Zhu ZM et al (2008) Increased apelin following bone marrow mononuclear cell transplantation contributes to the improvement of cardiac function in patients with severe heart failure. Cell Transplant 18:1311–1318
23. Bonaros N, Rauf R, Werner E, Schlechta B, Rohde E, Kocher A et al (2008) Neoangiogenesis after combined transplantation of skeletal myoblasts and angiopoietic progenitors leads to increased cell engraftment and lower apoptosis rates in ischemic heart failure. Interact Cardiovasc Thorac Surg 7:249–255
24. Stamm C, Westphal B, Kleine HD, Petzsch M, Kittner C, Klinge H et al (2003) Autologous bone-marrow stem-cell transplantation for myocardial regeneration. Lancet 361:45–46
25. Chachques JC (2009) Cellular cardiac regenerative therapy in which patients? Expert Rev Cardiovasc Ther 7:911–919
26. Kaminski A, Steinhoff G (2008) Current status of intramyocardial bone marrow stem cell transplantation. Semin Thorac Cardiovasc Surg 20:119–125
27. Soeki T, Tamura Y, Shinohara H, Tanaka H, Bando K, Fukuda N (2000) Serial changes in serum VEGF and HGF in patients with acute myocardial infarction. Cardiology 93:168–174
28. Freyman T, Polin G, Osman H, Crary J, Lu M, Cheng L et al (2006) A quantitative, randomized study evaluating three methods of mesenchymal stem cell delivery following myocardial infarction. Eur Heart J 27:1114–1122
29. Menasche P (2011) Cardiac cell therapy: lessons from clinical trials. J Mol Cell Cardiol 50:258–265
30. Larose E, Proulx G, Voisine P, Rodes-Cabau J, De Larochelliere R, Rossignol G et al (2010) Percutaneous versus surgical delivery of autologous myoblasts after chronic myocardial infarction: an in vivo cardiovascular magnetic resonance study. Catheter Cardiovasc Interv 75:120–127
31. Hu S, Liu S, Zheng Z, Yuan X, Li L, Lu M et al (2011) Isolated coronary artery bypass graft combined with bone marrow mononuclear cells delivered through a graft vessel for patients with previous myocardial infarction and chronic heart failure a single-center, randomized, double-blind, placebo-controlled clinical trial. J Am Coll Cardiol 57:2409–2415
32. Zhao Q, Sun Y, Xia L, Chen A, Wang Z (2008) Randomized study of mononuclear bone marrow cell transplantation in patients with coronary surgery. Ann Thorac Surg 86:1833–1840
33. Hendrikx M, Hensen K, Clijsters C, Jongen H, Koninckx R, Bijnens E et al (2006) Recovery of regional but not global contractile function by the direct intramyocardial autologous bone marrow transplantation: results from a randomized controlled clinical trial. Circulation 114:I101–I107
34. Patel AN, Geffner L, Vina RF, Saslavsky J, Urschel HC Jr, Kormos R et al (2005) Surgical treatment for congestive heart failure with autologous adult stem cell transplantation: a prospective randomized study. J Thorac Cardiovasc Surg 130:1631–1638

35. Mocini D, Staibano M, Mele L, Giannantoni P, Menichella G, Colivicchi F et al (2006) Autologous bone marrow mononuclear cell transplantation in patients undergoing coronary artery bypass grafting. Am Heart J 151:192–197
36. Ahmadi H, Baharvand H, Ashtiani SK, Soleimani M, Sadeghian H, Ardekani JM et al (2007) Safety analysis and improved cardiac function following local autologous transplantation of CD133(+) enriched bone marrow cells after myocardial infarction. Curr Neurovasc Res 4:153–160
37. Klein HM, Ghodsizad A, Marktanner R, Poll L, Voelkel T, Mohammad Hasani MR et al (2007) Intramyocardial implantation of CD133+ stem cells improved cardiac function without bypass surgery. Heart Surg Forum 10:E66–E69
38. Donndorf P, Kundt G, Kaminski A, Yerebakan C, Liebold A, Steinhoff G et al (2011) Intramyocardial bone marrow stem cell transplantation during coronary artery bypass surgery: a meta-analysis. J Thorac Cardiovasc Surg 142:911–920
39. Yerebakan C, Kaminski A, Westphal B, Donndorf P, Glass A, Liebold A et al (2011) Impact of preoperative left ventricular function and time from infarction on the long-term benefits after intramyocardial CD133(+) bone marrow stem cell transplant. J Thorac Cardiovasc Surg 142:1530–9.e3
40. Hofmann M, Wollert KC, Meyer GP, Menke A, Arseniev L, Hertenstein B et al (2005) Monitoring of bone marrow cell homing into the infarcted human myocardium. Circulation 111:2198–2202
41. Abdel-Latif A, Bolli R, Tleyjeh IM, Montori VM, Perin EC, Hornung CA et al (2007) Adult bone marrow-derived cells for cardiac repair: a systematic review and meta-analysis. Arch Intern Med 167:989–997
42. Lipinski MJ, Biondi-Zoccai GG, Abbate A, Khianey R, Sheiban I, Bartunek J et al (2007) Impact of intracoronary cell therapy on left ventricular function in the setting of acute myocardial infarction: a collaborative systematic review and meta-analysis of controlled clinical trials. J Am Coll Cardiol 50:1761–1767
43. Mansour S, Roy DC, Bouchard V, Nguyen BK, Stevens LM, Gobeil F et al (2010) COMPARE-AMI trial: comparison of intracoronary injection of CD133+ bone marrow stem cells to placebo in patients after acute myocardial infarction and left ventricular dysfunction: study rationale and design. J Cardiovasc Transl Res 3:153–159
44. Nasseri BA, Kukucka M, Dandel M, Knosalla C, Choi YH, Ebell W et al (2009) Two-dimensional speckle tracking strain analysis for efficacy assessment of myocardial cell therapy. Cell transplant 18:361–370
45. Pompillio G, Steinhoff G, Liebold A, Pesce M, Alemanni F, Capogrossi MC et al (2008) Direct minimally invasive intramyocardial injection of bone marrow-derived AC133+ stem cells in patients with refractory ischemia: preliminary results. Thorac Cardiovasc Surg 56:71–76
46. Bartunek J, Vanderheyden M, Vanderkerckhove B, Mansour S, De Bryne B, de Bondt P et al (2005) Intracoronary injection of CD133-positive enriched bone marrow progenitor cells promotes cardiac recovery after recent myocardial infarction: feasibility and safety. Circulation 112:I178–I183
47. Manginas A, Goussetis E, Koutelou M, Karatasakis G, Peristeri I, Theodorakos A et al (2007) Pilot study to evaluate the safety and feasibility of intracoronary CD133(+) and CD133(−) CD34(+) cell therapy in patients with nonviable anterior myocardial infarction. Catheter Cardiovasc Interv 69:773–781
48. Colombo A, Castellani M, Piccaluga E, Pusineri E, Palatresi S, Longari V et al (2011) Myocardial blood flow and infarct size after CD133+ cell injection in large myocardial infarction with good recanalization and poor reperfusion: results from a randomized controlled trial. J Cardiovasc Med 12:239–248

Chapter 15
$CD133^+$ Cells for the Treatment of Degenerative Diseases: Update and Perspectives

Mirella Meregalli, Andrea Farini, Marzia Belicchi, and Yvan Torrente

Abstract Stem cells are used in cell therapy for degenerative disorders. The main advantage of stem cells is that they can replenish their numbers for long periods through cell division and produce a progeny that can differentiate into multiple cell lineages with specific functions. CD133 is a member of a novel family of cell surface glycoproteins. The expression of human CD133 (AC133 antigen) was originally described in the hematopoietic $CD34^+$ stem cells, but now it becomes more and more evident that CD133 is a marker of stem and progenitor cell populations originating from various tissues and organs. The main objective of this chapter is to describe the potential sources of $CD133^+$ stem cells that harbor the ability to engraft, proliferate, and differentiate into functional cells. The characterization of such $CD133^+$ stem cells unlocks new opportunities in the treatment of degenerative diseases such as Duchenne muscular dystrophy.

Keywords $CD133^+$ stem cell • Degenerative disease • Heterogeneity • Muscular dystrophy • Therapeutic application

15.1 Introduction

Stem cells are interesting tools for studying the mechanisms of development and are clinically relevant in cell therapy of degenerative disorders [1, 2]. The biology of stem cells and their role in regeneration were the subject of intensive research in many

M. Meregalli • A. Farini • M. Belicchi • Y. Torrente, M.D., Ph.D. (✉)
Stem Cell Laboratory, Dipartimento di Fisiopatologia Medico-Chirurgica e dei Trapianti, Università degli Studi di Milano, Fondazione IRCCS Cà Granda Ospedale Maggiore Policlinico di Milano, Centro Dino Ferrari, via Francesco Sforza 35, 20122 Milan, Italy
e-mail: yvan.torrente@unimi.it

D. Corbeil (ed.), *Prominin-1 (CD133): New Insights on Stem & Cancer Stem Cell Biology*, Advances in Experimental Medicine and Biology 777, DOI 10.1007/978-1-4614-5894-4_15,

laboratories. Suitable stem cells for these purposes must be easily extracted, remain capable of efficient tissue regeneration, and upon transplantation integrate into host tissues allowing the functional recovery of the pathological phenotype [3]. The major advantage of stem cells is that they can proliferate for long periods and produce a progeny that can differentiate into various cell lineages with specialized functions [4]. Because of their interest in the medical field as they can be used to regenerate damaged tissues caused by disease or injury [5, 6], it is fundamental to assess specific and functional phenotypic markers for their unequivocal identification, prospective isolation, and eventually to manipulate them for genetic modification [7].

Several studies have demonstrated that the stem cell populations suitable for clinical demands might derive from multiple regions of the organism. Part of the regenerative capacity of stem cells resides in their intrinsic regulation as well as in their local microenvironment (called stem cell niche), which provides the appropriate factors and cues allowing their quiescence, proliferation, and differentiation. Besides the typical tissue-specific localization, which is often not fully documented, somatic stem cells are frequently identified by the expression of specific stemness marker antigens and evaluated by functional assays. In this context, various groups have described adult muscle side population [8], bone marrow-derived stem cells [9], muscle-derived stem cells [10], mesoangioblasts [11], and pericytes [12]. CD133 emerges as one of the major cell surface markers that labels various stem and progenitor cells originating from different sources of tissues or organs. This chapter focuses on the characterization of CD133$^+$ cells and their clinical potential particularly in degenerative conditions such as Duchenne macular dystrophy (DMD), a lethal X-linked recessive disorder caused by deficiency of the dystrophin protein.

15.2 CD133: Identification, Membrane Topology, and Gene

In 1997, murine prominin-1 was identified as a novel antigen (13A4) specifically localized in plasma membrane protrusions of neuroepithelial stem cells [13]. In the same year, the human CD133 homologue of mouse prominin-1 was revealed as a novel antigen (AC133) found on a subpopulation of CD34$^+$ hematopoietic stem cells derived from fetal liver, bone marrow, and cord blood [14].

Prominin-1 contains about 860 amino acids with a molecular weight of approximately 120 kDa. Note that its length and molecular mass are dependent of the splice variants (for more details, see Chap. 1). Structurally, it consists of an N-terminal extracellular domain, five transmembrane domains with two large extracellular loops, and a cytoplasmic tail with about 60 amino acids (see Fig. 1.1 in Chap. 1). Based on its amino acid sequence, human prominin-1 (hereafter CD133) possesses eight potential N-glycosylation sites: four on each extracellular loops [13, 15].

Prominin-1 *(PROM1)* genes are located on chromosome 5 (5B3) in mice and chromosome 4 (4p15.33) in humans and contain 34 and 37 exons, respectively, which span approximately 160 kb [16, 17]. Their transcripts are about 4.4 kb [13, 15, 17]. The promoter region of the human gene comprises at least 13 distinct

5'-untranslated region (5'-UTR) exons, alternatively spliced, giving rise to at least 16 distinct transcripts. Five promoters (P1, P2, P3, P4, and P5), all TATA-less, were identified. P1, P2, and P3 are located within a CpG island [18] (see Fig. 5.1 in Chap. 5), and DNA hypomethylation was shown to be associated with increased CD133 expression in cancer stem cells [19] (see Chap. 5 for more details concerning the *PROM1* gene regulation).

15.3 CD133: A Stem and Cancer Stem Cell Marker

The description of the expression of CD133 by hematopoietic stem and progenitor cells has directed the general interest toward its potential use as a cell surface marker of somatic stem cells [14]. In human hematopoietic lineages, CD133 expression as revealed by its particular AC133 epitope is restricted to a subpopulation of $CD34^+$ cells [15, 17, 20] (see also Chap. 1). Besides the hematopoietic system, CD133 (AC133) is found in various human subpopulation of cells harboring stem cell properties highlighting its clinical relevance [21]. CD133 is expressed on endothelial progenitor cells (EPCs), playing a role in angiogenesis and neovasculogenesis during both wound healing and tumor growth [22]. $CD133^+$ progenitor cells are identified in the central nervous system (CNS) (see Chap. 6). Neural $CD133^+$ cells possess the capacity to form neurospheres in vitro and to differentiate into neurons and CNS-resident cells in vivo [23, 24]. Remarkably, Bussolati and Richardson isolated rare $CD133^+$ cells from renal and prostate tissues, respectively, with high proliferative potential and stem cell properties [25, 26] (see Chaps. 8 and 11).

In cancerous tissues, other studies elegantly demonstrated that $CD133^+$ cells derived from human brain tumors displayed the properties of stem cells and could recapitulate the heterogeneity of the original tumor [27–29]. CD133 highlights stem and cancer stem cells.

15.4 Embryonic $CD133^+$ Cells

Embryonic stem cells (ESCs) are pluripotent cells derived from the early embryo. They proliferate over prolonged culture periods while remaining undifferentiated and maintaining a stable karyotype [30–32]. They could differentiate into cells present in all three germ layers [33]. Technically, an important step toward the advanced use of ESCs in cell-based therapies is understanding and eventually manipulating their regulation (proliferation versus differentiation into specific lineages) once they have been transferred into the patient.

Most established human (h) ESC lines have been characterized by the expression of cell surface antigens such as stage-specific embryonic antigen (SSEA)-3 and SSEA-4 [34], while CD133 was detected only in a subfraction of proliferating hESCs in culture, suggesting that its expression could identify a particular functional hESC subpopulation [31]. Indeed, King and colleagues demonstrated that a CD133$^+$ subpopulation of hESCs was predisposed to follow an ectodermal lineage fate [35]. They argued that CD133$^+$ hESCs represented one of many subpopulations of progenitor cells within an undifferentiated hESC [35]. Moreover, the authors demonstrated that stable CD133$^+$ hESC subpopulations existed within proliferating cultures and could be expanded without the loss of marker expression or lineage specificity [35]. At the terminal stages of differentiation, the expression of murine CD133 was observed only on cells co-expressing the neuroectodermal marker, nestin [36]. Thus, CD133 could be a useful marker to monitor the in vitro expansion of hESCs that will give rise specifically to neural progenitor cells, which might be suitable for the treatment of neurodegenerative disorders.

15.5 Somatic CD133$^+$ Stem Cells for Therapy

15.5.1 Bone Marrow-Derived CD133$^+$ Cells

In animal models, it has been established that circulating progenitor cells improved the repair of endothelial cells after injury and limited the progression of atherosclerosis [37, 38]. Moreover, it has been shown that intracoronary infusion of autologous bone marrow-derived stem cells (BMCs) is a safe and feasible procedure, which constructively influences the postinfarction remodeling in patients with myocardial infarction [39, 40]. Indeed, BMCs are to date the best-characterized and used cells in the majority of clinical trials. BMCs contain a complex assortment of progenitor cells, including hematopoietic and mesenchymal stem cells. Among them, bone marrow-derived circulating progenitor cells (BM-CPCs) expressed two hematopoietic stem cell markers: CD133 and CD34. Studies of tissue ischemia in animal models demonstrated that these cells were recruited into peripheral blood and participated in neovascularization [41]. Interestingly, Turan and colleagues assessed that the mobilization of CD133$^+$CD45$^+$ (and CD34$^+$CD45$^+$) BM-CPCs significantly increased with a peak on day 7 and decreased on day 8 following myocardial infarction [42]. The authors correlated this spontaneous mobilization response to myocardial repair after infarction. They also found a significant increase of CD133$^+$CD45$^+$ (and CD34$^+$CD45$^+$) BM-CPCs mobilization in the peripheral blood at 3, 6, and 12 months after intracoronary cell transplantation, with no significant change in the control group. The authors proposed that CD133$^+$ BM-CPCs could be considered for cell-based therapy to repair of cardiac tissue [42].

The teams of Steinhoff and Gams carried out similar preclinical and clinical trials and revealed that adult EPCs were able to stimulate angiogenesis in the postinfarct tissues [43, 44] (see also Chap. 14). Flores-Ramirez and colleagues showed that intracoronary infusion of a specific subpopulation of EPCs expressing CD133 into patients with moderate to severe postinfarct heart failure resulted in a net improvement of heart function [45]. Moreover, the authors assessed that the treatment was minimally invasive, safe, less expensive, and associated with minor adverse events. Of particular interest, they noticed that a single intracoronary infusion of CD133+ cells into the infracted area could improve heart function for at least 2 years resulting in a remarkable progress in quality of life [45].

15.5.1.1 CD133+ Cells and Inflammatory Myopathies

Among the autoimmune inflammatory myopathies (IMs), inclusion body myositis (IBM) is the most frequent and is characterized by chronic inflammation in conjunction with a degenerative process that causes a poor response to current immunotherapy and progressive disability of most affected individuals [46]. As it happens in the other muscle diseases, satellite cells fail to compensate the progressive loss of functional muscle tissue [46]. Hollemann and colleagues focused their efforts on the occurrence of hematopoietic progenitor cells (HPCs) with putative myoendothelial potential in human skeletal muscle in immune-mediated inflammation. In IMs, the authors detected several clusters of endomysial CD133+ cells, which were co-localized with CD45+ mononuclear infiltrates [47]. Given the close correlation between numbers of CD133+ HPCs and CD45+ leukocytes, they proposed that CD133+ HPCs enter skeletal muscle via the circulation along with inflammatory cells, as previously suggested by others [48, 49]. Moreover, they indicated that CD133+ HPCs were mobilized from the bone marrow into the circulation and recruited along with myelomonocytic cell populations to injured muscle, to allow the repair mechanism after inflammatory damage [47]. Interestingly, they detected in the muscles of IBM patients subpopulations of CD133+ cells co-expressing myogenic marker Pax7 with or without CD34 reactivity indicating endothelial or myogenic commitment of some HPCs. In agreement with previous studies, such information suggested that transplanted bone marrow-derived myogenic progenitors were able to fuse with preexisting myofibers [50]. In IM patients, circulating CD133+ myoendothelial progenitors are able to colonize skeletal muscle after injury and get in touch with resident satellite cells and EPCs so strictly that they participated in the permanent restoration of damaged muscle [47].

15.5.1.2 CD133+ Cells and Retinal Pigment Epithelium

Maw and colleagues demonstrated that murine CD133 was located at the base of the photoreceptor outer segments (OS) and regulated the formation of newly synthesized photoreceptive discs and, hence, the morphogenesis of these cells [15]. Moreover, they showed that patients affected by retinal degeneration shared a frame shift mutation in *PROM1* gene [15]. In the same line, Zacchigna and colleagues demonstrated using prominin-1 knockout mice that this molecule controlled OS morphogenesis and its maintenance [51] (for more details concerning the role of prominin-1/CD133 in photoreceptors, see Chap. 4).

In addition to photoreceptors, retinal pigment epithelium (RPE) plays an important role in the homeostasis of the retina. This epithelium is formed by a monolayer of cuboidal cells that are responsible for the movement of ions and water to maintain a proper state of dehydration for visual clarity; its pigmentation absorbs stray light that would otherwise degrade the visual image [52]. RPE dysfunction was linked to many devastating eye disorders, including age-related macular degeneration (AMD) [53], Stargardt's disease, and retinitis pigmentosa [54]. Unfortunately, current treatments for AMD can only slow down the progression of vision loss, primarily benefiting patients with advanced stages of the disease. Harris and colleagues isolated bone marrow-derived CD133$^+$ subpopulations and transplanted them into the intravitreal space [55]. Remarkably, transplanted CD133$^+$ cells incorporated into the damaged RPE layer, expressed key RPE-specific proteins, and assumed typical RPE hexanocuboidal morphology. Interestingly, the incorporation rate into the damaged RPE layer ameliorated with increasing numbers of transplanted CD133$^+$ cells [55]. These data provided encouraging evidence that bone marrow-derived CD133$^+$ cells could be suitable for the treatment of several degenerative disorders of the retina.

15.5.2 Blood-Derived CD133$^+$ Stem Cells

CD133$^+$ cells isolated from blood were considered to be hematopoietic and endothelial stem cells of bone marrow origin that could give rise to both endothelial cells and myoblasts [56]. As reported in 2004, CD133$^+$ cells isolated from peripheral blood exhibited stem cell behavior [48]. Phenotypically, more than 92% of the CD133$^+$ cells were positive for CD34; they co-expressed Thy-1 (95%) and CD45 (97%) and harbored a lineage-negative trait (CD33$^-$CD38$^-$). When cultured in proliferation medium, their myogenic commitment was demonstrated by the expression of several myogenic markers such as M-cadherin, MRF4, and myogenin. Once dissociated to single cells and cloned, these cells replicated as typical stem cells giving rise to small adherent colonies. In addition, a robust telomerase activity confirmed the stem cell-like activity of the blood-derived CD133$^+$ cells [48]. Different studies described how the cloned CD133-derived stem cells could differentiate not only into hematopoietic and endothelial tissues [48] but also into blast cells, macrophages, and granular polymorphonuclear cells [57]. Circulating CD133$^+$ cells differentiated into myotubes in vitro. Moreover, when cultured in the presence of a feeder layer consisting of mouse myogenic cells, they fused and formed myosin heavy chain (MyHC)-expressing heterozygous human/murine myotubes [48]. Interestingly, CD133$^+$ stem cells transplanted intramuscularly and intra-arterially in a murine model of DMD, the *scid/mdx* mouse, could restore dystrophin expression, participate in skeletal muscle regeneration, and regenerate the satellite cell pool. Histologically, the cells transplanted into the dystrophic skeletal muscles located beneath the basal lamina and distributed along freshly isolated fibers and co-expressed lamin A/C and M-cadherin [48], whereas after intra-arterial injection, they were detected in numerous vessels near regeneration areas and expressed VE-cadherin and CD31.

In a research program aiming to use human stem cells to repair muscle damage in muscular dystrophies, such as DMD, we revealed 6 years ago new insights into the mechanism implicated in the muscle homing of blood-derived CD133$^+$ stem cells, which allowed the systemic delivery of such stem cells. In this context, it is essential that injected CD133$^+$ stem cells migrate into the vasculature [58]. Interestingly, we showed that almost all CD133$^+$ cells co-expressed CD44 and lymphocyte function-associated antigen 1, while >40% of them expressed P-selectin glycoprotein ligand-1, very late antigen-4 (VLA-4), L-selectin, and C-C chemokine receptor type 7. This unique expression pattern of adhesion/receptor molecules gave to CD133$^+$ stem cells the potentiality to mediate the migration through the blood vessel wall [58]. The expression of vascular cell adhesion protein 1 (VCAM-1, CD106) by the dystrophic muscle vessels appeared also essential for the proper interaction and/or adhesion of injected CD133$^+$ cells. The elucidation of the mechanism including molecular players (e.g., VCAM-1/VLA-4 adhesion receptor complex) involved in muscle homing could thus improve potential therapy for such diseases.

It is also of special interest to determine whether cells are administered systemically rather than directly into the target tissue. In this perspective, we used very recently an innovative technology to study the distribution of the injected cells. We combined nanoparticles labeling of blood-derived CD133$^+$ cells and the X-ray computed microtomography to provide detailed information on the migration of injected stem cells in three-dimensional space within the tissues [59]. We demonstrated that the migration of intra-arterially delivered stem cells to dystrophic muscles was time dependent, and the number of migrating cells increased specifically in damaged muscle tissues. A better understanding of the kinetic of migration and distribution of the CD133$^+$ stem cells could be crucial to enhance their therapeutic potential in tissue repair.

Collectively, these data suggest that human blood-derived CD133$^+$ cells can migrate within muscle tissues and differentiated into endothelial, satellite cells, and human dystrophin-positive muscle fibers.

15.5.3 Muscle-Derived CD133$^+$ Stem Cells

Our laboratory isolated a similar CD133$^+$ cell population from human normal and dystrophic tissues. These cells co-expressed other markers such as CD34, CD45, and KDR. Muscle-derived CD133$^+$ cells represented 2% of the total dystrophic muscle-derived nucleated cells. They were present in both muscles, although dystrophic muscle biopsies of young patients (5–14 years old) contained a higher number of these cells than muscle biopsies of healthy donors matched for age [60]. Their hematopoietic commitment was confirmed by the detection of CD45. Their proliferation rate was approximately 20 populations doubling with a doubling time around 36 h [60]. Further in vitro analysis of normal and dystrophic muscle-derived CD133$^+$ cells revealed the expression of Pax-7, Myf-5, MyoD, M-cadherin, MRF-4, and myogenin after 24 days of culture under the proliferative conditions, suggesting a myogenic commitment and their differentiation into multinucleated myotubes expressing MyHCs [60].

In in vivo experiments, CD133$^+$ stem cells were transduced with a lentivirus carrying a construct designed to skip exon 51 of dystrophin transcript (see below) [61] and intramuscularly transplanted into dystrophic mice. We identified human spectrin$^+$ myofibers co-expressing the human lamin A/C, with no differences between muscles injected with normal or dystrophic cells. Similarly to the injection of blood-derived CD133$^+$ cells (see above), many muscle-derived CD133$^+$ cells were located near and within small vessels, co-expressing CD31 and VE-cadherin. Few transplanted cells were found around muscle capillaries, suggesting a pericyte commitment [62]. Because CD133$^+$ stem cells were recruited into muscle fibers and located near muscle vessels, we assessed them as myoendothelial progenitors [60]. Similarly, Negroni and colleagues identified recently a subpopulation of CD133$^+$ stem cells expressing the satellite cells marker CD56. They found that only 19% CD133$^+$CD34$^+$ cells and 23% CD133$^+$CD34$^-$ cells were CD56$^+$, but nevertheless both populations retained a high proliferation rate. The authors analyzed their myogenic potential and demonstrated that muscle-derived CD133$^+$CD34$^+$ cells formed more multinucleated myotubes expressing MyHCs in comparison to the CD133$^+$CD34$^-$ population [63].

15.6 Heterogeneity of CD133$^+$ Stem Cells

Exceptional efforts were made in the last decade in order to characterize stem cells. These investigations led to identification of their niche and the molecular signals that regulate their mobilization and homing. The mechanisms controlling self-renewal, commitment, and differentiation of stem cells were also partially described [64]. These findings produced invaluable tools for the development of rational cell therapy protocols that yielded positive results in preclinical models of genetic and acquired diseases [65] and, in many cases, entered clinical experimentation with positive outcomes [66]. Unfortunately, almost all studies regarding the stem cell behavior and development were strongly limited by the heterogeneity of stem cell populations [67]. For instance, CD133$^+$ stem cells represent a heterogeneous population, which include a common progenitor with round shape, low proliferative capacity, and potential to give rise to myogenic, endothelial, and pericyte cells. Negroni and colleagues demonstrated that CD133$^+$ stem cells are not all at the same stage of myogenic commitment, e.g., they identified a CD133$^+$ cell fraction already committed to myogenic differentiation, which gave rise to myonuclei and satellite cells, while another cell fraction with slower differentiation capacity secreted specific molecular factors stimulating the recovery of the degenerated muscle when injected in a dystrophic tissue [63]. As CD133$^+$ cells and mesoangioblasts displayed similar behaviors in experimental conditions and shared the ability to migrate through the vasculature, it was suggested a probable connection between the origin and the development of these two kind of progenitors cells. This paradox was dealt in the study of Grounds and colleagues, where it was highlighted how the number of resident satellite cells was much smaller than the number of committed myogenic precursors that populated the muscle tissue soon after injury [68]. As it was unexplainable why all progenitors were not able to defeat – or at least ameliorate – the symptoms of the diseases, it was proposed that these committed progenitors were the steps of differentiation of a common undifferentiated stem cell where the function of which is largely unknown.

15.7 Outlines of Clinical Trials

The utilization of CD133$^+$ stem cells for therapeutic purposes has recently opened a new promising field in the treatment of degenerative diseases. As described above, the data from our laboratory demonstrated that circulating and muscle-resident CD133$^+$ stem cells differentiated in myogenic cells and thus could be suitable for stem cell therapy of muscular dystrophy [48]. We extended these initial observations to a combination of cell- and gene-based approaches via the ex vivo introduction of corrective genes into dystrophic CD133$^+$ cells, which allowed their subsequent autologous transplantation [62]. To that end, we have used the exon skipping approach as a correction method. The specific antisense oligonucleotides mask the putative splicing sites of exons in the mutated region of the primary RNA transcript, and their removal would reestablish a proper reading frame. We treated DMD patients carrying a Δ49–50 deletion in dystrophin gene: theoretically skipping of exon 51 in this context should eliminate the original frame shift, revealing a correct reading frame between exon 48 and exon 52 [62]. We transduced CD133$^+$ stem cells isolated from both blood and muscle of these patients, and we demonstrated that genetically engineered DMD stem cells expressed a shortened dystrophin mRNA specifically deleted for the exon 51 [62]. Genetically engineered CD133$^+$ cells not only participated in myogenesis but also differentiated into satellite cells within the recipient dystrophic skeletal muscle [62]. Moreover, as DMD pathology affects the whole body musculature and its effective treatment requires the distribution of the injected cells to the whole body musculature, intra-arterial injections allow dispersed delivery of CD133$^+$ stem cells.

As highlighted in the previous sections, ongoing phase I/II studies are running in order to assess the efficacy and the safety of the injection of CD133$^+$ stem cells isolated from different tissues in application to different diseases. For instance, CD133$^+$ stem cells were obtained from peripheral blood of diabetic patients with critical ischemia in lower limbs. They were characterized in terms of capacity of differentiation, and the safety of their intramuscular administration was determined. The patient global health was evaluated, in particular, the ability of transplanted CD133$^+$ cells to increase the revascularization at lower limbs [42]. Another study evaluated whether the intracoronary infusion of autologous bone marrow-derived

$CD133^+$ EPCs was able to promote neovascularization and to improve myocardial perfusion and contractility in patients with refractory coronary heart disease. These patients were initially characterized by poor response to standard coronary interventions, severe impairment of the quality of life, and poor prognosis [45]. Recently, a young patient suffering from myocardial infarction was transplanted with bone marrow-derived $CD133^+CD34^+$ cells. As he/she showed improvement in functional capacity, it was suggested that the injected cells were able to ameliorate angiogenesis, promote myocyte recovery, and inhibit cellular apoptosis [57]. Moreover, it was shown that patients suffering from stroke mobilized an EPCs fraction expressing CD133, CD34, KDR, and CD45 that induced vasculogenesis and secreted angiogenic factors [69]. Now 15 ongoing phase I and II studies are started using $CD133^+$ stem cells (www.clinicaltrials.gov).

Although $CD133^+$ stem cells are already used in different areas to regenerate deteriorated tissues, several aspects need to be ameliorated in order to consider them fully suitable for cellular therapy particularly for muscular dystrophies. First of all, it is necessary to improve the strategy to deliver myogenic cells chronically to the various sites of sporadic degeneration. Unfortunately, it is not clearly established how long $CD133^+$ cells isolated from muscle or blood could contribute to host tissue regeneration. It was previously published that $CD133^+$ cells were able to exhibit myogenic properties 60 days after transplantation in scid/mdx mice [62]. It should also be determined how many times are necessary to perform cellular transplantation to have a net improvement in terms of efficiency of muscular development. Numerous ongoing works are now in progress trying to find answers to these important issues. Technically, the rate of proliferation of blood-derived $CD133^+$ cells in culture and their storage for repeated treatments need to be enhanced. Similarly, it would be of interest to understand why blood-derived $CD133^+$ cells are less efficient compared with muscle-derived $CD133^+$ cells toward their contribution to muscle nuclei.

15.8 Conclusion

Stem cell therapy is an appealing approach to treat pathologies such as muscular dystrophies because only a small number of cells, together with factors required to promote their expansion, are necessary to obtain a therapeutic outcome [3]. As prerequisites, candidate stem cells must be easily isolated, maintain their efficient myogenic conversion, and when transplanted in the host tissue, integrate into the muscles allowing the functional recovery of the dystrophic phenotype [3]. Survival and migration from the injection site to the compromised muscles remain unfortunately suboptimal for numerous cell populations, and these issues need to be improved. However, we demonstrated that human circulating and muscle-resident cells expressing CD133 behaved as a stem cell population capable of commitment to hematopoietic, endothelial, and myogenic lineages [48, 62], and the discovery of the molecular mechanism involved in the muscle homing allowed the systemic delivery of these cells into patients [56]. Remarkably, different ongoing and successful clinical trials demonstrated the possibility to use CD133+ stem cells in the treatment of ischemia, refractory coronary heart disease, cirrhosis, myocardial infarction, and muscular dystrophy [42, 45, 48, 57]. Although certain issues have to be clarified to deeply understand the behavior of CD133+ stem cells, they satisfied nonetheless most of the criteria for their use in the treatment of degenerative diseases. Finally, it will be interesting to elucidate the possible common origin of different stem cell populations, such as CD133+ cells, muscle-derived stem cells and mesoangioblasts, their relationship to satellite cells, and whether these myogenic cells had the potential to give rise to muscular cell lineages also in steady-state conditions [56]. In this sense, a method to isolate "pure" stem cell populations is desirable. The use of CD133 in combination with other markers might open new perspective in this field.

Acknowledgements Stem Cell Laboratory, Dipartimento di Fisiopatologia Medico-Chirurgica e dei Trapianti, Università degli Studi di Milano, Fondazione IRCCS Cà Granda, OMP Milano has been supported by the Association Monégasque contre les Myopathies (AMM), Associazione La Nostra Famiglia Fondo DMD Gli Amici di Emanuele, the Associazione Amici del Centro Dino Ferrari, EU's Framework programme 7 Optistem 223098, and Provincia di Trento Fondo 12-03-5277500-01.

References

1. Owonikoko T, Agha M, Balassanian R, Smith R, Raptis A (2007) Gemtuzumab therapy for isolated extramedullary AML relapse following allogeneic stem-cell transplant. Nat Clin Pract Oncol 4:491–495
2. Singh N, Pillay V, Choonara YE (2007) Advances in the treatment of Parkinson's disease. Prog Neurobiol 81:29–44
3. Price FD, Kuroda K, Rudnicki MA (2007) Stem cell based therapies to treat muscular dystrophy. Biochim Biophys Acta 1772:272–283
4. Bertoni C (2008) Clinical approaches in the treatment of Duchenne muscular dystrophy (DMD) using oligonucleotides. Front Biosci 13:517–527
5. Fuchs E, Segre JA (2000) Stem cells: a new lease on life. Cell 100:143–155
6. Weissman IL (2000) Stem cells: units of development, units of regeneration, and units in evolution. Cell 100:157–168
7. Tarnok A, Ulrich H, Bocsi J (2010) Phenotypes of stem cells from diverse origin. Cytometry A 77:6–10
8. Asakura A, Rudnicki MA (2002) Side population cells from diverse adult tissues are capable of in vitro hematopoietic differentiation. Exp Hematol 30:1339–1345
9. Dezawa M, Ishikawa H, Itokazu Y, Yoshihara T, Hoshino M, Takeda S, Ide C, Nabeshima Y (2005) Bone marrow stromal cells generate muscle cells and repair muscle degeneration. Science 309:314–317
10. Qu-Petersen Z, Deasy B, Jankowski R, Ikezawa M, Cummins J, Pruchnic R, Mytinger J, Cao B, Gates C, Wernig A, Huard J (2002) Identification of a novel population of muscle stem cells in mice: potential for muscle regeneration. J Cell Biol 157:851–864
11. Cossu G, Bianco P (2003) Mesoangioblasts-vascular progenitors for extravascular mesodermal tissues. Curr Opin Genet Dev 13:537–542
12. Doherty MJ, Ashton BA, Walsh S, Beresford JN, Grant ME, Canfield AE (1998) Vascular pericytes express osteogenic potential in vitro and in vivo. J Bone Miner Res 13:828–838

13. Weigmann A, Corbeil D, Hellwig A, Huttner WB (1997) Prominin, a novel microvilli-specific polytopic membrane protein of the apical surface of epithelial cells, is targeted to plasmalemmal protrusions of non-epithelial cells. Proc Natl Acad Sci USA 94:12425–12430
14. Yin AH, Miraglia S, Zanjani ED, Almeida-Porada G, Ogawa M, Leary AG, Olweus J, Kearney J, Buck DW (1997) AC133, a novel marker for human hematopoietic stem and progenitor cells. Blood 90:5002–5012
15. Miraglia S, Godfrey W, Yin AH, Atkins K, Warnke R, Holden JT, Bray RA, Waller EK, Buck DW (1997) A novel five-transmembrane hematopoietic stem cell antigen: isolation, characterization, and molecular cloning. Blood 90:5013–5021
16. Maw MA, Corbeil D, Koch J, Hellwig A, Wilson-Wheeler JC, Bridges RJ et al (2000) A frameshift mutation in prominin (mouse)-like 1 causes human retinal degeneration. Hum Mol Genet 9:27–34
17. Fargeas CA, Florek M, Huttner WB, Corbeil D (2003) Characterization of prominin-2, a new member of the prominin family of pentaspan membrane glycoproteins. J Biol Chem 278:8586–8596
18. Shmelkov SV, Jun L, St Clair R, McGarrigle D, Derderian CA, Usenko JK, Costa C, Zhang F, Guo X, Rafii S (2004) Alternative promoters regulate transcription of the gene that encodes stem cell surface protein AC133. Blood 103:2055–2061
19. Tabu K, Sasai K, Kimura T et al (2008) Promoter hypomethylation regulates CD133 expression in human gliomas. Cell Res 18:1037–1046
20. Florek M, Haase M, Marzesco AM, Freund D, Ehninger G, Huttner WB, Corbeil D (2005) Prominin-1/CD133, a neural and hematopoietic stem cell marker, is expressed in adult human differentiated cells and certain types of kidney cancer. Cell Tissue Res 319:15–26
21. Fargeas CA, Fonseca AV, Huttner WB, Corbeil D (2006) Prominin-1 (CD133): from progenitor cells to human diseases. Future Lipidology 1:213–225
22. Ribatti D (2004) The involvement of endothelial progenitor cells in tumor angiogenesis. J Cell Mol Med 8:294–300
23. Uchida N, Buck DW, He D, Reitsma MJ, Masek M, Phan TV, Tsukamoto AS, Gage FH, Weissman IL (2000) Direct isolation of human central nervous system stem cells. Proc Natl Acad Sci USA 97:14720–14725
24. Tamaki T, Okada Y, Uchiyama Y, Tono K, Masuda M, Wada M, Hoshi A, Ishikawa T, Akatsuka A (2007) Clonal multipotency of skeletal muscle-derived stem cells between mesodermal and ectodermal lineage. Stem Cells 25:2283–2290
25. Bussolati B, Bruno S, Grange C, Buttiglieri S, Deregibus MC, Cantino D, Camussi G (2005) Isolation of renal progenitor cells from adult human kidney. Am J Pathol 166:545–555
26. Richardson GD, Robson CN, Lang SH, Neal DE, Maitland NJ, Collins AT (2004) CD133, a novel marker for human prostatic epithelial stem cells. J Cell Sci 117:3539–3545
27. Hemmati HD, Nakano I, Lazareff JA, Masterman-Smith M, Geschwind DH, Bronner-Fraser M, Kornblum HI (2003) Cancerous stem cells can arise from pediatric brain tumors. Proc Natl Acad Sci USA 100:15178–15183
28. Singh SK, Hawkins C, Clarke ID, Squire JA, Bayani J, Hide T, Henkelman RM, Cusimano MD, Dirks PB (2004) Identification of human brain tumour initiating cells. Nature 432:396–401
29. Bao S, Wu Q, McLendon RE, Hao Y, Shi Q, Hjelmeland AB, Dewhirst MW, Bigner DD, Rich JN (2006) Glioma stem cells promote radioresistance by preferential activation of the DNA damage response. Nature 444:756–760
30. Amit M, Itskovitz-Eldor J (2002) Derivation and spontaneous differentiation of human embryonic stem cells. J Anat 200:225–232
31. Carpenter MK, Rosler E, Rao MS (2003) Characterization and differentiation of human embryonic stem cells. Cloning Stem Cells 5:79–88
32. Hoffman LM, Carpenter MK (2005) Characterization and culture of human embryonic stem cells. Nat Biotechnol 23:699–708

33. Strulovici Y, Leopold PL, O'Connor TP, Pergolizzi RG, Crystal RG (2007) Human embryonic stem cells and gene therapy. Mol Ther 15:850–866
34. Adewumi O, Aflatoonian B, Ahrlund-Richter L, Amit M, Andrews PW, Beighton G et al (2007) Characterization of human embryonic stem cell lines by the International Stem Cell Initiative. Nat Biotechnol 25:803–816
35. King FW, Ritner C, Liszewski W, Kwan HC, Pedersen A, Leavitt AD, Bernstein HS (2009) Subpopulations of human embryonic stem cells with distinct tissue-specific fates can be selected from pluripotent cultures. Stem Cells Dev 18:1441–1450
36. Kania G, Corbeil D, Fuchs J, Tarasov KV, Blyszczuk P, Huttner WB, Boheler KR, Wobus AM (2005) Somatic stem cell marker prominin-1/CD133 is expressed in embryonic stem cell-derived progenitors. Stem Cells 23:791–804
37. Walter DH, Rittig K, Bahlmann FH, Kirchmair R, Silver M, Murayama T, Nishimura H, Losordo DW, Asahara T, Isner JM (2002) Statin therapy accelerates reendothelialization: a novel effect involving mobilization and incorporation of bone marrow-derived endothelial progenitor cells. Circulation 105:3017–3024
38. Werner N, Priller J, Laufs U, Endres M, Bohm M, Dirnagl U, Nickenig G (2002) Bone marrow-derived progenitor cells modulate vascular reendothelialization and neointimal formation: effect of 3-hydroxy-3-methylglutaryl coenzyme a reductase inhibition. Arterioscler Thromb Vasc Biol 22:1567–1572
39. Assmus B, Schachinger V, Teupe C, Britten M, Lehmann R et al (2002) Transplantation of Progenitor Cells and Regeneration Enhancement in Acute Myocardial Infarction (TOPCARE-AMI). Circulation 106:3009–3017
40. Schachinger V, Erbs S, Elsasser A, Haberbosch W, Hambrecht R, Holschermann H et al (2006) Intracoronary bone marrow-derived progenitor cells in acute myocardial infarction. N Engl J Med 355:1210–1221
41. Asahara T, Masuda H, Takahashi T, Kalka C, Pastore C, Silver M, Kearne M, Magner M, Isner JM (1999) Bone marrow origin of endothelial progenitor cells responsible for postnatal vasculogenesis in physiological and pathological neovascularization. Circ Res 85:221–228
42. Turan RG, Bozdag-Turan I, Ortak J, Akin I, Kische S, Schneider H et al (2011) Improved mobilization of the CD34(+) and CD133(+) bone marrow-derived circulating progenitor cells by freshly isolated intracoronary bone marrow cell transplantation in patients with ischemic heart disease. Stem Cells Dev 20:1491–1501
43. Klein HM, Ghodsizad A, Marktanner R, Poll L, Voelkel T, Mohammad Hasani MR et al (2007) Intramyocardial implantation of CD133+ stem cells improved cardiac function without bypass surgery. Heart Surg Forum 10:E66–E69
44. Stamm C, Kleine HD, Choi YH, Dunkelmann S, Lauffs JA, Lorenzen B et al (2007) Intramyocardial delivery of CD133+ bone marrow cells and coronary artery bypass grafting for chronic ischemic heart disease: safety and efficacy studies. J Thorac Cardiovasc Surg 133:717–725
45. Flores-Ramirez R, Uribe-Longoria A, Rangel-Fuentes MM, Gutierrez-Fajardo P, Salazar-Riojas R, Cervantes-Garcia D et al (2010) Intracoronary infusion of CD133+ endothelial progenitor cells improves heart function and quality of life in patients with chronic post-infarct heart insufficiency. Cardiovasc Revasc Med 11:72–78
46. Buckingham M, Bajard L, Chang T, Daubas P, Hadchouel J, Meilhac S, Montarras D, Rocancourt D, Relaix F (2003) The formation of skeletal muscle: from somite to limb. J Anat 202:59–68
47. Hollemann D, Budka H, Loscher WN, Yanagida G, Fischer MB, Wanschitz JV (2008) Endothelial and myogenic differentiation of hematopoietic progenitor cells in inflammatory myopathies. J Neuropathol Exp Neurol 67:711–719
48. Torrente Y, Belicchi M, Sampaolesi M et al (2004) Human circulating AC133(+) stem cells restore dystrophin expression and ameliorate function in dystrophic skeletal muscle. J Clin Invest 114:182–195

49. Palermo AT, Labarge MA, Doyonnas R, Pomerantz J, Blau HM (2005) Bone marrow contribution to skeletal muscle: a physiological response to stress. Dev Biol 279:336–344
50. Dreyfus PA, Chretien F, Chazaud B, Kirova Y, Caramelle P, Garcia L, Butler-Browne G, Gherardi RK (2004) Adult bone marrow-derived stem cells in muscle connective tissue and satellite cell niches. Am J Pathol 164:773–779
51. Zacchigna S, Oh H, Wilsch-Bräuninger M, Missol-Kolka E, Jászai J, Jansen S et al (2009) Loss of the cholesterol-binding protein prominin-1/CD133 causes disk dysmorphogenesis and photoreceptor degeneration. J Neurosci 29:2297–2308
52. Bok D (1993) The retinal pigment epithelium: a versatile partner in vision. J Cell Sci Suppl 17:189–195
53. Zarbin MA (2004) Current concepts in the pathogenesis of age-related macular degeneration. Arch Ophthalmol 122:598–614
54. Margalit E, Sadda SR (2003) Retinal and optic nerve diseases. Artif Organs 27:963–974
55. Harris JR, Brown GA, Jorgensen M, Kaushal S, Ellis EA, Grant MB, Scott EW (2006) Bone marrow-derived cells home to and regenerate retinal pigment epithelium after injury. Invest Ophthalmol Vis Sci 47:2108–2113
56. Peault B, Rudnicki M, Torrente Y, Cossu G, Tremblay JP, Partridge T, Gussoni E, Kunkel LM, Huard J (2007) Stem and progenitor cells in skeletal muscle development, maintenance, and therapy. Mol Ther 15:867–877
57. Limsuwan A, Pienvichit P, Limpijankit T, Khowsathit P, Hongeng S, Pornkul R, Siripornpitak S, Boonbaichaiyapruk S (2010) Transcoronary bone marrow-derived progenitor cells in a child with myocardial infarction: first pediatric experience. Clin Cardiol 33:E7–E12
58. Gavina M, Belicchi M, Rossi B et al (2006) VCAM-1 expression on dystrophic muscle vessels has a critical role in the recruitment of human blood-derived $CD133^+$ stem cells after intra-arterial transplantation. Blood 108:2857–2866
59. Farini A, Villa C, Manescu A, Fiori F, Giuliani A, Razini P, Sitzia C, Del Fraro G, Belicchi M, Meregalli M, Rustichelli F, Torrente Y (2012) Novel insight into stem cell trafficking in dystrophic muscles. Int J Nanomedicine 7:3059–3067
60. Torrente Y, Belicchi M, Marchesi C, Dantona G, Cogiamanian F, Pisati F et al (2007) Autologous transplantation of muscle-derived $CD133^+$ stem cells in Duchenne muscle patients. Cell Transplant 16:563–577
61. Goyenvalle A, Vulin A, Fougerousse F, Leturcq F, Kaplan JC, Garcia L, Danos O (2004) Rescue of dystrophic muscle through U7 snRNA-mediated exon skipping. Science 306:1796–1799
62. Benchaouir R, Meregalli M, Farini A, D'Antona G, Belicchi M, Goyenvalle A, Battistelli M, Bresolin N, Bottinelli R, Garcia L, Torrente Y (2007) Restoration of human dystrophin following transplantation of exon-skipping-engineered DMD patient stem cells into dystrophic mice. Cell Stem Cell 1:646–657
63. Negroni E, Riederer I, Chaouch S, Belicchi M, Razini P, Di Santo J, Torrente Y, Butler-Browne GS, Mouly V (2009) In vivo myogenic potential of human $CD133^+$ muscle-derived stem cells: a quantitative study. Mol Ther 17:1771–1778
64. Smith S, Neaves W, Teitelbaum S (2006) Adult stem cell treatments for diseases? Science 313:439
65. Sampaolesi M, Blot S, D'Antona G, Granger N, Tonlorenzi R, Innocenzi A et al (2006) Mesoangioblast stem cells ameliorate muscle function in dystrophic dogs. Nature 444:574–579
66. Mavilio F, Pellegrini G, Ferrari S, Di Nunzio F, Di Iorio E, Recchia A et al (2006) Correction of junctional epidermolysis bullosa by transplantation of genetically modified epidermal stem cells. Nat Med 12:1397–1402
67. Graf T, Stadtfeld M (2008) Heterogeneity of embryonic and adult stem cells. Cell Stem Cell 3:480–483

68. Grounds MD, McGeachie JK (1987) A model of myogenesis in vivo, derived from detailed autoradiographic studies of regenerating skeletal muscle, challenges the concept of quantal mitosis. Cell Tissue Res 250:563–569
69. Navarro-Sobrino M, Rosell A, Hernandez-Guillamon M, Penalba A, Ribo M, Alvarez-Sabin J, Montaner J (2010) Mobilization, endothelial differentiation and functional capacity of endothelial progenitor cells after ischemic stroke. Microvasc Res 80:317–323

Index

D. Corbeil (ed.), *Prominin-1 (CD133): New Insights on Stem & Cancer Stem Cell Biology*, Advances in Experimental Medicine and Biology 777, DOI 10.1007/978-1-4614-5894-4,

MIX
Papier aus verantwortungsvollen Quellen
Paper from responsible sources
FSC® C105338

If you have any concerns about our products,
you can contact us on
ProductSafety@springernature.com

In case Publisher is established outside the EU,
the EU authorized representative is:
Springer Nature Customer Service Center GmbH
Europaplatz 3, 69115 Heidelberg, Germany

Printed by Libri Plureos GmbH
in Hamburg, Germany